T0325165

MEASURE THEORY AND FINE PROPERTIES OF FUNCTIONS

Revised Edition

TEXTBOOKS in MATHEMATICS

Series Editors: Al Boggess and Ken Rosen

PUBLISHED TITLES

ABSTRACT ALGEBRA: AN INQUIRY-BASED APPROACH
Jonathan K. Hodge, Steven Schlicker, and Ted Sundstrom

ABSTRACT ALGEBRA: AN INTERACTIVE APPROACH
William Paulsen

ADVANCED CALCULUS: THEORY AND PRACTICE
John Srdjan Petrovic

ADVANCED LINEAR ALGEBRA
Nicholas Loehr

ANALYSIS WITH ULTRASMALL NUMBERS
Karel Hrbacek, Olivier Lessmann, and Richard O'Donovan

APPLIED DIFFERENTIAL EQUATIONS: THE PRIMARY COURSE
Vladimir Dobrushkin

APPLYING ANALYTICS: A PRACTICAL APPROACH
Evan S. Levine

COMPUTATIONS OF IMPROPER REIMANN INTEGRALS
Ioannis Roussos

CONVEX ANALYSIS
Steven G. Krantz

COUNTEREXAMPLES: FROM ELEMENTARY CALCULUS TO THE BEGINNINGS OF ANALYSIS
Andrei Bourchtein and Ludmila Bourchtein

DIFFERENTIAL EQUATIONS: THEORY, TECHNIQUE, AND PRACTICE, SECOND EDITION
Steven G. Krantz

DIFFERENTIAL EQUATIONS WITH MATLAB®: EXPLORATION, APPLICATIONS, AND THEORY
Mark A. McKibben and Micah D. Webster

ELEMENTARY NUMBER THEORY
James S. Kraft and Lawrence C. Washington

ELEMENTS OF ADVANCED MATHEMATICS, THIRD EDITION
Steven G. Krantz

PUBLISHED TITLES CONTINUED

EXPLORING LINEAR ALGEBRA: LABS AND PROJECTS WITH MATHEMATICA®
Crista Arangala

AN INTRODUCTION TO NUMBER THEORY WITH CRYPTOGRAPHY
James Kraft and Larry Washington

AN INTRODUCTION TO PARTIAL DIFFERENTIAL EQUATIONS WITH MATLAB®, SECOND EDITION
Mathew Coleman

INTRODUCTION TO THE CALCULUS OF VARIATIONS AND CONTROL WITH MODERN APPLICATIONS
John T. Burns

INTRODUCTION TO MATHEMATICAL LOGIC, SIXTH EDITION
Elliott Mendelson

INTRODUCTION TO MATHEMATICAL PROOFS: A TRANSITION TO ADVANCED MATHEMATICS, SECOND EDITION
Charles E. Roberts, Jr.

LINEAR ALGEBRA, GEOMETRY AND TRANSFORMATION
Bruce Solomon

THE MATHEMATICS OF GAMES: AN INTRODUCTION TO PROBABILITY
David G. Taylor

MEASURE THEORY AND FINE PROPERTIES OF FUNCTIONS, REVISED EDITION
Lawrence C. Evans and Ronald F. Gariepy

QUADRACTIC IRRATIONALS: AN INTRODUCTION TO CLASSICAL NUMBER THEORY
Franz Holter-Koch

REAL ANALYSIS AND FOUNDATIONS, THIRD EDITION
Steven G. Krantz

RISK ANALYSIS IN ENGINEERING AND ECONOMICS, SECOND EDITION
Bilal M. Ayyub

RISK MANAGEMENT AND SIMULATION
Aparna Gupta

TRANSFORMATIONAL PLANE GEOMETRY
Ronald N. Umble and Zhigang Han

TEXTBOOKS in MATHEMATICS

MEASURE THEORY AND FINE PROPERTIES OF FUNCTIONS

Revised Edition

Lawrence C. Evans

University of California
Berkeley, USA

Ronald F. Gariepy

University of Kentucky
Lexington, USA

CRC Press
Taylor & Francis Group
Boca Raton   London   New York

CRC Press is an imprint of the
Taylor & Francis Group an **informa** business

A CHAPMAN & HALL BOOK

CRC Press
Taylor & Francis Group
6000 Broken Sound Parkway NW, Suite 300
Boca Raton, FL 33487-2742

© 2015 by Taylor & Francis Group, LLC
CRC Press is an imprint of Taylor & Francis Group, an Informa business

No claim to original U.S. Government works

Version Date: 20150317

ISBN 13: 978-1-4822-4238-6 (hbk)

This book contains information obtained from authentic and highly regarded sources. Reasonable efforts have been made to publish reliable data and information, but the author and publisher cannot assume responsibility for the validity of all materials or the consequences of their use. The authors and publishers have attempted to trace the copyright holders of all material reproduced in this publication and apologize to copyright holders if permission to publish in this form has not been obtained. If any copyright material has not been acknowledged please write and let us know so we may rectify in any future reprint.

Except as permitted under U.S. Copyright Law, no part of this book may be reprinted, reproduced, transmitted, or utilized in any form by any electronic, mechanical, or other means, now known or hereafter invented, including photocopying, microfilming, and recording, or in any information storage or retrieval system, without written permission from the publishers.

For permission to photocopy or use material electronically from this work, please access www.copyright.com (http://www.copyright.com/) or contact the Copyright Clearance Center, Inc. (CCC), 222 Rosewood Drive, Danvers, MA 01923, 978-750-8400. CCC is a not-for-profit organization that provides licenses and registration for a variety of users. For organizations that have been granted a photocopy license by the CCC, a separate system of payment has been arranged.

Trademark Notice: Product or corporate names may be trademarks or registered trademarks, and are used only for identification and explanation without intent to infringe.

Visit the Taylor & Francis Web site at
http://www.taylorandfrancis.com

and the CRC Press Web site at
http://www.crcpress.com

Contents

Preface to the Revised Edition   xi

Preface   xiii

1   General Measure Theory   1

   1.1   Measures and measurable functions   1
      1.1.1   Measures   1
      1.1.2   Systems of sets   5
      1.1.3   Approximation by open and compact sets   9
      1.1.4   Measurable functions   16
   1.2   Lusin's and Egoroff's Theorems   19
   1.3   Integrals and limit theorems   24
   1.4   Product measures, Fubini's Theorem, Lebesgue measure   29
   1.5   Covering theorems   35
      1.5.1   Vitali's Covering Theorem   35
      1.5.2   Besicovitch's Covering Theorem   39
   1.6   Differentiation of Radon measures   47
      1.6.1   Derivatives   47
      1.6.2   Integration of derivatives; Lebesgue decomposition   50
   1.7   Lebesgue points, approximate continuity   53
      1.7.1   Differentiation Theorem   53
      1.7.2   Approximate limits, approximate continuity . . .   56
   1.8   Riesz Representation Theorem   59
   1.9   Weak convergence   65
      1.9.1   Weak convergence of measures   65
      1.9.2   Weak convergence of functions   68
      1.9.3   Weak convergence in L^1   70
      1.9.4   Measures of oscillation   75
   1.10 References and notes   78

2 Hausdorff Measures 81

2.1 Definitions and elementary properties 81
2.2 Isodiametric inequality, $\mathcal{H}^n = \mathcal{L}^n$ 87
2.3 Densities 92
2.4 Functions and Hausdorff measure 96
    2.4.1 Hausdorff measure and Lipschitz mappings . . . 96
    2.4.2 Graphs of Lipschitz functions 97
    2.4.3 Integrals over balls 98
2.5 References and notes 100

3 Area and Coarea Formulas 101

3.1 Lipschitz functions, Rademacher's Theorem 101
    3.1.1 Lipschitz continuous functions 101
    3.1.2 Rademacher's Theorem 103
3.2 Linear maps and Jacobians 108
    3.2.1 Linear mappings 108
    3.2.2 Jacobians 114
3.3 The area formula 114
    3.3.1 Preliminaries 114
    3.3.2 Proof of the area formula 119
    3.3.3 Change of variables formula 122
    3.3.4 Applications 123
3.4 The coarea formula 126
    3.4.1 Preliminaries 126
    3.4.2 Proof of the coarea formula 134
    3.4.3 Change of variables formula 139
    3.4.4 Applications 140
3.5 References and notes 142

4 Sobolev Functions 143

4.1 Definitions and elementary properties 143
4.2 Approximation 145
    4.2.1 Approximation by smooth functions 145
    4.2.2 Product and chain rules 153
    4.2.3 $W^{1,\infty}$ and Lipschitz continuous functions 155
4.3 Traces 156
4.4 Extensions 158

4.5 Sobolev inequalities 162
    4.5.1 Gagliardo–Nirenberg–Sobolev inequality 162
    4.5.2 Poincaré's inequality on balls 164
    4.5.3 Morrey's inequality 167
4.6 Compactness 168
4.7 Capacity 170
    4.7.1 Definitions and elementary properties 171
    4.7.2 Capacity and Hausdorff dimension 179
4.8 Quasicontinuity, precise representatives of Sobolev functions 183
4.9 Differentiability on lines 187
    4.9.1 Sobolev functions of one variable 188
    4.9.2 Differentiability on a.e. line 189
4.10 References and notes 190

5 Functions of Bounded Variation, Sets of Finite Perimeter   **193**

5.1 Definitions, Structure Theorem 193
5.2 Approximation and compactness 199
    5.2.1 Lower semicontinuity 199
    5.2.2 Approximation by smooth functions 199
    5.2.3 Compactness 203
5.3 Traces 204
5.4 Extensions 210
5.5 Coarea formula for BV functions 212
5.6 Isoperimetric inequalities 215
    5.6.1 Sobolev's and Poincaré's inequalities for BV . . . 216
    5.6.2 Isoperimetric inequalities 217
    5.6.3 $\mathcal{H}^{n-1}$ and Cap_1 220
5.7 The reduced boundary 221
    5.7.1 Estimates 221
    5.7.2 Blow-up 225
    5.7.3 Structure Theorem for sets of finite perimeter . . 231
5.8 Gauss–Green Theorem 235
5.9 Pointwise properties of BV functions 236
5.10 Essential variation on lines 244
    5.10.1 BV functions of one variable 245
    5.10.2 Essential variation on almost all lines 247
5.11 A criterion for finite perimeter 249

5.12 References and notes 255

6 Differentiability, Approximation by C^1 Functions   257

6.1 L^p differentiability, approximate differentiability 257
    6.1.1 L^{1^*} differentiability for BV 257
    6.1.2 L^{p*} differentiability a.e. for $W^{1,p}$ 260
    6.1.3 Approximate differentiability 262
6.2 Differentiability a.e. for $W^{1,p}$ $(p > n)$ 265
6.3 Convex functions 266
6.4 Second derivatives a.e. for convex functions 273
6.5 Whitney's Extension Theorem 276
6.6 Approximation by C^1 functions 282
    6.6.1 Approximation of Lipschitz continuous functions 283
    6.6.2 Approximation of BV functions 283
    6.6.3 Approximation of Sobolev functions 286
6.7 References and notes 288

Bibliography   289

Notation   293

Index   297

Preface to the Revised Edition

We published the original edition of this book in 1992 and have been extremely gratified with its popularity for now over 20 years. The publisher recently asked us to write an update, and we have agreed to do so in return for a promise that the future price be kept reasonable.

For this revised edition the entire book has been retyped into LaTeX and we have accordingly been able to set up better cross-references with page numbers. There have been countless improvements in notation, format and clarity of exposition, and the bibliography has been updated. We have also added several new sections, describing the π-λ Theorem, weak compactness criteria in L^1 and Young measure methods for weak convergence.

We will post any future corrections or comments at LCE's homepage, accessible through the math.berkeley.edu website. We remain very grateful to the many readers who have written us over the years, suggesting improvements and error fixes.

LCE has been supported during the writing of the revised edition by the National Science Foundation (under the grant DMS-1301661), by the Miller Institute for Basic Research in Science and by the Class of 1961 Collegium Chair at UC Berkeley.

Best wishes to our readers, past and future.

LCE/RFG
November, 2014
Berkeley/Lexington

Preface

These notes gather together what we regard as the essentials of real analysis on $\mathbb{R}^n$.

There are of course many good texts describing, on the one hand, Lebesgue measure for the real line and, on the other, general measures for abstract spaces. But we believe there is still a need for a source book documenting the rich structure of measure theory on $\mathbb{R}^n$, with particular emphasis on integration and differentiation. And so we packed into these notes all sorts of interesting topics that working mathematical analysts need to know, but are mostly not taught. These include Hausdorff measures and capacities (for classifying "negligible" sets for various fine properties of functions), Rademacher's Theorem (asserting the differentiability of Lipschitz continuous functions almost everywhere), Aleksandrov's Theorem (asserting the twice differentiability of convex functions almost everywhere), the area and coarea formulas (yielding change-of-variables rules for Lipschitz continuous maps between $\mathbb{R}^n$ and $\mathbb{R}^m$), and the Lebesgue–Besicovitch Differentiation Theorem (amounting to the fundamental theorem of calculus for real analysis).

This book is definitely not for beginners. We explicitly assume our readers are at least fairly conversant with both Lebesgue measure and abstract measure theory. The expository style reflects this expectation. We do not offer lengthy heuristics or motivation, but as compensation have tried to present *all* the technicalities of the proofs: "God is in the details."

Chapter 1 comprises a quick review of mostly standard real analysis, Chapter 2 introduces Hausdorff measures, and Chapter 3 discusses the area and coarea formulas. In Chapters 4 through 6 we analyze the fine properties of functions possessing weak derivatives of various sorts. Sobolev functions, which is to say functions having weak first partial derivatives in an L^p space, are the subject of Chapter 4; functions of bounded variation, that is, functions having measures as weak first partial derivatives, the subject of Chapter 5. Finally, Chapter 6 discusses the approximation of Lipschitz continuous, Sobolev and BV functions by C^1 functions, and several related subjects.

We have listed in the references the primary sources we have relied upon for these notes. In addition many colleagues, in particular S. Antman, J.-A. Cohen, M. Crandall, A. Damlamian, H. Ishii, N.V. Krylov, N. Owen, P. Souganidis, S. Spector, and W. Strauss, have suggested improvements and detected errors. We have also made use of S. Katzenburger's class notes. Early drafts of the manuscript were typed by E. Hampton, M. Hourihan, B. Kaufman, and J. Slack.

LCE was partially supported by NSF Grants DMS-83-01265, 86-01532, and 89-03328, and by the Institute for Physical Science and Technology at the University of Maryland. RFG was partially supported by NSF Grant DMS-87-04111 and by NSF Grant RII-86-10671 and the Commonwealth of Kentucky through the Kentucky EPSCoR program.

Warnings

Our terminology is occasionally at variance with standard usage. The principal changes are these:

- What we call a *measure* is usually called an *outer measure*.

- For us a function is *integrable* if it has an integral (which may equal $\pm\infty$).

- We call a function f *summable* if $|f|$ has a finite integral.

- We do *not* identify two L^p, BV or Sobolev functions that agree almost everywhere.

Chapter 1

General Measure Theory

This chapter is mostly a review of standard measure theory, with particular attention paid to Radon measures on $\mathbb{R}^n$.

Sections 1.1 through 1.4 are a rapid recounting of abstract measure theory. In Section 1.5 we establish Vitali's and Besicovitch's Covering Theorems, the latter being the key for the Lebesgue–Besicovitch Differentiation Theorem for Radon measures in Sections 1.6 and 1.7. Section 1.8 provides a vector-valued version of Riesz's Representation Theorem. In Section 1.9 we study weak compactness for sequences of measures and functions.

The reader should as necessary consult the Appendix for a summary of our notation.

1.1 Measures and measurable functions

1.1.1 Measures

Although we intend later to work almost exclusively in $\mathbb{R}^n$, it is most convenient to start abstractly.

Let X denote a nonempty set, and 2^X the collection of all subsets of X.

DEFINITION 1.1. *A mapping $\mu : 2^X \to [0, \infty]$ is called a* **measure** *on X provided*

(i) $\mu(\emptyset) = 0$, *and*

(ii) *if*

$$A \subseteq \bigcup_{k=1}^{\infty} A_k,$$

then

$$\mu(A) \le \sum_{k=1}^{\infty} \mu(A_k).$$

Condition (ii) is called *subadditivity*.

Warning: Most texts call such a mapping μ an *outer measure*, reserving the name *measure* for μ restricted to the collection of μ-measurable subsets of X (see below). We will see, however, that there are definite advantages to being able to "measure" even nonmeasurable sets.

DEFINITION 1.2. *Let μ be a measure on X and $C \subseteq X$. Then μ* **restricted to** *C, written*

$$\mu \llcorner C,$$

is the measure defined by

$$(\mu \llcorner C)(A) := \mu(A \cap C) \quad \text{for all } A \subseteq X.$$

DEFINITION 1.3. *A set $A \subseteq X$ is* **μ-measurable** *if for each set $B \subseteq X$ we have*

$$\mu(B) = \mu(B \cap A) + \mu(B - A).$$

THEOREM 1.1 (Elementary properties of measures). *Let μ be a measure on X.*

(i) *If $A \subseteq B \subseteq X$, then*

$$\mu(A) \le \mu(B).$$

(ii) *A set A is μ-measurable if and only if $X - A$ is μ-measurable.*

(iii) *The sets $\emptyset$ and X are μ-measurable. More generally, if $\mu(A) = 0$, then A is μ-measurable.*

(iv) *If C is any subset of X, then each μ-measurable set is also $\mu \llcorner C$-measurable.*

Proof. 1. Assertion (i) follows at once from the definition. To show (ii), assume A is μ-measurable and $B \subseteq X$. Then

$$\mu(B) = \mu(B \cap A) + \mu(B - A) = \mu(B - (X - A)) + \mu(B \cap (X - A));$$

and so $X - A$ is μ-measurable.

2. Suppose now $\mu(A) = 0$, $B \subseteq X$. Then $\mu(B \cap A) = 0$, and consequently

$$\mu(B) \geq \mu(B - A) = \mu(B \cap A) + \mu(B - A).$$

The opposite inequality is clear from subadditivity.

3. Assume A is μ-measurable, $B \subseteq X$. Then

$$
\begin{aligned}
\mu \llcorner C(B) &= \mu(B \cap C) \\
&= \mu((B \cap C) \cap A) + \mu((B \cap C) - A) \\
&= \mu((B \cap A) \cap C) + \mu((B - A) \cap C) \\
&= \mu \llcorner C(B \cap A) + \mu \llcorner C(B - A).
\end{aligned}
$$

Hence A is $\mu \llcorner C$-measurable. ☐

THEOREM 1.2 (Sequences of measurable sets). *Let $\{A_k\}_{k=1}^{\infty}$ be a sequence of μ-measurable sets.*

(i) *The sets*

$$\bigcup_{k=1}^{\infty} A_k \quad and \quad \bigcap_{k=1}^{\infty} A_k$$

are μ-measurable.

(ii) *If the sets $\{A_k\}_{k=1}^{\infty}$ are disjoint, then*

$$\mu\left(\bigcup_{k=1}^{\infty} A_k\right) = \sum_{k=1}^{\infty} \mu(A_k).$$

(iii) *If $A_1 \subseteq \ldots A_k \subseteq A_{k+1} \ldots$, then*

$$\lim_{k \to \infty} \mu(A_k) = \mu\left(\bigcup_{k=1}^{\infty} A_k\right).$$

(iv) *If $A_1 \supset \ldots A_k \supset A_{k+1} \ldots$ and $\mu(A_1) < \infty$, then*

$$\lim_{k \to \infty} \mu(A_k) = \mu\left(\bigcap_{k=1}^{\infty} A_k\right).$$

Proof. 1. Since subadditivity implies

$$\mu(B) \le \mu(B \cap A) + \mu(B - A)$$

for all $A, B \subseteq \mathbb{R}^n$, it suffices to show the opposite inequality in order to prove the set A is μ-measurable.

For each set $B \subseteq \mathbb{R}^n$,

$$\begin{aligned}
\mu(B) &= \mu(B \cap A_1) + \mu(B - A_1) \\
&= \mu(B \cap A_1) + \mu((B - A_1) \cap A_2) + \mu((B - A_1) - A_2) \\
&\ge \mu(B \cap (A_1 \cup A_2)) + \mu(B - (A_1 \cup A_2)),
\end{aligned}$$

and thus $A_1 \cup A_2$ is μ-measurable. By induction, the union of finitely many μ-measurable sets is μ-measurable.

2. Because

$$X - (A_1 \cap A_2) = (X - A_1) \cup (X - A_2),$$

the intersection of two, and thus of finitely many, μ-measurable sets is μ-measurable.

3. Assume now the sets $\{A_k\}_{k=1}^{\infty}$ are disjoint, and write

$$B_j := \bigcup_{k=1}^{j} A_k \quad (j = 1, 2, \dots).$$

Then

$$\begin{aligned}
\mu(B_{j+1}) &= \mu(B_{j+1} \cap A_{j+1}) + \mu(B_{j+1} - A_{j+1}) \\
&= \mu(A_{j+1}) + \mu(B_j) \quad (j = 1, \dots);
\end{aligned}$$

whence

$$\mu\left(\bigcup_{k=1}^{j} A_k\right) = \sum_{k=1}^{j} \mu(A_k) \quad (j = 1, \dots).$$

It follows that

$$\sum_{k=1}^{\infty} \mu(A_k) \le \mu\left(\bigcup_{k=1}^{\infty} A_k\right),$$

from which assertion (ii) follows.

4. To prove (iii), we note from (ii) that

$$\lim_{k \to \infty} \mu(A_k) = \mu(A_1) + \sum_{k=1}^{\infty} \mu(A_{k+1} - A_k) = \mu \left(\bigcup_{k=1}^{\infty} A_k \right).$$

Assertion (iv) follows from (iii), since

$$\mu(A_1) - \lim_{k \to \infty} \mu(A_k) = \lim_{k \to \infty} \mu(A_1 - A_k) = \mu \left(\bigcup_{k=1}^{\infty} (A_1 - A_k) \right)$$

$$\geq \mu(A_1) - \mu \left(\bigcap_{k=1}^{\infty} A_k \right).$$

5. Recall that if B is any subset of X, then each μ-measurable set is also $\mu \, \llcorner \, B$-measurable. Since $B_j := \cup_{k=1}^{j} A_k$ is μ-measurable by Step 1, for each $B \subseteq X$ with $\mu(B) < \infty$ we have

$$\mu \left(B \cap \bigcup_{k=1}^{\infty} A_k \right) + \mu \left(B - \bigcup_{k=1}^{\infty} A_k \right)$$

$$= (\mu \, \llcorner \, B) \left(\bigcup_{k=1}^{\infty} B_k \right) + (\mu \, \llcorner \, B) \left(\bigcap_{k=1}^{\infty} (X - B_k) \right)$$

$$= \lim_{k \to \infty} (\mu \, \llcorner \, B)(B_k) + \lim_{k \to \infty} (\mu \, \llcorner \, B)(X - B_k)$$

$$= \mu(B).$$

Thus $\cup_{k=1}^{\infty} A_k$ is μ-measurable, as is $\cap_{k=1}^{\infty} A_k$, since

$$X - \bigcap_{k=1}^{\infty} A_k = \bigcup_{k=1}^{\infty} (X - A_k).$$

This proves (i). $\qquad \square$

1.1.2 Systems of sets

We introduce next certain important classes of subsets of X.

DEFINITION 1.4. *A collection of subsets $\mathcal{A} \subseteq 2^X$ is a σ-algebra provided*

(i) $\emptyset, X \in \mathcal{A}$;

(ii) $A \in \mathcal{A}$ implies $X - A \in \mathcal{A}$;

(iii) $A_k \in \mathcal{A}$ $(k = 1, \dots)$ implies

$$\bigcup_{k=1}^{\infty} A_k \in \mathcal{A};$$

(iv) $A_k \in \mathcal{A}$ $(k = 1, \dots)$ implies

$$\bigcap_{k=1}^{\infty} A_k \in \mathcal{A}.$$

Remark. Since

$$X - \left(\bigcap_{k=1}^{\infty} A_k \right) = \bigcup_{k=1}^{\infty} (X - A_k),$$

(iv) in fact follows from (ii) and (iii). Similarly, (ii) and (iv) imply (iii). □

THEOREM 1.3 (Measurable sets as a σ-algebra). *If μ is a measure on a nonempty set X, then the collection of all μ-measurable subsets of X is a σ-algebra.*

Proof. This follows at once from Theorems 1.1 and 1.2. □

The intersection of any collection of σ-algebras is a σ-algebra, and consequently the following makes sense:

DEFINITION 1.5. *If $\mathcal{C} \subseteq 2^X$ is any nonempty collection of subsets of X, the σ-algebra generated by $\mathcal{C}$, denoted*

$$\sigma(\mathcal{C}),$$

is the smallest σ-algebra containing $\mathcal{C}$.

An important special case is when $\mathcal{C}$ is the collection of all open subsets of $\mathbb{R}^n$:

DEFINITION 1.6.

(i) *The **Borel σ-algebra** of $\mathbb{R}^n$ is the smallest σ-algebra of $\mathbb{R}^n$ containing the open subsets of $\mathbb{R}^n$.*

(ii) *A measure μ on $\mathbb{R}^n$ is called* **Borel** *if each Borel set is μ-measurable.*

Caratheodory's criterion (Theorem 1.9) will provide us with a convenient way to check that a measure is Borel.

For various applications it is convenient to introduce as well certain classes of subsets having less structure than σ-algebras.

DEFINITION 1.7. *A nonempty collection of subsets $\mathcal{P} \subseteq 2^X$ is a* **π-system** *provided*

$$A, B \in \mathcal{P} \quad \text{implies} \quad A \cap B \in \mathcal{P}.$$

So a π-system is simply a collection of subsets closed under finite intersections.

DEFINITION 1.8. *A collection of subsets $\mathcal{L} \subseteq 2^X$ is a* **λ-system** *provided*

(i) $X \in \mathcal{L}$;

(ii) $A, B \in \mathcal{L}$ and $B \subseteq A$ implies $A - B \in \mathcal{L}$;

(iii) if $A_k \in \mathcal{L}$ and $A_k \subseteq A_{k+1}$ for $k = 1, \dots$, then

$$\bigcup_{k=1}^{\infty} A_k \in \mathcal{L}.$$

Since both π-systems and λ-systems have less stringent properties than σ-algebras, it will be easy in applications to check that various interesting collections of sets are indeed π- or λ-systems. The following will then provide a link back to σ-algebras:

THEOREM 1.4 (π–λ Theorem). *If $\mathcal{P}$ is a π-system and $\mathcal{L}$ is a λ-system with*

$$\mathcal{P} \subseteq \mathcal{L},$$

then

$$\sigma(\mathcal{P}) \subseteq \mathcal{L}.$$

Proof. 1. Define

$$\mathcal{S} := \bigcap_{\mathcal{L}' \supseteq \mathcal{P}} \mathcal{L}',$$

the intersection of all λ-systems $\mathcal{L}'$ containing $\mathcal{P}$. Clearly $\mathcal{P} \subseteq \mathcal{S} \subseteq \mathcal{L}$, and it is easy to check that $\mathcal{S}$ is itself a λ-system.

2. *Claim #1*: $\mathcal{S}$ is a π-system.

Proof of claim: Select any $A, B \in \mathcal{S}$; we must show $A \cap B \in \mathcal{S}$. Define

$$\mathcal{A} := \{C \subseteq X \mid A \cap C \in \mathcal{S}\}.$$

Since $\mathcal{S}$ is a λ-system, it follows that $\mathcal{A}$ is a λ-system. Therefore $\mathcal{S} \subseteq \mathcal{A}$. But then since $B \in \mathcal{S}$, we see that $A \cap B \in \mathcal{S}$.

3. *Claim #2*: $\mathcal{S}$ is a σ-algebra.

Proof of claim: This will follow since $\mathcal{S}$ is both a λ- and a π-system. Since $X \in \mathcal{S}$, it follows that $\emptyset = X - X \in \mathcal{S}$. Clearly $A \in \mathcal{S}$ implies $X - A \in \mathcal{S}$. Since $\mathcal{S}$ is closed under complements and under finite intersections, it is closed under finite unions. Hence if $A_1, A_2, \cdots \in \mathcal{S}$, then $B_n := \cup_{k=1}^{n} A_k \in \mathcal{S}$. As $\mathcal{S}$ is a λ-system, we see that therefore $\cup_{k=1}^{\infty} A_k \in \mathcal{S}$. Thus $\mathcal{S}$ is a σ-algebra.

4. Since $\mathcal{S} \supseteq \mathcal{P}$ is a σ-algebra, it follows that

$$\sigma(\mathcal{P}) \subseteq \mathcal{S} \subseteq \mathcal{L}. \qquad \qquad \square$$

As a first application, we show that finite Borel measures in $\mathbb{R}^n$ are uniquely determined by their values on closed "rectangles" with sides parallel to the coordinate axes:

THEOREM 1.5 (Borel measures and rectangles). *Let μ and ν be two finite Borel measures on $\mathbb{R}^n$ such that*

$$\mu(R) = \nu(R)$$

for all closed "rectangles"

$$R := \{x \in \mathbb{R}^n \mid -\infty \leq a_i \leq x_i \leq b_i \leq \infty \ (i = 1, \ldots, n)\}.$$

Then

$$\mu(B) = \nu(B)$$

for all Borel sets $B \subseteq \mathbb{R}^n$.

Proof. We apply the π-λ Theorem with

$$\mathcal{P} := \{R \subseteq \mathbb{R}^n \mid R \text{ is a rectangle}\}$$

and
$$\mathcal{L} := \{B \subseteq \mathbb{R}^n \mid B \text{ is Borel}, \mu(B) = \nu(B)\}.$$
Then $\mathcal{P} \subseteq \mathcal{L}$, $\mathcal{P}$ is clearly a π-system, and we check that $\mathcal{L}$ is a λ-system. Consequently, the π-λ Theorem implies $\sigma(\mathcal{P}) \subseteq \mathcal{L}$. But $\sigma(\mathcal{P})$ comprises the Borel sets, since each open subset of $\mathbb{R}^n$ can be written as a countable union of closed rectangles. $\qquad\square$

Remark. This proof illustrates the usefulness of λ-systems. It is not so clear that $\{B \text{ Borel} \mid \mu(B) = \nu(B)\}$ is a σ-algebra, since it is not obviously closed under intersections. $\qquad\square$

1.1.3 Approximation by open and compact sets

Next we introduce certain classes of measures that admit good approximations of various types.

DEFINITION 1.9.

(i) *A measure μ on X is **regular** if for each set $A \subseteq X$ there exists a μ-measurable set B such that $A \subseteq B$ and $\mu(A) = \mu(B)$.*

(ii) *A measure μ on $\mathbb{R}^n$ is **Borel regular** if μ is Borel and for each $A \subseteq \mathbb{R}^n$ there exists a Borel set B such that $A \subseteq B$ and $\mu(A) = \mu(B)$.*

(iii) *A measure μ on $\mathbb{R}^n$ is a **Radon** measure if μ is Borel regular and $\mu(K) < \infty$ for each compact set $K \subset \mathbb{R}^n$.*

THEOREM 1.6 (Increasing sets). *Let μ be a regular measure on X. If $A_1 \subseteq \ldots A_k \subseteq A_{k+1} \ldots$, then*

$$\lim_{k \to \infty} \mu(A_k) = \mu\left(\bigcup_{k=1}^{\infty} A_k\right).$$

Remark. An important point is that the sets $\{A_k\}_{k=1}^{\infty}$ need not be μ-measurable here. $\qquad\square$

Proof. Since μ is regular, there exist measurable sets $\{C_k\}_{k=1}^{\infty}$, with $A_k \subseteq C_k$ and $\mu(A_k) = \mu(C_k)$ for each k. Set $B_k := \cap_{j \geq k} C_j$. Then $A_k \subseteq B_k$, each B_k is μ-measurable, and $\mu(A_k) = \mu(B_k)$. Thus

$$\lim_{k \to \infty} \mu(A_k) = \lim_{k \to \infty} \mu(B_k) = \mu\left(\bigcup_{k=1}^{\infty} B_k\right) \geq \mu\left(\bigcup_{k=1}^{\infty} A_k\right).$$

But $A_k \subseteq \cup_{j=1}^{\infty} A_j$, and so also

$$\lim_{k \to \infty} \mu(A_k) \leq \mu \left(\bigcup_{j=1}^{\infty} A_j \right). \qquad \Box$$

We demonstrate next that if μ is Borel regular, we can create a Radon measure by restricting μ to a measurable set of finite measure.

THEOREM 1.7 (Restriction and Radon measures). *Let μ be a Borel regular measure on $\mathbb{R}^n$. Suppose $A \subseteq \mathbb{R}^n$ is μ-measurable and $\mu(A) < \infty$.*

Then $\mu \llcorner A$ is a Radon measure.

Remark. If A is a Borel set, then $\mu \llcorner A$ is Borel regular, even if $\mu(A) = \infty$. $\qquad \Box$

Proof. 1. Let $\nu := \mu \llcorner A$. Clearly $\nu(K) < \infty$ for each compact K. Since Theorem 1.1, (iv) asserts that every μ-measurable set is ν-measurable, ν is a Borel measure.

2. *Claim: ν is Borel regular.*

Proof of claim: Since μ is Borel regular, there exists a Borel set B such that $A \subseteq B$ and $\mu(A) = \mu(B) < \infty$. Then, since A is μ-measurable,

$$\mu(B - A) = \mu(B) - \mu(A) = 0.$$

Choose $C \subseteq \mathbb{R}^n$. Then

$$\begin{aligned}
(\mu \llcorner B)(C) &= \mu(C \cap B) \\
&= \mu(C \cap B \cap A) + \mu((C \cap B) - A) \\
&\leq \mu(C \cap A) + \mu(B - A) \\
&= (\mu \llcorner A)(C).
\end{aligned}$$

Thus $\mu \llcorner B = \mu \llcorner A$, and so we may as well assume A is a Borel set.

3. Now let $C \subseteq \mathbb{R}^n$ We must show that there exists a Borel set D such that $C \subseteq D$ and $\nu(C) = \nu(D)$. Since μ is a Borel regular measure, there exists a Borel set E such that $A \cap C \subseteq E$ and $\mu(E) = \mu(A \cap C)$. Let $D := E \cup (\mathbb{R}^n - A)$. Since A and E are Borel sets, so is D. Moreover, $C \subseteq (A \cap C) \cup (\mathbb{R}^n - A) \subseteq D$. Finally, since $D \cap A = E \cap A$,

$$\nu(D) = \mu(D \cap A) = \mu(E \cap A) \leq \mu(E) = \mu(A \cap C) = \nu(C). \qquad \Box$$

We consider next the possibility of measure theoretically approximating by open, closed or compact sets.

LEMMA 1.1. *Let μ be a Borel measure on $\mathbb{R}^n$ and let B be a Borel set.*

(i) *If $\mu(B) < \infty$, there exists for each $\epsilon > 0$ a closed set C such that*

$$C \subseteq B, \; \mu(B - C) < \epsilon.$$

(ii) *If μ is a Radon measure, then there exists for each $\epsilon > 0$ an open set U such that*

$$B \subseteq U, \; \mu(U - B) < \epsilon.$$

Proof. 1. Let $\nu := \mu \mathbin{\llcorner} B$. Since μ is Borel and $\mu(B) < \infty$, ν is a finite Borel measure. Let

$$\mathcal{F} := \{A \subseteq \mathbb{R}^n \mid A \text{ is } \mu \text{ -measurable and for each } \epsilon > 0$$
$$\text{there exists a closed set } C \subseteq A \text{ with } \nu(A - C) < \epsilon\}.$$

Obviously, $\mathcal{F}$ contains all closed sets.

2. *Claim #1:* If $\{A_i\}_{i=1}^{\infty} \subseteq \mathcal{F}$, then $A := \cap_{i=1}^{\infty} A_i \in \mathcal{F}$.

Proof of claim: Fix $\epsilon > 0$. Since $A_i \in \mathcal{F}$, there exists a closed set $C_i \subseteq A_i$ with $\nu(A_i - C_i) < \frac{\epsilon}{2^i}$ $(i = 1, 2, \dots)$. Let $C := \cap_{i=1}^{\infty} C_i$. Then C is closed and

$$\nu(A - C) = \nu \left(\bigcap_{i=1}^{\infty} A_i - \bigcap_{i=1}^{\infty} C_i \right)$$
$$\leq \nu \left(\bigcup_{i=1}^{\infty} (A_i - C_i) \right)$$
$$\leq \sum_{i=1}^{\infty} \nu(A_i - C_i) < \epsilon.$$

Thus $A \in \mathcal{F}$.

3. *Claim #2:* If $\{A_i\}_{i=1}^{\infty} \subseteq \mathcal{F}$, then $A := \cup_{i=1}^{\infty} A_i \in \mathcal{F}$.

Proof of claim: Fix $\epsilon > 0$ and choose C_i as above. Since $\nu(A) < \infty$, we have

$$\lim_{x \to \infty} \nu \left(A - \bigcup_{i=1}^{m} C_i \right) = \nu \left(\bigcup_{i=1}^{\infty} A_i - \bigcup_{i=1}^{\infty} C_i \right)$$

$$\leq \nu \left(\bigcup_{i=1}^{\infty} (A_i - C_i) \right)$$

$$\leq \sum_{i=1}^{\infty} \nu(A_i - C_i) < \epsilon.$$

Consequently, there exists an integer m such that

$$\nu \left(A - \bigcup_{i=1}^{m} C_i \right) < \epsilon.$$

But $\cup_{i=1}^{m} C_i$ is closed, and so $A \in \mathcal{F}$.

4. Since every open subset of $\mathbb{R}^n$ can be written as a countable union of closed sets, Claim #2 shows that $\mathcal{F}$ contains all open sets. Consider next

$$\mathcal{G} := \{ A \in \mathcal{F} \mid \mathbb{R}^n - A \in \mathcal{F} \}.$$

Trivially, if $A \in \mathcal{G}$, then $\mathbb{R}^n - A \in \mathcal{G}$. Note also that $\mathcal{G}$ contains all open sets.

5. *Claim #3*: If $\{A_i\}_{i=1}^{\infty} \subseteq \mathcal{G}$, then $A = \cup_{i=1}^{\infty} A_i \in \mathcal{G}$.

Proof of claim: By Claim #2, $A \in \mathcal{F}$. Since also $\{\mathbb{R}^n - A_i\}_{i=1}^{\infty} \subseteq \mathcal{F}$, Claim #1 implies $\mathbb{R}^n - A = \cap_{i=1}^{\infty} (\mathbb{R}^n - A_i) \in \mathcal{F}$.

6. Thus $\mathcal{G}$ is a σ-algebra containing the open sets and therefore also the Borel sets. In particular, $B \in \mathcal{G}$; and hence, given $\epsilon > 0$, there is a closed set $C \subseteq B$ such that

$$\mu(B - C) = \nu(B - C) < \epsilon.$$

This establishes (i).

7. Write $U_m := B^0(0, m)$, the open ball with center 0, radius m. Then $U_m - B$ is a Borel set with $\mu(U_m - B) < \infty$, and so we can apply (i) to find a closed set $C_m \subseteq U_m - B$ such that $\mu((U_m - C_m) - B) = \mu((U_m - B) - C_m) < \frac{\epsilon}{2^m}$.

Let $U := \cup_{m=1}^{\infty} (U_m - C_m)$; U is open. Now $B \subseteq \mathbb{R}^n - C_m$ and thus $U_m \cap B \subseteq U_m - C_m$. Consequently,

$$B = \bigcup_{m=1}^{\infty} (U_m \cap B) \subseteq \bigcup_{m=1}^{\infty} (U_m - C_m) = U.$$

Furthermore,

$$\mu(U-B) = \mu\left(\bigcup_{m=1}^{\infty}(U_m - C_m) - B\right) \leq \sum_{m=1}^{\infty}\mu((U_m - C_m) - B) < \epsilon.$$

$\square$

THEOREM 1.8 (Approximation by open and by compact sets). *Let μ be a Radon measure on $\mathbb{R}^n$. Then*

(i) *for each set $A \subseteq \mathbb{R}^n$,*

$$\mu(A) = \inf\{\mu(U) \mid A \subseteq U, U\, open\},$$

and

(ii) *for each μ-measurable set $A \subseteq \mathbb{R}^n$,*

$$\mu(A) = \sup\{\mu(K) \mid K \subseteq A, K\, compact\}.$$

Remark. Assertion (i) does not require A to be μ-measurable. $\square$

Proof. 1. If $\mu(A) = \infty$, (i) is obvious, and so let us suppose $\mu(A) < \infty$. Assume first A is a Borel set. Fix $\epsilon > 0$. Then by Lemma 1.1, there exists an open set $U \supset A$ with $\mu(U - A) < \epsilon$. Since $\mu(U) = \mu(A) + \mu(U - A) < \infty$, (i) holds.

Now, let A be an arbitrary set. Since μ is Borel regular, there exists a Borel set $B \supset A$ with $\mu(A) = \mu(B)$. Then

$$\mu(A) = \mu(B) = \inf\{\mu(U) \mid B \subseteq U, U \text{ open}\}$$
$$\geq \inf\{\mu(U) \mid A \subseteq U, U \text{ open}\}.$$

The reverse inequality is clear, and so assertion (i) is proved.

2. Now let A be μ-measurable, with $\mu(A) < \infty$. Set $\nu := \mu \llcorner A$; then ν is a Radon measure according to Theorem 1.7. Fix $\epsilon > 0$. Applying (i) to ν and $\mathbb{R}^n - A$, we obtain an open set U with $\mathbb{R}^n - A \subseteq U$ and $\nu(U) \leq \epsilon$. Let $C := \mathbb{R}^n - U$. Then C is closed and $C \subseteq A$. Moreover,

$$\mu(A - C) = \nu(\mathbb{R}^n - C) = \nu(U) \leq \epsilon.$$

Thus

$$0 \leq \mu(A) - \mu(C) \leq \epsilon,$$

and so

$$\mu(A) = \sup\{\mu(C) \mid C \subseteq A, C \text{ closed}\}. \qquad (\star)$$

3. Suppose that $\mu(A) = \infty$. Define $D_k := \{x \mid k - 1 \leq |x| < k\}$. Then $A = \cup_{k=1}^{\infty}(D_k \cap A)$; so $\infty = \mu(A) = \sum_{k=1}^{\infty} \mu(A \cap D_k)$. Since μ is a Radon measure, $\mu(D_k \cap A) < \infty$. Then by the above, there exists a closed set $C_k \subseteq D_k \cap A$ with $\mu(C_k) \geq \mu(D_k \cap A) - \frac{1}{2^k}$. Now $\cup_{k=1}^{\infty} C_k \subseteq A$ and

$$\lim_{n \to \infty} \mu\left(\bigcup_{k=1}^{n} C_k\right) = \mu\left(\bigcup_{k=1}^{\infty} C_k\right)$$

$$= \sum_{k=1}^{\infty} \mu(C_k) \geq \sum_{k=1}^{\infty} \left(\mu(D_k \cap A) - \frac{1}{2^k}\right) = \infty.$$

But $\cup_{k=1}^{n} C_k$ is closed for each n, whence in this case we also have assertion $(\star)$.

4. Finally, let $B(m)$ denote the closed ball with center 0, radius m. Let C be closed, $C_m := C \cap B(m)$. Each set C_m is compact and $\mu(C) = \lim_{m \to \infty} \mu(C_m)$. Hence for each μ-measurable set A,

$$\sup\{\mu(K) \mid K \subseteq A, K \text{ compact}\} = \sup\{\mu(C) \mid C \subseteq A, C \text{ closed}\}. \quad \square$$

We introduce next a simple and very useful way to verify that a measure is Borel.

THEOREM 1.9 (Caratheodory's criterion). *Let μ be a measure on $\mathbb{R}^n$. If for all sets A, $B \subseteq \mathbb{R}^n$, we have*

$$\mu(A \cup B) = \mu(A) + \mu(B) \quad \text{whenever } \operatorname{dist}(A, B) > 0,$$

then μ is a Borel measure.

Proof. 1. Suppose $A, C \subseteq \mathbb{R}^n$ and C is closed. We must show

$$\mu(A) \geq \mu(A \cap C) + \mu(A - C), \qquad (\star)$$

the opposite inequality following from subadditivity.

If $\mu(A) = \infty$, then $(\star)$ is obvious. Assume instead $\mu(A) < \infty$. Define

$$C_n := \left\{x \in \mathbb{R}^n \mid \operatorname{dist}(x, C) \leq \frac{1}{n}\right\} \quad (n = 1, 2, \dots).$$

Then $\operatorname{dist}(A - C_n, A \cap C) \geq \frac{1}{n} > 0$. Therefore, by hypothesis,

$$\mu(A - C_n) + \mu(A \cap C) = \mu((A - C_n) \cup (A \cap C)) \leq \mu(A). \qquad (\star\star)$$

2. *Claim*: $\lim_{n \to \infty} \mu(A - C_n) = \mu(A - C)$.

Proof of claim: Set

$$R_k := \left\{ x \in A \mid \frac{1}{k+1} < \operatorname{dist}(x, C) \leq \frac{1}{k} \right\} \qquad (k = 1, \ldots).$$

Since C is closed, $A - C = (A - C_n) \cup \cup_{k=n}^{\infty} R_k$; consequently,

$$\mu(A - C_n) \leq \mu(A - C) \leq \mu(A - C_n) + \sum_{k=n}^{\infty} \mu(R_k).$$

If we can show $\sum_{k=1}^{\infty} \mu(R_k) < \infty$, we will then have

$$\lim_{n \to \infty} \mu(A - C_n) \leq \mu(A - C)$$

$$\leq \lim_{n \to \infty} \mu(A - C_n) + \lim_{n \to \infty} \sum_{k=n}^{\infty} \mu(R_k)$$

$$= \lim_{n \to \infty} \mu(A - C_n),$$

thereby establishing the claim.

3. Now $\operatorname{dist}(R_i, R_j) > 0$ if $j \geq i + 2$. Hence by induction we find

$$\sum_{k=1}^{m} \mu(R_{2k}) = \mu \left(\bigcup_{k=1}^{m} R_{2k} \right) \leq \mu(A),$$

and likewise

$$\sum_{k=0}^{m} \mu(R_{2k+1}) = \mu \left(\bigcup_{k=0}^{m} R_{2k+1} \right) \leq \mu(A).$$

Combining these results and letting $m \to \infty$, we discover

$$\sum_{k=1}^{\infty} \mu(R_k) \leq 2\mu(A) < \infty.$$

4. We therefore have

$$\mu(A - C) + \mu(A \cap C) = \lim_{n \to \infty} \mu(A - C_n) + \mu(A \cap C) \leq \mu(A),$$

according to $(\star\star)$. This proves $(\star)$ and thus the closed set C is μ-measurable. $\qquad \square$

1.1.4   Measurable functions

We now extend the notion of measurability from sets to functions.

Let X be a set and Y a topological space. Assume μ is a measure on X.

DEFINITION 1.10.

(i) *A function $f : X \to Y$ is called μ-measurable if for each open set $U \subseteq Y$, the set*

$$f^{-1}(U)$$

is μ-measurable.

(ii) *A function $f : X \to Y$ is Borel measurable if for each open set $U \subseteq Y$, the set*

$$f^{-1}(U)$$

is Borel measurable.

EXAMPLE. If $f : \mathbb{R}^n \to Y$ is continuous, then f is Borel-measurable. This follows since $f^{-1}(U)$ is open, and therefore μ-measurable, for each open set $U \subseteq Y$.   □

THEOREM 1.10 (Inverse images).

(i) *If $f : X \to Y$ is μ-measurable, then $f^{-1}(B)$ is μ-measurable for each Borel set $B \subseteq Y$.*

(ii) *A function $f : X \to [-\infty, \infty]$ is μ-measurable if and only if $f^{-1}([-\infty, a))$ is μ-measurable for each $a \in \mathbb{R}$.*

(iii) *If $f : X \to \mathbb{R}^n$ and $g : X \to \mathbb{R}^m$ are μ-measurable, then*

$$(f, g) : X \to \mathbb{R}^{n+m}$$

is μ-measurable.

Proof. 1. We check that

$$\{A \subseteq Y \mid f^{-1}(A) \text{ is } \mu\text{-measurable}\}$$

is a σ-algebra containing the open sets and hence the Borel sets.

2. Likewise,

$$\{A \subseteq [-\infty, \infty] \mid f^{-1}(A) \text{ is } \mu\text{-measurable}\}$$

is a σ-algebra containing $[-\infty, a)$ for each $a \in \mathbb{R}$, and therefore containing the Borel subsets of $\mathbb{R}$.

3. Let $h := (f, g)$. Then

$$\{A \subseteq \mathbb{R}^{n+m} \mid h^{-1}(A) \text{ is } \mu\text{-measurable}\}$$

is a σ-algebra containing all open sets of the form $U \times V$, where $U \subseteq \mathbb{R}^n$ and $V \subseteq \mathbb{R}^m$ are open. $\qquad\qquad\qquad\qquad\qquad\qquad\square$

Measurable functions inherit the good properties of measurable sets:

THEOREM 1.11 (Properties of measurable functions).

(i) *If $f, g : X \to [-\infty, \infty]$ are μ-measurable, then so are*

$$f + g, \ fg, \ |f|, \ \min(f, g) \ and \ \max(f, g).$$

The function $\frac{f}{g}$ is also μ-measurable, provided $g \neq 0$ on X.

(ii) *If the functions $f_k : X \to [-\infty, \infty]$ are μ-measurable ($k = 1, 2, \ldots$), then*

$$\inf_{k \geq 1} f_k, \ \sup_{k \geq 1} f_k, \ \liminf_{k \to \infty} f_k, \ and \ \limsup_{k \to \infty} f_k$$

are also μ-measurable.

Proof. 1. As noted above, $f : X \to [-\infty, \infty]$ is μ-measurable if and only if $f^{-1}[-\infty, a]$ is μ-measurable for each $a \in \mathbb{R}$.

2. Suppose $f, g : X \to \mathbb{R}$ are μ-measurable, Then

$$(f + g)^{-1}(-\infty, a) = \bigcup_{\substack{r, s \text{ rational} \\ r + s < a}} (f^{-1}(-\infty, r) \cap g^{-1}(-\infty, s)),$$

and so $f + g$ is μ-measurable. Since

$$(f^2)^{-1}(-\infty, a) = f^{-1}(-\infty, a^{\frac{1}{2}}) - f^{-1}(-\infty, -a^{\frac{1}{2}}],$$

for $a \geq 0$, f^2 is μ-measurable. Consequently,

$$fg = \frac{1}{2}[(f + g)^2 - f^2 - g^2]$$

is μ-measurable as well. Next observe that if $g \neq 0$,

$$\left(\tfrac{1}{g}\right)^{-1}(-\infty, a) = \begin{cases} g^{-1}(a^{-1}, 0) & \text{if } a < 0 \\ g^{-1}(-\infty, 0) & \text{if } a = 0 \\ g^{-1}(-\infty, 0) \cup g^{-1}(a^{-1}, \infty) & \text{if } a > 0; \end{cases}$$

thus $\frac{1}{g}$ and so also $\frac{f}{g}$ are μ-measurable.

3. Finally,

$$f^+ = f\chi_{\{f \geq 0\}} = \max(f, 0), \quad f^- = -f\chi_{\{f < 0\}} = \max(-f, 0)$$

are μ-measurable, and consequently so are

$$|f| = f^+ + f^-,$$
$$\max(f, g) = (f - g)^+ + g,$$
$$\min(f, g) = -(f - g)^- + g.$$

4. Suppose next the functions $f_k \colon X \to [-\infty, \infty]$ $(k = 1, 2, \dots)$ are μ-measurable. Then

$$\left(\inf_{k \geq 1} f_k\right)^{-1} [-\infty, a] = \bigcup_{k=1}^{\infty} f_k^{-1}[-\infty, a)$$

and

$$\left(\sup_{k \geq 1} f_k\right)^{-1} [-\infty, a] = \bigcap_{k=1}^{\infty} f_k^{-1}[-\infty, a].$$

Therefore

$$\inf_{k \geq 1} f_k, \quad \sup_{k \geq 1} f_k$$

are μ- measurable.

5. We complete the proof by noting that

$$\liminf_{k \to \infty} f_k = \sup_{m \geq 1} \inf_{k \geq m} f_k, \quad \limsup_{k \to \infty} f_k = \inf_{m \geq 1} \sup_{k \geq m} f_k. \qquad \square$$

Next is an elegant and quite useful way to rewrite a nonnegative measurable function.

THEOREM 1.12 (Decomposition of nonnegative measurable functions). *Assume that $f : X \to [0, \infty]$ is μ-measurable. Then there exist μ-measurable sets $\{A_k\}_{k=1}^{\infty}$ in X such that*

$$f = \sum_{k=1}^{\infty} \frac{1}{k} \chi_{A_k}.$$

Observe that the sets $\{A_k\}_{k=1}^{\infty}$ need not be disjoint and that this assertion is valid even if f is not a simple function.

Proof. Set

$$A_1 := \{x \in X \mid f(x) \geq 1\},$$

and inductively define for $k = 2, 3, \ldots$

$$A_k := \left\{ x \in X \mid f(x) \geq \frac{1}{k} + \sum_{j=1}^{k-1} \frac{1}{j} \chi_{A_j} \right\}.$$

An induction argument shows that

$$f \geq \sum_{k=1}^{m} \frac{1}{k} \chi_{A_k} \quad (m = 1, \ldots);$$

and therefore

$$f \geq \sum_{k=1}^{\infty} \frac{1}{k} \chi_{A_k}.$$

If $f(x) = \infty$, then $x \in A_k$ for all k. If instead $0 \leq f(x) < \infty$, then for infinitely many n, $x \notin A_n$. Hence for infinitely many n,

$$0 \leq f(x) - \sum_{k=1}^{n-1} \frac{1}{k} \chi_{A_k} \leq \frac{1}{n}. \qquad \square$$

1.2 Lusin's and Egoroff's Theorems

THEOREM 1.13 (Extending continuous functions). *Suppose $K \subseteq \mathbb{R}^n$ is compact and $f : K \to \mathbb{R}^m$ is continuous. Then there exists a continuous mapping $\bar{f} : \mathbb{R}^n \to \mathbb{R}^m$ such that*

$$\bar{f} = f \text{ on } K.$$

Remark. Extension theorems preserving more of the structure of f will be presented in Sections 3.1, 4.4, 5.4, and 6.5.   □

Proof. 1. The assertion for $m > 1$ follows easily from the case $m = 1$, and so we may assume $f : K \to \mathbb{R}$.

Let $U := \mathbb{R}^n - K$. For $x \in U$ and $s \in K$, set

$$u_s(x) := \max\left\{2 - \frac{|x - s|}{\text{dist}(x, K)}, 0\right\}.$$

Then

$$\begin{cases} x \mapsto u_s(x) \text{ is continuous on } U, \\ 0 \le u_s(x) \le 1, \\ u_s(x) = 0 \text{ if } |x - s| \ge 2\,\text{dist}(x, K). \end{cases}$$

Now let $\{s_j\}_{j=1}^\infty$ be a countable dense subset of K, and define

$$\sigma(x) := \sum_{j=1}^\infty 2^{-j} u_{s_j}(x) \quad \text{for } x \in U.$$

Observe $0 < \sigma(x) \le 1$ for $x \in U$. Next, set

$$v_k(x) := \frac{2^{-k} u_{s_k}(x)}{\sigma(x)}$$

for $x \in U$, $k = 1, 2, \ldots$. The functions $\{v_k\}_{k=1}^\infty$ form a partition of unity on U. Define

$$\bar{f}(x) := \begin{cases} f(x) & \text{if } x \in K \\ \sum_{k=1}^\infty v_k(x) f(s_k) & \text{if } x \in U. \end{cases}$$

According to the Weierstrass M-test, $\bar{f}$ is continuous on U.

2. We must show

$$\lim_{x \to a, x \in U} \bar{f}(x) = f(a)$$

for each $a \in K$. Fix $\epsilon > 0$. There exists $\delta > 0$ such that

$$|f(a) - f(s_k)| < \epsilon$$

for all s_k such that $|a - s_k| < \delta$. Suppose $x \in U$ with $|x - a| < \frac{\delta}{4}$. If $|a - s_k| \geq \delta$, then

$$\delta \leq |a - s_k| \leq |a - x| + |x - s_k| < \frac{\delta}{4} + |x - s_k|,$$

so that

$$|x - s_k| \geq \frac{3}{4}\delta > 2|x - a| \geq 2\,\mathrm{dist}(x, K).$$

Thus, $v_k(x) = 0$ whenever $|x - a| < \frac{\delta}{4}$ and $|a - s_k| \geq \delta$. Since

$$\sum_{k=1}^{\infty} v_k(x) = 1$$

if $x \in U$, we calculate for $|x - a| < \frac{\delta}{4}$, $x \in U$, that

$$|\bar{f}(x) - f(a)| \leq \sum_{k=1}^{\infty} v_k(x)|f(s_k) - f(x)| < \epsilon. \qquad \square$$

We now show that a measurable function can measure theoretically approximated by a continuous function.

THEOREM 1.14 (Lusin's Theorem). *Let μ be a Borel regular measure on $\mathbb{R}^n$ and $f : \mathbb{R}^n \to \mathbb{R}^m$ be μ-measurable. Assume that $A \subseteq \mathbb{R}^n$ is μ-measurable and $\mu(A) < \infty$. Fix $\epsilon > 0$.*

Then there exists a compact set $K \subseteq A$ such that

(i) *$\mu(A - K) < \epsilon$, and*

(ii) *$f|_K$ is continuous.*

Proof. 1. For each positive integer i, let $B_{ij}{}_{j=1}^{\infty} \subset \mathbb{R}^m$ be disjoint Borel sets such that $\mathbb{R}^m = \cup_{j=1}^{\infty} B_{ij}$ and $\mathrm{diam}\, B_{ij} < \frac{1}{i}$. Define $A_{ij} := A \cap f^{-1}(B_{ij})$. Then A_{ij} is μ-measurable and $A = \cup_{j=1}^{\infty} A_{ij}$.

2. Write $\nu := \mu \llcorner A$; ν is a Radon measure. Theorem 1.8 implies the existence of a compact set $K_{ij} \subseteq A_{ij}$ with $\nu(A_{ij} - K_{ij}) < \frac{\epsilon}{2^{i+j}}$. Then

$$\mu\left(A - \bigcup_{j=1}^{\infty} K_{ij}\right) = \nu\left(A - \bigcup_{j=1}^{\infty} K_{ij}\right)$$

$$= \nu \left(\bigcup_{j=1}^{\infty} A_{ij} - \bigcup_{j=1}^{\infty} K_{ij} \right)$$

$$\leq \nu \left(\bigcup_{j=1}^{\infty} (A_{ij} - K_{ij}) \right) < \frac{\epsilon}{2^i}.$$

Since $\lim_{N \to \infty} \mu \left(A - \cup_{j=1}^{N} K_{ij} \right) = \mu \left(A - \cup_{j=1}^{\infty} K_{ij} \right)$, there exists a number $N(i)$ such that

$$\mu \left(A - \bigcup_{j=1}^{N(i)} K_{ij} \right) < \frac{\epsilon}{2^i}.$$

3. Set $D_i := \cup_{j=1}^{N(i)} K_{ij}$; then D_i is compact. For each i and j, we fix $b_{ij} \in B_{ij}$ and we then define $g_i : D_i \to \mathbb{R}^m$ by setting $g_i(x) = b_{ij}$ for $x \in K_{ij}$ $(j \leq N(i))$. Since $K_{i1}, \ldots, K_{iN(i)}$ are compact, disjoint sets, and thus are a positive distance apart, g_i is continuous. Furthermore, $|f(x) - g_i(x)| < \frac{1}{i}$ for all $x \in D_i$. Set $K := \cap_{i=1}^{\infty} D_i$. Then K is compact and

$$\mu(A - K) \leq \sum_{i=1}^{\infty} \mu(A - D_i) < \epsilon.$$

Since $|f(x) - g_i(x)| < \frac{1}{i}$ for each $x \in D_i$, we see $g_i \to f$ uniformly on K. Thus $f|_K$ is continuous, as required. $\qquad\square$

THEOREM 1.15 (Approximation by continuous functions).
Let μ be a Borel regular measure on $\mathbb{R}^n$ and suppose that $f : \mathbb{R}^n \to \mathbb{R}^m$ is μ-measurable. Assume $A \subset \mathbb{R}^n$ is μ-measurable and $\mu(A) < \infty$. Fix $\epsilon > 0$.

Then there exists a continuous function $\bar{f} : \mathbb{R}^n \to \mathbb{R}^m$ such that

$$\mu(\{x \in A \mid \bar{f}(x) \neq f(x)\}) < \epsilon.$$

Proof. According to Lusin's Theorem, there exists a compact set $K \subseteq A$ such that $\mu(A - K) < \epsilon$ and $f|_K$ is continuous. Then by Theorem 1.13 there exists a continuous function $\bar{f} : \mathbb{R}^n \to \mathbb{R}^m$ such that $\bar{f}|_K = f|_K$ and

$$\mu\{x \in A \mid \bar{f}(x) \neq f(x)\} \leq \mu(A - K) < \epsilon. \qquad\square$$

Remark. Compare this with Whitney's Extension Theorem 6.10, which identifies conditions ensuring the existence of a C^1 extension $\bar{f}$. □

NOTATION The expression

$$\mu\text{-a.e.}$$

means "almost everywhere with respect the measure μ," that is, except possibly on a set A with $\mu(A) = 0$.

THEOREM 1.16 (Egoroff's Theorem). *Let μ be a measure on $\mathbb{R}^n$ and suppose $f_k : \mathbb{R}^n \to \mathbb{R}^m$ $(k = 1, 2, \dots)$ are μ-measurable. Assume also $A \subset \mathbb{R}^n$ is μ-measurable, with $\mu(A) < \infty$, and*

$$f_k \to f \quad \mu\text{-a.e. on } A.$$

Then for each $\epsilon > 0$ there exists a μ-measurable set $B \subseteq A$ such that

(i) $\mu(A - B) < \epsilon$, *and*

(ii) $f_k \to f$ *uniformly on B.*

Proof. For $i, j = 1, 2, \dots$ define

$$C_{ij} := \bigcup_{k=j}^{\infty} \{x \mid |f_k(x) - f(x)| > 2^{-i}\}.$$

Then $C_{i,j+1} \subseteq C_{ij}$ for all i, j; and so, since $\mu(A) < \infty$,

$$\lim_{j \to \infty} \mu(A \cap C_{ij}) = \mu\left(A \cap \bigcap_{j=1}^{\infty} C_{ij}\right) = 0.$$

Hence there exists an integer $N(i)$ such that $\mu(A \cap C_{i,N(i)}) < \epsilon 2^{-i}$. Let $B := A - \cup_{i=1}^{\infty} C_{i,N(i)}$. Then

$$\mu(A - B) \leq \sum_{i=1}^{\infty} \mu\left(A \cap C_{i,N(i)}\right) < \epsilon.$$

Then for each i, each $x \in B$, and all $n \geq N(i)$, we have $|f_n(x) - f(x)| \leq 2^{-i}$. Consequently $f_n \to f$ uniformly on B. □

1.3   Integrals and limit theorems

Now we want to extend calculus to the measure theoretic setting. This section presents integration theory; differentiation theory is harder and will be set forth later in Section 1.6.

NOTATION $f^+ = \max(f, 0)$, $f^- = \max(-f, 0)$,

$$f = f^+ - f^-.$$

Let μ be a measure on a nonempty set X.

DEFINITION 1.11. *A function* $g : X \to [-\infty, \infty]$ *is called a* **simple function** *if the image of* g *is countable.*

DEFINITION 1.12.

(i) *If* g *is a nonnegative, simple, μ-measurable function, we define its* **integral**

$$\int g \, d\mu := \sum_{0 \le y \le \infty} y\mu(g^{-1}\{y\}).$$

(ii) *If* g *is a simple μ-measurable function and either* $\int g^+ \, d\mu < \infty$ *or* $\int g^- \, d\mu < \infty$, *we call* g *a* **μ-integrable simple function** *and define*

$$\int g \, d\mu := \int g^+ \, d\mu - \int g^- \, d\mu.$$

This expression may equal $\pm\infty$.

Thus if g is a μ-integrable simple function,

$$\int g \, d\mu := \sum_{-\infty \le y \le \infty} y\mu(g^{-1}\{y\}).$$

DEFINITION 1.13.

(i) *Let* $f \colon X \to [-\infty, \infty]$. *We define the* **upper integral**

$$\int^* f \, d\mu :=$$

$$\inf \left\{ \int g \, d\mu \mid g \ \mu\text{-integrable, simple, } g \ge f \ \mu\text{-a.e.} \right\}$$

and the **lower integral**

$$\int_* f \, d\mu :=$$

$$\sup \left\{ \int g \, d\mu \mid g \text{ } \mu\text{-integrable, simple, } g \leq f \text{ } \mu\text{-}a.e. \right\}.$$

(ii) *A μ-measurable function $f : X \to [-\infty, \infty]$ is called μ-* **integrable** *if $\int^* f \, d\mu = \int_* f \, d\mu$, in which case we write*

$$\int f \, d\mu := \int^* f \, d\mu = \int_* f \, d\mu.$$

Warning: Our use of the term "integrable" differs from most texts. For us, a function is "integrable" provided it has an integral, even if this integral equals $+\infty$ or $-\infty$.

Note that a nonnegative μ-measurable function is always μ-integrable.   □

We assume the reader to be familiar with all the usual properties of integrals.

DEFINITION 1.14.

(i) *A function $f : X \to [-\infty, \infty]$ is μ-**summable** if f is μ-integrable and*

$$\int |f| \, d\mu < \infty.$$

(ii) *We say a function $f : \mathbb{R}^n \to [-\infty, \infty]$ is **locally μ-summable** if $f|_K$ is μ-summable for each compact set $K \subset \mathbb{R}^n$.*

DEFINITION 1.15. *We say ν is a **signed measure** on $\mathbb{R}^n$ if there exists a Radon measure μ on $\mathbb{R}^n$ and a locally μ-summable function $f : \mathbb{R}^n \to [-\infty, \infty]$ such that*

$$\nu(K) = \int_K f \, d\mu \qquad\qquad (\star)$$

for all compact sets $K \subseteq \mathbb{R}^n$.

NOTATION

(i) We write
$$\nu = \mu \llcorner f$$
provided $(\star)$ holds for all compact sets K. Note that therefore
$$\mu \llcorner A = \mu \llcorner \chi_A.$$

(ii) We denote by
$$L^1(X, \mu)$$
the set of all μ-summable functions on X, and
$$L^1_{\text{loc}}(\mathbb{R}^n, \mu)$$
the set of all locally μ-summable functions.

(iii) Likewise, if $1 < p < \infty$,
$$L^p(X, \mu)$$
denotes the set of all μ-measurable functions f on X such that $|f|^p$ is μ-summable , and
$$L^p_{\text{loc}}(\mathbb{R}^n, \mu)$$
the set of μ-measurable functions f such that $|f|^p$ is locally μ-summable.

(iv) *We do not identify two L^p (or L^p_{loc}) functions that agree μ-a.e.*

The following three limit theorems for integrals are among the most important assertions in all of analysis.

THEOREM 1.17 (Fatou's Lemma). *Let $f_k : X \to [0, \infty]$ be μ-measurable for $k = 1, \ldots$. Then*
$$\int \liminf_{k \to \infty} f_k \, d\mu \le \liminf_{k \to \infty} \int f_k \, d\mu.$$

Proof. Take $g := \sum_{j=1}^{\infty} a_j \chi_{A_j}$ to be a nonnegative simple function less than or equal to $\liminf_{k \to \infty} f_k$. Suppose the μ-measurable sets $\{A_j\}_{j=1}^{\infty}$ are disjoint and $a_j > 0$ for $j = 1, \ldots$.

Fix $0 < t < 1$. Then

$$A_j = \bigcup_{k=1}^{\infty} B_{j,k},$$

where

$$B_{j,k} := A_j \cap \{x \mid f_l(x) > t a_j \text{ for all } l \geq k\}.$$

Note

$$A_j \supseteq B_{j,k+1} \supseteq B_{j,k} \quad (k = 1, \dots).$$

Thus

$$\int f_k \, d\mu \geq \sum_{j=1}^{\infty} \int_{A_j} f_k \, d\mu \geq \sum_{j=1}^{\infty} \int_{B_{j,k}} f_k \, d\mu \geq t \sum_{j=1}^{\infty} a_j \mu(B_{j,k});$$

and so

$$\liminf_{k \to \infty} \int f_k \, d\mu \geq t \sum_{j=1}^{\infty} a_j \mu(A_j) = t \int g \, d\mu.$$

This inequality holds for each $0 < t < 1$ and each simple function g less than or equal to $\liminf_{k \to \infty} f_k$. Consequently,

$$\liminf_{k \to \infty} \int f_k \, d\mu \geq \int_* \liminf_{k \to \infty} f_k \, d\mu = \int \liminf_{k \to \infty} f_k \, d\mu. \qquad \square$$

THEOREM 1.18 (Monotone Convergence Theorem). *Let $f_k : X \to [0, \infty]$ be μ-measurable $(k = 1, \dots)$, with*

$$f_1 \leq \cdots \leq f_k \leq f_{k+1} \leq \cdots.$$

Then

$$\lim_{k \to \infty} \int f_k \, d\mu = \int \lim_{k \to \infty} f_k \, d\mu.$$

Proof. Clearly,

$$\int f_j \, d\mu \leq \int \lim_{k \to \infty} f_k \, d\mu \quad (j = 1, \dots);$$

and therefore

$$\lim_{k \to \infty} \int f_k \, d\mu \leq \int \lim_{k \to \infty} f_k \, d\mu.$$

The opposite inequality follows from Fatou's Lemma. $\qquad \square$

THEOREM 1.19 (Dominated Convergence Theorem). *Assume*
$g \geq 0$ *is* μ*-summable and* $f, \{f_k\}_{k=1}^{\infty}$ *are* μ*-measurable. Suppose*

$$f_k \to f \quad \mu\text{-}a.e.$$

as $k \to \infty$*, and*

$$|f_k| \leq g \quad (k = 1, \dots).$$

Then

$$\lim_{k \to \infty} \int |f_k - f| \, d\mu = 0.$$

Proof. By Fatou's Lemma,

$$\int 2g \, d\mu = \int \liminf_{k \to \infty} (2g - |f - f_k|) \, d\mu \leq \liminf_{k \to \infty} \int 2g - |f - f_k| \, d\mu;$$

whence

$$\limsup_{k \to \infty} \int |f - f_k| \, d\mu \leq 0. \qquad \square$$

THEOREM 1.20 (Variant of Dominated Convergence Theorem). *Assume* $g, \{g_k\}_{k=1}^{\infty}$ *are* μ*-summable and* $f, \{f_k\}_{k=1}^{\infty}$ *are* μ*-measurable.*

Suppose $f_k \to f$ μ*-a.e. and*

$$|f_k| \leq g_k \quad (k = 1, \dots),$$

If also

$$g_k \to g \quad \mu\text{-}a.e.$$

and

$$\lim_{k \to \infty} \int g_k \, d\mu = \int g \, d\mu,$$

then

$$\lim_{k \to \infty} \int |f_k - f| \, d\mu = 0.$$

Proof. Similar to proof of the Dominated Convergence Theorem. $\quad \square$

It is easy to see that $\lim_{k \to \infty} \int |f_k - f| \, d\mu = 0$ does not necessarily imply $f_k \to f$ μ-a.e. But if we pass to an appropriate subsequence, we can obtain a.e. convergence.

THEOREM 1.21 (Almost everywhere convergent subsequence). *Assume* $f, \{f_k\}_{k=1}^{\infty}$ *are* μ-*summable and*

$$\lim_{k \to \infty} \int |f_k - f| \, d\mu = 0.$$

Then there exists a subsequence $\{f_{k_j}\}_{j=1}^{\infty}$ *for which*

$$f_{k_j} \to f \quad \mu\text{-}a.e.$$

Proof. We select a subsequence $\{f_{k_j}\}_{j=1}^{\infty}$ of the functions $\{f_k\}_{k=1}^{\infty}$ satisfying

$$\sum_{j=1}^{\infty} \int |f_{k_j} - f| \, d\mu < \infty.$$

In view of the Monotone Convergence Theorem, this implies

$$\int \sum_{j=1}^{\infty} |f_{k_j} - f| \, d\mu < \infty;$$

and thus

$$\sum_{j=1}^{\infty} |f_{k_j} - f| < \infty \quad \mu\text{-a.e.}$$

Consequently, $f_{k_j} \to f$ at μ-a.e. point. $\qquad\square$

1.4 Product measures, Fubini's Theorem, Lebesgue measure

Let X and Y be nonempty sets.

DEFINITION 1.16. *Let* μ *be a measure on* X *and* ν *a measure on* Y. *We define the measure* $\mu \times \nu : 2^{X \times Y} \to [0, \infty]$ *by setting*

$$(\mu \times \nu)(S) := \inf \left\{ \sum_{i=1}^{\infty} \mu(A_i)\nu(B_i) \right\},$$

for each $S \subseteq X \times Y$, *where the infimum is taken over all collections of*

μ-measurable sets $A_i \subseteq X$ and ν-measurable sets $B_i \subseteq Y (i = 1, \dots)$ such that

$$S \subseteq \bigcup_{i=1}^{\infty} (A_i \times B_i).$$

The measure $\mu \times \nu$ is called the **product measure** of μ and ν.

DEFINITION 1.17.

(i) A subset $A \subseteq X$ is **σ-finite** with respect to μ if we can write

$$A = \bigcup_{k=1}^{\infty} B_k,$$

where each B_k is μ-measurable and $\mu(B_k) < \infty$ for $k = 1, 2, \dots$.

(ii) A function $f : X \to [-\infty, \infty]$ is **σ-finite** with respect to μ if f is μ-measurable and $\{x \mid f(x) \neq 0\}$ is σ-finite with respect to μ.

THEOREM 1.22 (Fubini's Theorem). Let μ be a measure on X and ν a measure on Y.

(i) Then $\mu \times \nu$ is a regular measure on $X \times Y$, even if μ and ν are not regular.

(ii) If $A \subseteq X$ is μ-measurable and $B \subseteq Y$ is ν-measurable, then $A \times B$ is $(\mu \times \nu)$-measurable and

$$(\mu \times \nu)(A \times B) = \mu(A)\nu(B).$$

(iii) If $S \subseteq X \times Y$ is σ-finite with respect to $\mu \times \nu$, then the cross section

$$S_y := \{x \mid (x, y) \in S\}$$

is μ-measurable for ν-a.e. y,

$$S_x := \{y \mid (x, y) \in S\}$$

is ν-measurable for μ-a.e. x, $\mu(S_y)$ is ν-integrable, and $\nu(S_x)$ is μ-integrable. Moreover,

$$(\mu \times \nu)(S) = \int_Y \mu(S_y) \, d\nu(y) = \int_X \nu(S_x) \, d\mu(x).$$

(iv) *If f is $(\mu \times \nu)$-integrable and f is σ-finite with respect to $\mu \times \nu$ (in particular, if f is $(\mu \times \nu)$-summable), then the mapping*

$$y \mapsto \int_X f(x, y) \, d\mu(x)$$

is ν-integrable, the mapping

$$x \mapsto \int_Y f(x, y) \, d\nu(y)$$

is μ-integrable, and

$$\int_{X \times Y} f \, d(\mu \times \nu) = \int_Y \left[\int_X f(x, y) \, d\mu(x) \right] d\nu(y)$$

$$= \int_X \left[\int_Y f(x, y) \, d\nu(y) \right] d\mu(x).$$

Remark. We will later study the coarea formula (Theorem 3.10), which is a kind of "curvilinear" version of Fubini's Theorem. □

Proof. 1. Let $\mathcal{F}$ denote the collection of all sets $S \subseteq X \times Y$ for which the mapping

$$x \mapsto \chi_S(x, y)$$

is μ-integrable for each $y \in Y$ and the mapping

$$y \mapsto \int_X \chi_S(x, y) \, d\mu(x)$$

is ν-integrable. If $S \in \mathcal{F}$, we write

$$\rho(S) := \int_Y \left[\int_X \chi_S(x, y) \, d\mu(x) \right] d\nu(y).$$

2. Define

$$\mathcal{P}_0 := \{A \times B \mid A \ \mu\text{-measurable}, B \ \nu\text{-measurable}\},$$

$$\mathcal{P}_1 := \{\cup_{j=1}^{\infty} S_j \mid S_j \in \mathcal{P}_0 \ (j = 1, \ldots)\},$$

$$\mathcal{P}_2 := \{\cap_{j=1}^{\infty} S_j \mid S_j \in \mathcal{P}_1 \ (j = 1, \ldots)\}.$$

Note $\mathcal{P}_0 \subseteq \mathcal{F}$ and

$$\rho(A \times B) = \mu(A)\nu(B)$$

when $A \times B \in \mathcal{P}_0$. If $A_1 \times B_1, A_2 \times B_2 \in \mathcal{P}_0$, then

$$(A_1 \times B_1) \cap (A_2 \times B_2) = (A_1 \cap A_2) \times (B_1 \cap B_2) \in \mathcal{P}_0,$$

and

$$(A_1 \times B_1) - (A_2 \times B_2) = ((A_1 - A_2) \times B_1) \cup ((A_1 \cap A_2) \times (B_1 - B_2))$$

is a disjoint union of members of $\mathcal{P}_0$. It follows that each set in $\mathcal{P}_1$ is a countable disjoint union of sets in $\mathcal{P}_0$. Hence $\mathcal{P}_1 \subseteq \mathcal{F}$.

3. *Claim #1*: For each $S \subseteq X \times Y$,

$$(\mu \times \nu)(S) = \inf\{\rho(R) \mid S \subseteq R \in \mathcal{P}_1\}.$$

Proof of claim: First we note that if $S \subseteq R = \cup_{i=1}^{\infty} (A_i \times B_i)$, then

$$\rho(R) \le \sum_{i=1}^{\infty} \rho(A_i \times B_i) = \sum_{i=1}^{\infty} \mu(A_i)\nu(B_i).$$

Thus

$$\inf\{\rho(R) \mid S \subseteq R \in \mathcal{P}_1\} \le (\mu \times \nu)(S).$$

Moreover, there exists a disjoint collection of sets $\{A_j' \times B_j'\}_{j=1}^{\infty}$ in $\mathcal{P}_0$ such that

$$R = \bigcup_{j=1}^{\infty} (A_j' \times B_j').$$

Thus

$$\rho(R) = \sum_{j=1}^{\infty} \mu(A_j')\nu(B_j') \ge (\mu \times \nu)(S).$$

4. Fix $A \times B \in \mathcal{P}_0$. Then

$$(\mu \times \nu)(A \times B) \le \mu(A)\nu(B) = \rho(A \times B) \le \rho(R)$$

for all $R \in \mathcal{P}_1$ such that $A \times B \subseteq R$. Thus Claim #1 implies

$$(\mu \times \nu)(A \times B) = \mu(A)\nu(B).$$

5. Next we must prove $A \times B$ is $(\mu \times \nu)$-measurable. So suppose $T \subseteq X \times Y$ and $T \subseteq R \in \mathcal{P}_1$. Then $R - (A \times B)$ and $R \cap (A \times B)$ are disjoint and belong to $\mathcal{P}_1$. Consequently,

$$(\mu \times \nu)(T - (A \times B)) + (\mu \times \nu)(T \cap (A \times B))$$
$$\leq \rho(R - (A \times B)) + \rho(R \cap (A \times B))$$
$$= \rho(R),$$

and so, according to Claim #1,

$$(\mu \times \nu)(T - (A \times B)) + (\mu \times \nu)(T \cap (A \times B)) \leq (\mu \times \nu)(T).$$

Thus $(A \times B)$ is $(\mu \times \nu)$-measurable. This proves (ii).

6. *Claim #2:* For each $S \subseteq X \times Y$ there is a set $R \in \mathcal{P}_2$ such that $S \subseteq R$ and

$$\rho(R) = (\mu \times \nu)(S).$$

Proof of claim: If $(\mu \times \nu)(S) = \infty$, set $R = X \times Y$. If $(\mu \times \nu)(S) < \infty$, then for each $j = 1, 2, \ldots$ there is according to Claim #1 a set $R_j \in \mathcal{P}_1$ such that $S \subseteq R_j$ and

$$\rho(R_j) < (\mu \times \nu)(S) + \frac{1}{j}.$$

Define

$$R := \bigcap_{j=1}^{\infty} R_j \in \mathcal{P}_2.$$

Then $R \in \mathcal{F}$, and by the Dominated Convergence Theorem,

$$(\mu \times \nu)(S) \leq \rho(R) = \lim_{k \to \infty} \rho\left(\cap_{j=1}^{k} R_j\right) \leq (\mu \times \nu)(S).$$

7. From (ii) we see that every member of $\mathcal{P}_2$ is $(\mu \times \nu)$-measurable and thus (i) follows from Claim #2.

8. If $S \subseteq X \times Y$ and $(\mu \times \nu)(S) = 0$, then there is a set $R \in \mathcal{P}_2$ such that $S \subseteq R$ and $\rho(R) = 0$; thus $S \in \mathcal{F}$ and $\rho(S) = 0$.

Now suppose that $S \subseteq X \times Y$ is $(\mu \times \nu)$-measurable and $(\mu \times \nu)(S) < \infty$. Then there is a $R \in \mathcal{P}_2$ such that $S \subseteq R$ and

$$(\mu \times \nu)(R - S) = 0;$$

hence

$$\rho(R - S) = 0.$$

Thus

$$\mu(\{x \mid (x,y) \in S\}) = \mu(\{x \mid (x,y) \in R\})$$

for ν-a.e. $y \in Y$, and

$$(\mu \times \nu)(S) = \rho(R) = \int \mu(\{x \mid (x,y) \in S\}) \, d\nu(y).$$

Assertion (iii) follows, provided $(\mu \times \nu)(S) < \infty$. If S is σ-finite with respect to $\mu \times \nu$, we decompose S into countably many sets with finite measure.

9. Assertion (iv) reduces to (iii) when $f = \chi_S$. If f is $(\mu \times \nu)$-integrable, is nonnegative and is σ-finite with respect to $\mu \times \nu$, we use Theorem 1.12 to write

$$f = \sum_{k=1}^{\infty} \frac{1}{k} \chi_{A_k}.$$

Then assertion (iv) follows for f from the Monotone Convergence Theorem. Finally, for general f we write

$$f = f^+ - f^-,$$

to deduce (iv) in general. $\qquad\square$

DEFINITION 1.18.

(i) **One-dimensional Lebesgue measure** *on* $\mathbb{R}^1$ *is*

$$\mathcal{L}^1(A) := \inf \left\{ \sum_{i=1}^{\infty} \operatorname{diam} C_i \mid A \subseteq \bigcup_{i=1}^{\infty} C_i, C_i \subseteq \mathbb{R} \right\}$$

for all $A \subseteq \mathbb{R}$.

(ii) *We inductively define* **n-dimensional Lebesgue measure** $\mathcal{L}^n$ *on* $\mathbb{R}^n$ *by*

$$\mathcal{L}^n := \mathcal{L}^{n-1} \times \mathcal{L}^1 = \mathcal{L}^1 \times \cdots \times \mathcal{L}^1 \quad (n \text{ times})$$

THEOREM 1.23 (Another characterization of Lebesgue measure). *We have*

$$\mathcal{L}^n = \mathcal{L}^{n-k} \times \mathcal{L}^k$$

for each $k \in \{1, \ldots, n-1\}$.

Proof. Let $Q := [-L, L]^n$ denote a closed cube with sides parallel to the coordinate axes and define

$$\mu := \mathcal{L}^n \llcorner Q, \ \nu = (\mathcal{L}^{n-k} \times \mathcal{L}^k) \llcorner Q.$$

Then $\mu(R) = \nu(R) < \infty$ for each "rectangle" $R := \{x \mid -\infty \leq a_i \leq x_i \leq b_i \leq \infty \ (i = 1, \ldots, n)\}$. According then to Theorem 1.5, μ and ν agree on all Borel sets.

This conclusion is valid for each cube Q as above, and thus $\mathcal{L}^n$ and $\mathcal{L}^{n-k} \times \mathcal{L}^k$ agree on Borel subsets of $\mathbb{R}^n$. Since both are Radon measures, they thus agree on all subsets of $\mathbb{R}^n$. □

We hereafter assume the reader's familiarity with all the usual facts about $\mathcal{L}^n$.

NOTATION We will write "dx," "dy," etc. rather than "$d\mathcal{L}^n$" in integrals taken with respect to $\mathcal{L}^n$.

We also write $L^1(\mathbb{R}^n)$ for $L^1(\mathbb{R}^n, \mathcal{L}^n)$, etc.

1.5 Covering theorems

We present in this section the fundamental covering theorems of Vitali and of Besicovitch. Vitali's Covering Theorem is easier and is most useful for investigating $\mathcal{L}^n$ on $\mathbb{R}^n$. Besicovitch's Covering Theorem is much harder to prove, but it is necessary for studying arbitrary Radon measures on $\mathbb{R}^n$. The crucial geometric difference is that Vitali's Covering Theorem provides a cover of enlarged balls, whereas Besicovitch's Covering Theorem yields a cover out of the original balls, at the price of a certain controlled amount of overlap.

These covering theorems will be employed throughout the rest of this book, the first and most important applications being the differentiation theorems in Section 1.6.

1.5.1 Vitali's Covering Theorem

NOTATION If $B = B(x, r)$ is a closed ball in $\mathbb{R}^n$, we write

$$\hat{B} = B(x, 5r)$$

to denote the concentric closed ball with radius 5 times the radius of B.

DEFINITION 1.19.

(i) *A collection $\mathcal{F}$ of closed balls in $\mathbb{R}^n$ is a* **cover** *of a set $A \subset \mathbb{R}^n$ if*

$$A \subseteq \bigcup_{B \in \mathcal{F}} B.$$

(ii) *$\mathcal{F}$ is a* **fine cover** *of A if, in addition,*

$$\inf\{\operatorname{diam} B \mid x \in B, B \in \mathcal{F}\} = 0$$

for each $x \in A$.

THEOREM 1.24 (Vitali's Covering Theorem). *Let $\mathcal{F}$ be any collection of nondegenerate closed balls in $\mathbb{R}^n$ with*

$$\sup\{\operatorname{diam} B \mid B \in \mathcal{F}\} < \infty.$$

Then there exists a countable family $\mathcal{G}$ of disjoint balls in $\mathcal{F}$ such that

$$\bigcup_{B \in \mathcal{F}} B \subseteq \bigcup_{B \in \mathcal{G}} \hat{B}.$$

Proof. 1. Write $D := \sup\{\operatorname{diam} B \mid B \in \mathcal{F}\}$. Set

$$\mathcal{F}_j := \{B \in \mathcal{F} \mid \frac{D}{2^j} < \operatorname{diam} B \leq \frac{D^{j-1}}{2}\} \quad (j = 1, 2, \dots).$$

We define $\mathcal{G}_j \subseteq \mathcal{F}_j$ as follows:

(a) Let $\mathcal{G}_1$ be any maximal disjoint collection of balls in $\mathcal{F}_1$.

(b) Assuming $\mathcal{G}_1 \dots, \mathcal{G}_{k-1}$ have been selected, we choose $\mathcal{G}_k$ to be any maximal disjoint subcollection of

$$\{B \in \mathcal{F}_k \mid B \cap B' = \emptyset \text{ for all } B' \in \cup_{j=1}^{k-1}\mathcal{G}_j\}.$$

Finally, define

$$\mathcal{G} := \cup_{j=1}^{\infty}\mathcal{G}_j.$$

Clearly $\mathcal{G}$ is a collection of disjoint balls and $\mathcal{G} \subseteq \mathcal{F}$.

2. *Claim:* For each ball $B \in \mathcal{F}$, there exists a ball $B' \in \mathcal{G}$ such that $B \cap B' \neq \emptyset$ and $B \subseteq \hat{B}'$.

Proof of claim: Fix $B \in \mathcal{F}$. There then exists an index j such that $B \in \mathcal{F}_j$. By the maximality of $\mathcal{G}_j$, there exists a ball $B' \in \cup_{k=1}^{j}\mathcal{G}_k$ with $B \cap B' \neq \emptyset$. But $\operatorname{diam} B' \geq \frac{D}{2^j}$ and $\operatorname{diam} B \leq \frac{D}{2^{j-1}}$; so that $\operatorname{diam} B \leq 2\operatorname{diam} B'$. Thus $B \subseteq \hat{B}'$, as claimed. $\qquad\square$

A technical consequence we will need later is this:

THEOREM 1.25 (Variant of Vitali Covering Theorem). *Assume that $\mathcal{F}$ is a fine cover of A by closed balls and*

$$\sup\{\operatorname{diam} B \mid B \in F\} < \infty.$$

Then there exists a countable family $\mathcal{G}$ of disjoint balls in $\mathcal{F}$ such that for each finite subset $\{B_1, \ldots, B_m\} \subseteq \mathcal{F}$, we have

$$A - \bigcup_{k=1}^{m} B_k \subseteq \bigcup_{B \in \mathcal{G} - \{B_1, \ldots, B_m\}} \hat{B}.$$

Proof. Choose $\mathcal{G}$ as in the proof of the Vitali Covering Theorem and select $\{B_1, \ldots, B_m\} \subseteq \mathcal{F}$.

If $A \subseteq \cup_{k=1}^{m} B_k$, we are done. Otherwise, let $x \in A - \cup_{k=1}^{m} B_k$. Since the balls in $\mathcal{F}$ are closed and $\mathcal{F}$ is a fine cover, there exists $B \in \mathcal{F}$ with $x \in B$ and $B \cap B_k = \emptyset$ for $k = 1, \ldots, m$. But then, from the claim in the proof above, there exists a ball $B' \in \mathcal{G}$ such that $B \cap B' \neq \emptyset$ and so $B \subseteq \hat{B}'$. $\qquad\square$

Next we show we can measure and theoretically "fill up" an arbitrary open set with many countably disjoint closed balls.

THEOREM 1.26 (Filling open sets with balls). *Let $U \subset \mathbb{R}^n$ be open, $\delta > 0$. There exists a countable collection $\mathcal{G}$ of disjoint closed balls in U such that $\operatorname{diam} B < \delta$ for all $B \in \mathcal{G}$ and*

$$\mathcal{L}^n \left(U - \bigcup_{B \in \mathcal{G}} B \right) = 0.$$

Proof. 1. Fix $1 - \frac{1}{5^n} < \theta < 1$. Assume first $\mathcal{L}^n(U) < \infty$.

2. *Claim*: There exists a finite collection $\{B_i\}_{i=1}^{M_1}$ of disjoint closed balls in U such that $\operatorname{diam} B_i < \delta$ for $i = 1, \ldots, M_1$, and

$$\mathcal{L}^n \left(U - \bigcup_{i=1}^{M_1} B_i \right) \leq \theta \mathcal{L}^n(U). \tag{$\star$}$$

Proof of claim: Let $\mathcal{F}_1 := \{B \subseteq U \mid \operatorname{diam} B < \delta\}$. By the Vitali Covering Theorem there exists a countable disjoint family $\mathcal{G}_1 \subseteq \mathcal{F}_1$ such that

$$U \subseteq \bigcup_{B \in \mathcal{G}_1} \hat{B}.$$

Thus

$$\mathcal{L}^n(U) \leq \sum_{B \in \mathcal{G}_1} \mathcal{L}^n(\hat{B}) = 5^n \sum_{B \in \mathcal{G}_1} \mathcal{L}^n(B) = 5^n \mathcal{L}^n \left(\bigcup_{B \in \mathcal{G}_1} B \right).$$

Hence

$$\mathcal{L}^n \left(\bigcup_{B \in \mathcal{G}_1} B \right) \geq \frac{1}{5^n} \mathcal{L}^n(U),$$

and consequently

$$\mathcal{L}^n \left(U - \bigcup_{B \in \mathcal{G}_1} B \right) \leq \left(1 - \frac{1}{5^n} \right) \mathcal{L}^n(U).$$

Since $\mathcal{G}_1$ is countable and since $1 - \frac{1}{5^n} < \theta < 1$, there exist finitely many balls $B_1, \ldots, B_{M_1}$ in $\mathcal{G}_1$ satisfying $(\star)$.

3. Now let

$$U_2 := U - \bigcup_{i=1}^{M_1} B_i,$$

$$\mathcal{F}_2 := \{B \mid B \subseteq U_2, \operatorname{diam} B < \delta\},$$

and, as above, find finitely many disjoint balls $B_{M_1+1}, \ldots, B_{M_2}$ in $\mathcal{F}_2$ such that

$$\mathcal{L}^n \left(U - \bigcup_{i=1}^{M_2} B_i \right) = \mathcal{L}^n \left(U_2 - \bigcup_{i=M_1+1}^{M_2} B_i \right)$$

$$\leq \theta \mathcal{L}^n(U_2) \leq \theta^2 \mathcal{L}^n(U).$$

4. Continue this process to obtain a countable collection of disjoint balls such that

$$\mathcal{L}^n \left(U - \bigcup_{i=1}^{M_k} B_i \right) \leq \theta^k \mathcal{L}^n(U) \quad (k = 1, \ldots).$$

Since $\theta^k \to 0$, the theorem is proved if $\mathcal{L}^n(U) < \infty$.

Should $\mathcal{L}^n(U) = \infty$, we apply the above construction to each of the open sets

$$U_m := \{x \in U \mid m < |x| < m + 1\} \quad (m = 0, 1, \dots). \qquad \square$$

Remark. See also Theorem 1.28 in the next section, which replaces $\mathcal{L}^n$ in the preceding proof by an arbitrary Radon measure. $\qquad \square$

1.5.2 Besicovitch's Covering Theorem

If μ is an arbitrary Radon measure on $\mathbb{R}^n$, there is no systematic way to control $\mu(\hat{B})$ in terms of $\mu(B)$. Vitali's Covering Theorem is consequently not so useful for studying such a measure; we need instead a covering theorem that does not require us to enlarge balls.

THEOREM 1.27 (Besicovitch's Covering Theorem).

There exists a constant N_n, depending only on the dimension n, with the following property:

If $\mathcal{F}$ is any collection of nondegenerate closed balls in $\mathbb{R}^n$ with

$$\sup\{\operatorname{diam} B \mid B \in \mathcal{F}\} < \infty$$

and if A is the set of centers of balls in $\mathcal{F}$, then there exist N_n countable collections $\mathcal{G}_1, \dots, \mathcal{G}_{N_n}$ of disjoint balls in $\mathcal{F}$ such that

$$A \subseteq \bigcup_{i=1}^{N_n} \bigcup_{B \in \mathcal{G}_i} B.$$

Proof. 1. First suppose that A is bounded. Write

$$D := \sup\{\operatorname{diam} B \mid B \in \mathcal{F}\}.$$

Choose any ball $B_1 = B(a_1, r_1) \in \mathcal{F}$ such that $r_1 \geq \frac{3}{4}\frac{D}{2}$. Inductively choose B_j for $j \geq 2$, as follows. Let $A_j := A - \cup_{i=1}^{j-1} B_i$. If $A_j = \emptyset$, stop and set $J := j - 1$. If $A_j \neq \emptyset$, choose $B_j = B(a_j, r_j) \in \mathcal{F}$ such that $a_j \in A_j$ and

$$r_j \geq \frac{3}{4}\sup\{r \mid B(a, r) \in \mathcal{F}, a \in A_j\}.$$

If $A_j \neq \emptyset$ for all j, set $J := \infty$.

2 *Claim #1*: If $j > i$, then $r_j \leq \frac{4}{3}r_i$.

Proof of claim: Suppose $j > i$. Then $a_j \in A_i$ and so

$$r_i \geq \frac{3}{4} \sup\{r \mid B(A, r) \in \mathcal{F}, a \in A_i\} \geq \frac{3}{4} r_j.$$

3. *Claim #2:* The balls $\{B(a_j, \frac{r_j}{3})\}_{j=1}^{J}$ are disjoint.

Proof of claim: Let $j > i$. Then $a_j \notin B_i$; hence

$$|a_i - a_j| > r_i = \frac{r_i}{3} + \frac{2r_i}{3} \geq \frac{r_i}{3} + \frac{2}{3}\frac{3}{4}r_j > \frac{r_i}{3} + \frac{r_j}{3}.$$

4. *Claim #3:* If $J = \infty$, then $\lim_{j \to \infty} r_j = 0$.

Proof of claim: By Claim #2 the balls $\{B(a_j, \frac{r_j}{3})\}_{j=1}^{J}$ are disjoint. Since $a_j \in A$ and A is bounded, $r_j \to 0$.

5. *Claim #4:* $A \subseteq \cup_{j=1}^{J} B_j$.

Proof of claim: If $J < \infty$, this is trivial. Suppose $J = \infty$. If $a \in A$, there exists an $r > 0$ such that $B(a, r) \in \mathcal{F}$. Then by Claim #3, there exists an r_j with $r_j < \frac{3}{4}r$, a contradiction to the choice of r_j, if $a \notin \cup_{i=1}^{j-1} B_i$.

6. Fix $k > 1$ and let $I := \{j \mid 1 \leq j < k, B_j \cap B_k \neq \emptyset\}$. We need to estimate the cardinality of I. Set

$$K := I \cap \{j \mid r_j \leq 3r_k\}.$$

7. *Claim #5:* $\operatorname{Card}(K) \leq 20^n$.

Proof of claim: Let $j \in K$. Then $B_j \cap B_k \neq \emptyset$ and $r_j \leq 3r_k$. Choose any $x \in B(a_j, \frac{r_j}{3})$. Then

$$|x - a_k| \leq |x - a_j| + |a_j - a_k| \leq \frac{r_j}{3} + r_j + r_k$$

$$= \frac{4}{3}r_j + r_k \leq 4r_k + r_k = 5r_k.$$

Thus $B(a_j, \frac{r_j}{3}) \subseteq B(a_k, 5r_k)$. Recall from Claim #2 that the balls $B(a_i, \frac{r_i}{3})$ are disjoint. Thus Claim #1 implies

$$\alpha(n)5^n r_k^n = \mathcal{L}^n(B(a_k, 5r_k)) \geq \sum_{j \in K} \mathcal{L}^n\left(B\left(a_j, \frac{r_j}{3}\right)\right)$$

$$= \sum_{j \in K} \alpha(n)\left(\frac{r_j}{3}\right)^n \geq \sum_{j \in K} \alpha(n)\left(\frac{r_k}{4}\right)^n$$

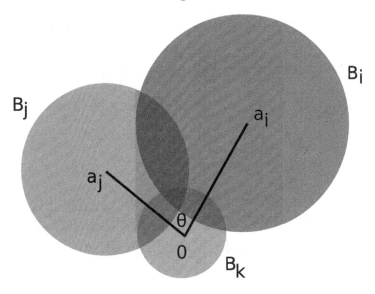

$$= \operatorname{Card}(K)\alpha(n)\frac{r_k^n}{4^n}.$$

Consequently,

$$5^n \geq \operatorname{Card}(K)\frac{1}{4^n}.$$

8. We must now estimate $\operatorname{Card}(I - K)$.

Let $i, j \in I - K$, with $i \neq j$. Then $1 \leq i,j < k$, $B_i \cap B_k \neq \emptyset$, $B_j \cap B_k \neq \emptyset$, $r_i > 3r_k$, and $r_j > 3r_k$. For simplicity of notation, we assume $a_k = 0$.

Let $0 \leq \theta \leq \pi$ be the angle between the vectors a_i and a_j. We want to find a lower bound on θ, and to this end we first assemble some geometric facts:

Since $i, j < k, 0 = a_k \notin B_i \cup B_j$. Thus $r_i < |a_i|$ and $r_j < |a_j|$. Since $B_i \cap B_k \neq \emptyset$ and $B_j \cap B_k \neq \emptyset$, $|a_i| \leq r_i + r_k$ and $|a_j| \leq r_j + r_k$. Finally, without loss of generality we can assume $|a_i| \leq |a_j|$. In summary,

$$\begin{cases} 3r_k < r_i < |a_i| \leq r_i + r_k \\ 3r_k < r_j < |a_j| \leq r_j + r_k \\ |a_i| \leq |a_j|. \end{cases}$$

9. *Claim #6a:* If $\cos\theta > \frac{5}{6}$, then $a_i \in B_j$.

Proof of claim: Suppose $|a_i - a_j| \geq |a_j|$; then the law of cosines gives

$$\cos\theta = \frac{|a_i|^2 + |a_j|^2 - |a_i - a_j|^2}{2|a_i||a_j|} \leq \frac{|a_i|^2}{2|a_i||a_j|} = \frac{|a_i|}{2|a_j|} \leq \frac{1}{2} < \frac{5}{6}.$$

Suppose instead that $|a_i - a_j| \leq |a_j|$ and $a_i \notin B_j$. Then $r_j < |a_i - a_j|$ and

$$\begin{aligned}
\cos\theta &= \frac{|a_i|^2 + |a_j|^2 - |a_i - a_j|^2}{2|a_i||a_j|} \\
&= \frac{|a_i|}{2|a_j|} + \frac{(|a_j| - |a_i - a_j|)(|a_j| + |a_i - a_j|)}{2|a_i||a_j|} \\
&\leq \frac{1}{2} + \frac{(|a_j| - |a_i - a_j|)(2|a_j|)}{2|a_i||a_j|} \\
&\leq \frac{1}{2} + \frac{r_j + r_k - r_j}{r_i} = \frac{1}{2} + \frac{r_k}{r_i} \leq \frac{5}{6}.
\end{aligned}$$

10. *Claim #6b:* If $a_i \in B_j$, then

$$0 \leq |a_i - a_j| + |a_i| - |a_j| \leq |a_j|\epsilon(\theta),$$

for

$$\epsilon(\theta) := \frac{8}{3}(1 - \cos\theta).$$

Proof of claim: Since $a_i \in B_j$, we must have $i < j$; hence $a_j \notin B_i$ and so $|a_i - a_j| > r_i$. Thus

$$\begin{aligned}
0 &\leq \frac{|a_i - a_j| + |a_i| - |a_j|}{|a_j|} \\
&\leq \frac{|a_i - a_j| + |a_i| - |a_j|}{|a_j|} \cdot \frac{|a_i - a_j| - |a_i| + |a_j|}{|a_i - a_j|} \\
&= \frac{|a_i - a_j|^2 - (|a_j| - |a_i|)^2}{|a_j||a_i - a_j|} \\
&= \frac{|a_i|^2 + |a_j|^2 - 2|a_i||a_j|\cos\theta|a_i|^2 - |a_j|^2 + 2|a_i||a_j|}{|a_j||a_i - a_j|} \\
&= \frac{2|a_i|(1 - \cos\theta)}{|a_i - a_j|}
\end{aligned}$$

$$\leq \frac{2(r_i + r_k)(1 - \cos\theta)}{r_i}$$

$$\leq \frac{2(1 + \frac{1}{3})r_i(1 - \cos\theta)}{r_i} = \epsilon(\theta).$$

11. *Claim #6c:* If $a_i \in B_j$, then $\cos\theta \leq \frac{61}{64}$.

Proof of claim: Since $a_i \in B_j$ and $a_j \notin B_i$, we have $r_i < |a_i - a_j| \leq r_j$. Since $i < j, r_j \leq \frac{4}{3}r_i$. Therefore,

$$
\begin{aligned}
|a_i - a_j| + |a_i| - |a_j| &\geq r_i + r_i - r_j - r_k \\
&\geq \frac{3}{2}r_j - r_j - r_k \\
&= \frac{1}{2}r_j - r_k \geq \frac{1}{6}r_j \\
&= \frac{1}{6}\frac{3}{4}(r_j + \frac{1}{3}r_j) \geq \frac{1}{8}(r_j + r_k) \\
&\geq \frac{1}{8}|a_j|.
\end{aligned}
$$

Then, by Claim #6b,

$$\frac{1}{8}|a_j| \leq |a_i - a_j| + |a_i| - |a_j| \leq |a_j|\epsilon(\theta).$$

Hence $\cos\theta \leq \frac{61}{64}$.

12. *Claim #6:* For all $i, j \in I - K$ with $i \neq j$, let θ denote the angle between $a_i - a_k$ and $a_j - a_k$. Then

$$\theta \geq \arccos\frac{61}{64} =: \theta_0 > 0.$$

This follows from Claims #6a–c.

13. *Claim #7:* There exists a constant L_n depending only on n such that $\mathrm{Card}(I - K) \leq L_n$.

Proof of claim: First, fix $r_0 > 0$ such that if $x \in \partial B(1)$ and $y, z \in B(x, r_0)$, then the angle between y and z is less than the constant θ_0 from Claim #6. Choose L_n so that $\partial B(1)$ can be covered by L_n balls with radius r_0 and centers on $\partial B(1)$, but cannot be covered by $L_n - 1$ such balls.

Then ∂B_k can be covered by L_n balls of radius $r_0 r_k$, with centers on ∂B_k. By Claim #6, if $i, j \in I - K$ with $i \neq j$, then the angle between $a_i - a_j$ and $a_j - a_k$ exceeds θ_0. Thus by the construction of r_0, the rays $a_j - a_k$ and $a_i - a_k$ cannot both go through the same ball on ∂B_k. Consequently, $\mathrm{Card}(I - K) \leq L_n$.

14. Finally, set $M_n := 20^n + L_n + 1$. Then by Claims #5 and #7,

$$\mathrm{Card}(I) = \mathrm{Card}(K) + \mathrm{Card}(I - K) \leq 20^n + L_n < M_n.$$

15. We next define the families of disjoint balls $\mathcal{G}_1, \ldots, \mathcal{G}_{M_n}$.

First, we define $\sigma : \{1, 2, \ldots\} \rightarrow \{1, \ldots, M_n\}$:

(a) Let $\sigma(i) = i$ for $1 \leq i \leq M_n$.

(b) For $k \geq M_n$ inductively define $\sigma(k+1)$ as follows. According to the calculations above,

$$\mathrm{Card}\{j \mid 1 \leq j \leq k, B_j \cap B_{k+1} \neq \emptyset\} < M_n,$$

so there exists $l \in \{1, \ldots, M_n\}$ such that $B_{k+1} \cap B_j = \emptyset$ for all j such that $\sigma(j) = l$ $(j = 1, \ldots, k)$. Set $\sigma(k+1) = l$.

Now, let

$$\mathcal{G}_j = \{B_i \mid \sigma(i) = j\}$$

for $1 \leq j \leq M_n$. From the definition of $\sigma(i)$ it follows that each $\mathcal{G}_j$ consists of disjoint balls from $\mathcal{F}$. Moreover, each B_i is in some $\mathcal{G}_j$; so that

$$A \subseteq \bigcup_{i=1}^{J} B_i = \bigcup_{i=1}^{M_n} \bigcup_{B \in \mathcal{G}_i} B.$$

16. Next, we extend the result to unbounded sets A.

For $l \geq 1$, let $A_l = A \cap \{x \mid 3D(l-1) \leq |x| < 3Dl\}$ and set $\mathcal{F}^l := \{B(a, r) \in \mathcal{F} \mid a \in A_l\}$. Then by Step 15, there exist countable collections $\mathcal{G}_1^l, \ldots, \mathcal{G}_{Mn}^l$ of disjoint closed balls in $\mathcal{F}^l$ such that

$$A_l \subseteq \bigcup_{i=1}^{Mn} \bigcup_{B \in \mathcal{G}_i^l} B.$$

Let

$$\mathcal{G}_j := \bigcup_{l=1}^{\infty} \mathcal{G}_j^{2l-1} \quad \text{for } 1 \le j \le M_n,$$

$$\mathcal{G}_{j+M_n} = \bigcup_{l=1}^{\infty} \mathcal{G}_j^{2l} \quad \text{for } 1 \le j \le M_n.$$

Set $N_n := 2M_n$. $\qquad\qquad\qquad\qquad\qquad\qquad\qquad\qquad \square$

We now see as a consequence of Besicovitch's Theorem that we can "fill up" an arbitrary open set with a countable collection of disjoint balls in such a way that the remainder has μ-measure zero.

THEOREM 1.28 (More on filling open sets with balls). *Let μ be a Borel measure on $\mathbb{R}^n$, and $\mathcal{F}$ any collection of nondegenerate closed balls. Let A denote the set of centers of the balls in $\mathcal{F}$. Assume*

$$\mu(A) < \infty$$

and

$$\inf\{r \mid B(a,r) \in \mathcal{F}\} = 0$$

for each $a \in A$.

Then for each open set $U \subseteq \mathbb{R}^n$, there exists a countable collection $\mathcal{G}$ of disjoint balls in $\mathcal{F}$ such that

$$\bigcup_{B \in \mathcal{G}} B \subseteq U$$

and

$$\mu\left((A \cap U) - \bigcup_{B \in \mathcal{G}} B \right) = 0.$$

Remark. The set A need not be μ-measurable here. Compare this assertion with Theorem 1.26 based on Vitali's Covering Theorem, above. $\qquad\qquad\qquad\qquad\qquad\qquad\qquad\qquad\qquad\qquad \square$

Proof. 1. Fix $1 - \frac{1}{N_n} < \theta < 1$.

Claim: There exists a finite collection $\{B_1, \dots B_{M_1}\}$ of disjoint closed balls in U such that

$$\mu\left((A \cap U) - \bigcup_{i=1}^{M_1} B_i \right) \le \theta\mu(A \cap U). \qquad (\star)$$

Proof of claim: Let $\mathcal{F}_1 = \{B \in \mathcal{F} \mid \text{diam } B \leq 1, B \subset U\}$. By Bescovitch's Theorem there exist families $\mathcal{G}_1, \ldots, \mathcal{G}_{N_n}$ of disjoint balls in $\mathcal{F}_1$ such that

$$A \cap U \subseteq \bigcup_{i=1}^{N_n} \bigcup_{B \in \mathcal{G}_i} B.$$

Thus

$$\mu(A \cap U) \leq \sum_{i=1}^{N_n} \mu\left(A \cap U \cap \bigcup_{B \in \mathcal{G}_i} B\right).$$

Consequently, there exists an integer j between 1 and N_n for which

$$\mu\left(A \cap U \cap \bigcup_{B \in \mathcal{G}_j} B\right) \geq \frac{1}{N_n}\mu(A \cap U).$$

By Theorem 1.6, there exist balls $B_1, \ldots, B_{M_1} \in \mathcal{G}_j$ such that

$$\mu\left(A \cap U \cap \bigcup_{i=1}^{M_1} B_i\right) \geq (1 - \theta)\mu(A \cap U).$$

But

$$\mu(A \cap U) = \mu\left(A \cap U \cap \bigcup_{i=1}^{M_1} B_i\right) + \mu\left((A \cap U) - \bigcup_{i=1}^{M_1} B_i\right),$$

since $\bigcup_{i=1}^{M_1} B_i$ is μ-measurable. Therefore $(\star)$ holds.

2. Now let $U_2 := U - \bigcup_{i=1}^{M_1} B_i$ and $\mathcal{F}_2 := \{B \mid B \in \mathcal{F}, \text{diam } B \leq 1, B \subset U_2\}$. As above, we find finitely many disjoint balls $B_{M_1+1}, \ldots, B_{M_2}$ such that

$$\mu\left((A \cap U) - \bigcup_{i=1}^{M_2} B_i\right) = \mu\left((A \cap U_2) - \bigcup_{i=M_1+1}^{M_2} B_i\right)$$
$$\leq \theta\mu(A \cap U_2)$$
$$\leq \theta^2\mu(A \cap U).$$

3. Continue this process to obtain a countable collection of disjoint balls from F and within U such that

$$\mu\left((A \cap U) - \bigcup_{i=1}^{M_k} B_i\right) \leq \theta^k \mu(A \cap U).$$

Since $\theta^k \to 0$ and $\mu(A) < \infty$, the theorem is proved. $\qquad\square$

1.6 Differentiation of Radon measures

We now utilize the covering theorems of the previous section to study the differentiation of Radon measures on $\mathbb{R}^n$.

1.6.1 Derivatives

Let μ and ν be Radon measures on $\mathbb{R}^n$.

DEFINITION 1.20. *For each point $x \in \mathbb{R}^n$, define*

$$
\overline{D}_\mu \nu(x) := \begin{cases} \limsup_{r \to 0} \dfrac{\nu(B(x,r))}{\mu(B(x,r))} & \text{if } \mu(B(x,r)) > 0 \text{ for all } r > 0 \\ +\infty & \text{if } \mu(B(x,r)) = 0 \text{ for some } r > 0 \end{cases}
$$

and

$$
\underline{D}_\mu \nu(x) := \begin{cases} \liminf_{r \to 0} \dfrac{\nu(B(x,r))}{\mu(B(x,r))} & \text{if } \mu(B(x,r)) > 0 \text{ for all } r > 0 \\ +\infty & \text{if } \mu(B(x,r)) = 0 \text{ for some } r > 0. \end{cases}
$$

DEFINITION 1.21. *If $\overline{D}_\mu \nu(x) = \underline{D}_\mu \nu(x) < +\infty$, we say ν is **differentiable** with respect to μ at x and write*

$$
D_\mu \nu(x) := \overline{D}_\mu \nu(x) = \underline{D}_\mu \nu(x).
$$

$D_\mu \nu$ *is the **derivative** of ν with respect to μ. We also call $D_\mu \nu$ the **density** of ν with respect to μ.*

Our goals are to learn when $D_\mu \nu$ exists and when ν can be recovered by integrating $D_\mu \nu$.

LEMMA 1.2. *Fix $0 < \alpha < \infty$. Then*

(i) $A \subseteq \{x \in \mathbb{R}^n \mid \underline{D}_\mu \nu(x) \leq \alpha\}$ *implies $\nu(A) \leq \alpha \mu(A)$.*

(ii) $A \subseteq \{x \in \mathbb{R}^n \mid \overline{D}_\mu \nu(x) \geq \alpha\}$ *implies $\nu(A) \geq \alpha \mu(A)$.*

Remark. The set A need not be μ- nor ν-measurable here. □

Proof. We may assume $\mu(\mathbb{R}^n), \nu(\mathbb{R}^n) < \infty$, since we could otherwise consider μ and ν restricted to compact subsets of $\mathbb{R}^n$.

Fix $\epsilon > 0$. Let U be open, $A \subseteq U$, where A satisfies the hypothesis of (i). Set

$$\mathcal{F} := \{B \mid B = B(a,r), a \in A, B \subseteq U, \nu(B) \leq (\alpha + \epsilon)\mu(B)\}.$$

Then $\inf\{r \mid B(a,r) \in \mathcal{F}\} = 0$ for each $a \in A$, and so Theorem 1.28 provides us with a countable collection $\mathcal{G}$ of disjoint balls in $\mathcal{F}$ such that

$$\nu\left(A - \bigcup_{B \in \mathcal{G}} B\right) = 0.$$

Then

$$\nu(A) \leq \sum_{B \in \mathcal{G}} \nu(B) \leq (\alpha + \epsilon) \sum_{B \in \mathcal{G}} \mu(B) \leq (\alpha + \epsilon)\mu(U).$$

This estimate is valid for each open set $U \supseteq A$, and hence Theorem 1.8 implies $\nu(A) \leq (\alpha + \epsilon)\mu(A)$. This proves (i). The proof of (ii) is similar. $\square$

THEOREM 1.29 (Differentiating measures). *Let μ and ν be Radon measures on $\mathbb{R}^n$. Then*

(i) *$D_\mu \nu$ exists and is finite μ-a.e., and*

(ii) *$D_\mu \nu$ is μ-measurable.*

Proof. We may assume $\nu(\mathbb{R}^n), \mu(\mathbb{R}^n) < \infty$, as we could otherwise consider μ and ν restricted to compact subsets of $\mathbb{R}^n$.

1. *Claim #1: $D_\mu \nu$ exists and is finite μ-a.e.*

Proof of claim: Let $I := \{x \mid \overline{D}_\mu \nu(x) = +\infty\}$. Observe that for each $\alpha > 0, I \subseteq \{x \mid \overline{D}_\mu \nu(x) \geq \alpha\}$. Thus by Lemma 1.2,

$$\mu(I) \leq \frac{1}{\alpha}\nu(I).$$

Send $\alpha \to \infty$ to conclude $\mu(I) = 0$, and so $\overline{D}_\mu \nu$ is finite μ-a.e.

For each $0 < a < b$, define

$$R(a,b) := \{x \mid \underline{D}_\mu \nu(x) < a < b < \overline{D}_\mu \nu(x) < \infty\}.$$

Again using Lemma 1.2, we see that

$$b\mu(R(a,b)) \le \nu(R(a,b)) \le a\mu(R(a,b));$$

whence $\mu(R(a,b)) = 0$, since $b > a$. Furthermore,

$$\{x \mid \underline{D}_\mu\nu(x) < \overline{D}_\mu\nu(x) < \infty\} = \bigcup_{\substack{0<a<b \\ a,b \text{ rational}}} R(a,b);$$

and consequently $D_\mu\nu$ exists and is finite μ-a.e.

2. *Claim #2*: For each $x \in \mathbb{R}^n$ and $r > 0$,

$$\limsup_{y \to x} \mu(B(y,r)) \le \mu(B(x,r)).$$

A similar assertion holds for ν.

Proof of claim: Choose $y_k \in \mathbb{R}^n$ with $y_k \to x$. Set $f_k := \chi_{B(y_k,r)}$ and $f = \chi_{B(x,r)}$. Then

$$\limsup_{k \to \infty} f_k \le f$$

and so

$$\liminf_{k \to \infty}(1 - f_k) \ge (1 - f).$$

Thus by Fatou's Lemma,

$$\int_{B(x,2r)} (1-f)\,d\mu \le \int_{B(x,2r)} \liminf_{k \to \infty}(1-f_k)\,d\mu$$

$$\le \liminf_{k \to \infty} \int_{B(x,2r)} (1-f_k)\,d\mu;$$

that is,

$$\mu(B(x,2r)) - \mu(B(x,r)) \le \liminf_{k \to \infty}(\mu(B(x,2r)) - \mu(B(y_k,r))).$$

Now since μ is a Radon measure, $\mu(B(x,2r)) < \infty$; the claim follows.

3. *Claim #3*: $D_\mu\nu$ is μ-measurable.

Proof of claim: According to Claim #2, for all $r > 0$, the functions $x \mapsto \mu(B(x,r))$ and $x \mapsto \nu(B(x,r))$ are upper semicontinuous and thus Borel measurable. Consequently, for every $r > 0$,

$$f_r(x) := \begin{cases} \dfrac{\nu(B(x,r))}{\mu(B(x,r))} & \text{if } \mu(B(x,r)) > 0 \\ +\infty & \text{if } \mu(B(x,r)) = 0 \end{cases}$$

is μ-measurable. But

$$D_\mu \nu = \lim_{r \to 0} f_r = \lim_{k \to \infty} f_{\frac{1}{k}} \quad \mu\text{-a.e.}$$

and so $D_\mu \nu$ is μ-measurable. $\qquad\qquad\qquad\qquad\qquad\qquad\square$

1.6.2 Integration of derivatives; Lebesgue decomposition

DEFINITION 1.22. *Assume μ and ν are Borel measures on $\mathbb{R}^n$.*

(i) *The measure ν is* **absolutely continuous** *with respect to μ, written*

$$\nu << \mu,$$

provided $\mu(A) = 0$ implies $\nu(A) = 0$ for all $A \subseteq \mathbb{R}^n$.

(ii) *The measures ν and μ are* **mutually singular**, *written*

$$\nu \perp \mu,$$

if there exists a Borel subset $B \subseteq \mathbb{R}^n$ such that

$$\mu(\mathbb{R}^n - B) = \nu(B) = 0.$$

THEOREM 1.30 (Differentiation of Radon measures). *Let ν, μ be Radon measures on $\mathbb{R}^n$, with $\nu << \mu$. Then*

$$\nu(A) = \int_A D_\mu \nu \, d\mu$$

for all μ-measurable sets $A \subseteq \mathbb{R}^n$.

Remark. This is a version of the **Radon–Nikodym Theorem.** Observe we prove not only that ν has a density with respect to μ, but also that this density $D_\mu \nu$ can be computed by "differentiating" ν with respect to μ. These assertions comprise in effect the fundamental theorem of calculus for Radon measures on $\mathbb{R}^n$. $\qquad\qquad\square$

Proof. 1. Let A be μ-measurable. Then there exists a Borel set B with $A \subseteq B, \mu(B - A) = 0$. Thus $\nu(B - A) = 0$ and so A is ν-measurable. Hence each μ-measurable set is also ν-measurable.

2. Set

$$Z := \{x \in \mathbb{R}^n \mid D_\mu \nu(x) = 0\}, \ I := \{x \in \mathbb{R}^n \mid D_\mu \nu(x) = +\infty\};$$

Z and I are μ-measurable. By Theorem 1.29, $\mu(I) = 0$ and so $\nu(I) = 0$. Also, Lemma 1.2 implies $\nu(Z) \leq \alpha\mu(Z)$ for all $\alpha > 0$; thus $\nu(Z) = 0$. Hence

$$\nu(Z) = 0 = \int_Z D_\mu\nu \, d\mu$$

and

$$\nu(I) = 0 = \int_I D_\mu\nu \, d\mu.$$

3. Now let A be μ-measurable and fix $1 < t < \infty$. Define for each integer m

$$A_m := A \cap \{x \in \mathbb{R}^n \mid t^m \leq D_\mu\nu(x) < t^{m+1}\}.$$

Then A_m is μ-measurable, and so also ν-measurable. Moreover,

$$A - \bigcup_{m=-\infty}^{\infty} A_m \subseteq Z \cup I \cup \{x \mid \overline{D}_\mu\nu(x) \neq \underline{D}_\mu\nu(x)\};$$

and hence

$$\nu\left(A - \bigcup_{m=-\infty}^{\infty} A_m\right) = 0.$$

Consequently, Lemma 1.2 implies

$$\nu(A) = \sum_{m=-\infty}^{\infty} \nu(A_m) \leq \sum_m t^{m+1}\mu(A_m)$$

$$= t\sum_m t^m \mu(A_m) \leq t \int_A D_\mu\nu \, d\mu.$$

Similarly, Lemma 1.2 gives

$$\nu(A) = \sum_m \nu(A_m) \geq \sum_m t^m \mu(A_m)$$

$$= t^{-1}\sum_m t^{m+1}\mu(A_m) \geq t^{-1}\sum_m \int_{A_m} D_\mu\nu \, d\mu$$

$$= t^{-1}\int_{A_m} D_\mu\nu \, d\mu.$$

Thus $\frac{1}{t}\int_A D_\mu\nu \, d\mu \leq \nu(A) \leq t\int_A D_\mu\nu \, d\mu$ for all $1 < t < \infty$. Now send $t \to 1^+$. $\qquad\square$

THEOREM 1.31 (Lebesgue Decomposition Theorem). *Let ν and μ be Radon measures on $\mathbb{R}^n$.*

(i) *Then*
$$\nu = \nu_{\mathrm{ac}} + \nu_{\mathrm{s}},$$

where $\nu_{\mathrm{ac}}, \nu_{\mathrm{s}}$ are Radon measures on $\mathbb{R}^n$ with
$$\nu_{\mathrm{ac}} << \mu, \ \nu_{\mathrm{s}} \perp \mu.$$

(ii) *Furthermore,*
$$D_\mu \nu = D_\mu \nu_{\mathrm{ac}}, \ D_\mu \nu_{\mathrm{s}} = 0 \quad \mu\text{-}a.e.;$$

and consequently
$$\nu(A) = \int_A D_\mu \nu \, d\mu + \nu_{\mathrm{s}}(A)$$

for each Borel set $A \subseteq \mathbb{R}^n$.

DEFINITION 1.23. *We call ν_{ac} the* **absolutely continuous part** *and ν_{s} the* **singular part** *of ν with respect to μ.*

Proof. 1. As before, we may as well assume $\mu(\mathbb{R}^n), \nu(\mathbb{R}^n) < \infty$.

Define
$$\mathcal{E} := \{A \subseteq \mathbb{R}^n \mid A \text{ Borel}, \ \mu(\mathbb{R}^n - A) = 0\},$$

and choose $B_k \in \mathcal{E}$ such that
$$\nu(B_k) \leq \inf_{A \in \mathcal{E}} \nu(A) + \frac{1}{k} \quad (k = 1\ldots).$$

Write $B := \cap_{k=1}^\infty B_k$. Since
$$\mu(\mathbb{R}^n - B) \leq \sum_{k=1}^\infty \mu(\mathbb{R}^n - B_k) = 0.$$

we have $B \in \mathcal{E}$, and so
$$\nu(B) = \inf_{A \in \mathcal{E}} \nu(A). \tag{$\star$}$$

Define
$$\nu_{\mathrm{ac}} := \nu \, \llcorner \, B, \ \nu_{\mathrm{s}} := \nu \, \llcorner \, (\mathbb{R}^n - B);$$

these are Radon measures according to Theorem 1.7.

3. Now suppose $A \subseteq B$, A is a Borel set, $\mu(A) = 0$, but $\nu(A) > 0$. Then $B - A \in \mathcal{E}$ and $\nu(B - A) < \nu(B)$, a contradiction to $(\star)$. Consequently, $\nu_{\mathrm{ac}} << \mu$. On the other hand, $\mu(\mathbb{R}^n - B) = 0$, and thus $\nu_{\mathrm{s}} \perp \mu$.

Finally, fix $\alpha > 0$ and set

$$C := \{x \in B \mid D_\mu \nu_{\mathrm{s}}(x) \geq \alpha\}.$$

According to Lemma 1.2,

$$\alpha\mu(C) \leq \nu_{\mathrm{s}}(C) = 0,$$

and therefore $D_\mu \nu_{\mathrm{s}} = 0$ μ-a.e. This implies

$$D_\mu \nu_{\mathrm{ac}} = D_\mu \nu \quad \mu\text{-a.e.} \qquad \square$$

1.7 Lebesgue points, approximate continuity

1.7.1 Differentiation Theorem

NOTATION We denote the **average** of f over the set E with respect μ by

$$\fint_E f \, d\mu := \frac{1}{\mu(E)} \int_E f \, d\mu,$$

provided $0 < \mu(E) < \infty$ and the integral is defined.

THEOREM 1.32 (Lebesgue–Besicovitch Differentiation Theorem).
  Let μ be a Radon measure on $\mathbb{R}^n$ and $f \in L^1_{\mathrm{loc}}(\mathbb{R}^n, \mu)$. Then

$$\lim_{r \to 0} \fint_{B(x,r)} f \, d\mu = f(x)$$

for μ-a.e. $x \in \mathbb{R}^n$.

Proof. For Borel sets $B \subseteq \mathbb{R}^n$, define $\nu^\pm(B) := \int_B f^\pm \, d\mu$, and for arbitrary $A \subseteq \mathbb{R}^n$, let $\nu^\pm(A) := \inf\{\nu^\pm(B) \mid A \subseteq B, B \text{ Borel}\}$. Then ν^+ and ν^- are Radon measures, and so, according to Theorem 1.30,

$$\nu^+(A) = \int_A D_\mu \nu^+ \, d\mu = \int_A f^+ \, d\mu$$

and

$$\nu^-(A) = \int_A D_\mu \nu^- \, d\mu = \int_A f^- \, d\mu$$

for all μ-measurable A. Thus $D_\mu \nu^\pm = f^\pm$ μ-a.e. Consequently,

$$\lim_{r \to 0} \fint_{B(x,r)} f \, d\mu = \lim_{r \to 0} \frac{1}{\mu(B(x,r))} [\nu^+(B(x,r)) - \nu^-(B(x,r))]$$

$$= D_\mu \nu^+(x) - D_\mu \nu^-(x)$$

$$= f^+(x) - f^-(x) = f(x)$$

for μ-a.e. point x.   □

THEOREM 1.33 (Lebesgue points for Radon measures). *Let μ be a Radon measure on $\mathbb{R}^n$, and suppose $f \in L^p_{loc}(\mathbb{R}^n, \mu)$ for some $1 \le p < \infty$. Then*

$$\lim_{r \to 0} \fint_{B(x,r)} |f - f(x)|^p \, d\mu = 0 \qquad (\star)$$

for μ-a.e. point x.

DEFINITION 1.24. *A point x for which $(\star)$ holds is called a **Lebesgue point** of f with respect to μ.*

Proof. Let $\{r_i\}_{i=1}^\infty$ be a countable dense subset of $\mathbb{R}$. By the Lebesgue–Besicovitch Differentiation Theorem 1.32,

$$\lim_{r \to 0} \fint_{B(x,r)} |f - r_i|^p \, d\mu = |f(x) - r_i|^p$$

for μ-a.e. x and $i = 1, 2, \ldots$. Thus there exists a set $A \subseteq \mathbb{R}^n$ such that $\mu(A) = 0$, for which $x \in \mathbb{R}^n - A$ implies

$$\lim_{r \to 0} \fint_{B(x,r)} |f - r_i|^p \, d\mu = |f(x) - r_i|^p$$

for all i. Fix $x \in \mathbb{R}^n - A$ and $\epsilon > 0$. Choose r_i such that $|f(x) - r_i|^p < \frac{\epsilon}{2^p}$. Then

$$\limsup_{r \to 0} \fint_{B(x,r)} |f - f(x)|^p \, d\mu$$

$$\le 2^{p-1} \left[\limsup_{r \to 0} \fint_{B(x,r)} |f - r_i|^p \, d\mu \right.$$

$$+ \fint_{B(x,r)} |f(x) - r_i|^p \, d\mu \Bigg]$$

$$= 2^{p-1}[|f(x) - r_i|^p + |f(x) - r_i|^p] < \epsilon. \qquad \square$$

For the case $\mu = \mathcal{L}^n$, this stronger assertion holds:

THEOREM 1.34 (Differentiation with noncentered balls). *As-sume that $f \in L^p_{\text{loc}}$ for some $1 \le p < \infty$. Then*

$$\lim_{B \to \{x\}} \fint_B |f - f(x)|^p \, dy = 0 \ \text{for } \mathcal{L}^n\text{-a.e. } x.,$$

*where the limit is taken over all closed balls B **containing** x, as* diam $B \to 0$.

The point is that the balls need not be centered at x.

Proof. We show that for each sequence of closed balls $\{B_k\}_{k=1}^\infty$ with $x \in B_k$ and $d_k := \text{diam } B_k \to 0$,

$$\fint_{B_k} |f - f(x)|^p \, dy \to 0$$

as $k \to \infty$, at each Lebesgue point of f.

Choose balls $\{B_k\}_{k=1}^\infty$ as above. Then $B_k \subseteq B(x, d_k)$, and conse-quently,

$$\fint_{B_k} |f - f(x)|^p \, dy \le 2^n \fint_{B(x,d_k)} |f - f(x)|^p \, dy.$$

The right-hand side goes to zero if x is a Lebesgue point. $\qquad \square$

THEOREM 1.35 (Points of density 1 and density 0). *Let $E \subseteq \mathbb{R}^n$ be $\mathcal{L}^n$-measurable. Then*

$$\lim_{r \to 0} \frac{\mathcal{L}^n(B(x,r) \cap E)}{\mathcal{L}^n(B(x,r))} = 1 \quad \text{for } \mathcal{L}^n\text{-a.e. } x \in E$$

and

$$\lim_{x \to 0} \frac{\mathcal{L}^n(B(x,r) \cap E)}{\mathcal{L}^n(B(x,r))} = 0 \quad \text{for } \mathcal{L}^n\text{-a.e. } x \in \mathbb{R}^n - E.$$

Proof. Set $f = \chi_E$, $\mu = \mathcal{L}^n$ in the Lebesgue–Besicovitch Differentia-tion Theorem. $\qquad \square$

DEFINITION 1.25. *Let $E \subseteq \mathbb{R}^n$. A point $x \in \mathbb{R}^n$ is a **point of density 1** for E if*

$$\lim_{r \to 0} \frac{\mathcal{L}^n(B(x,r) \cap E)}{\mathcal{L}^n(B(x,r))} = 1$$

*and a **point of density 0** for E if*

$$\lim_{r \to 0} \frac{\mathcal{L}^n(B(x,r) \cap E)}{\mathcal{L}^n(B(x,r))} = 0.$$

Remark. We regard the set of points of density 1 of E as comprising the *measure theoretic interior* of E; according to Theorem 1.35, $\mathcal{L}^n$-a.e. point in an $\mathcal{L}^n$-measurable set E belongs to its measure theoretic interior. Similarly, the points of density 0 for E make up the *measure theoretic exterior* of E. In Section 5.8 we will define and investigate the *measure theoretic boundary* of certain sets E. □

DEFINITION 1.26. *Assume $f \in L^1_{\mathrm{loc}}(\mathbb{R}^n)$. Then*

$$f^*(x) := \begin{cases} \lim_{r \to 0} \fint_{B(x,r)} f\,dy & \text{if this limit exists} \\ 0 & \text{otherwise} \end{cases}$$

*is the **precise representative** of f.*

Remark. Note that if $f, g \in L^1_{\mathrm{loc}}(\mathbb{R}^n)$, with $f = g$ $\mathcal{L}^n$-a.e., then $f^* = g^*$ for *all* points $x \in \mathbb{R}^n$. In view of the Lebesgue–Besicovitch Differentiation Theorem with $\mu = \mathcal{L}^n$, $\lim_{r \to 0} \fint_{B(x,r)} f\,dy$ exists $\mathcal{L}^n$-a.e. In Chapters 4 and 5, we will prove that if f is a Sobolev or BV function, then $f^* = f$, except possibly on a "very small" set of appropriate capacity or Hausdorff measure zero.

Observe also that it is possible for the above limit to exist even if x is not a Lebesgue point of f; cf. Theorem 5.19 in Section 5.9. □

1.7.2 Approximate limits, approximate continuity

DEFINITION 1.27. *Let $f : \mathbb{R}^n \to \mathbb{R}^m$. We say $l \in \mathbb{R}^m$ is the **approximate limit** of f as $y \to x$, written*

$$\mathrm{ap}\lim_{y \to x} f(y) = l,$$

if for each $\epsilon > 0$,

$$\lim_{r \to 0} \frac{\mathcal{L}^n(B(x,r) \cap \{|f - l| \geq \epsilon\})}{\mathcal{L}^n(B(x,r))} = 0.$$

So if l is the approximate limit of f at x, then for each $\epsilon > 0$ the set $\{|f - l| \geq \epsilon\}$ has density 0 at x.

THEOREM 1.36 (Uniqueness of approximate limits). *An approximate limit, if it exists, is unique.*

Proof. Assume for each $\epsilon > 0$ that both

$$\frac{\mathcal{L}^n(B(x,r) \cap \{|f - l| \geq \epsilon\})}{\mathcal{L}^n(B(x,r))} \to 0 \tag{$\star$}$$

and

$$\frac{\mathcal{L}^n(B(x,r) \cap \{|f - l'| \geq \epsilon\})}{\mathcal{L}^n(B(x,r))} \to 0 \tag{$\star\star$}$$

as $r \to 0$. Then if $l \neq l'$, we set $\epsilon := \frac{|l - l'|}{3}$ and observe for each $y \in B(x,r)$ that

$$3\epsilon = |l - l'| \leq |f(y) - l| + |f(y) - l'|.$$

Thus

$$B(x,r) \subseteq \{|f - l| \geq \epsilon\} \cup \{|f - l'| \geq \epsilon\}.$$

Therefore

$$\mathcal{L}^n(B(x,r)) \leq \mathcal{L}^n(B(x,r) \cap \{|f - l| \geq \epsilon\})$$
$$+ \mathcal{L}^n(B(x,r) \cap \{|f - l'| \geq \epsilon\}),$$

a contradiction to $(\star)$, $(\star\star)$. $\qquad\square$

DEFINITION 1.28. *Let $f : \mathbb{R}^n \to \mathbb{R}$.*

(i) *We say l is the **approximate lim sup** of f as $y \to x$, written*

$$\operatorname{ap} \limsup_{y \to x} f(y) = l,$$

if l is the infimum of the real numbers t such that

$$\lim_{r \to 0} \frac{\mathcal{L}^n(B(x,r) \cap \{f > t\})}{\mathcal{L}^n(B(x,r))} = 0.$$

(ii) *Similarly, l is the **approximate lim inf** of f as $y \to x$, written*

$$\operatorname{ap} \liminf_{y \to x} f(y) = l,$$

if l is the supremum of the real numbers t such that

$$\lim_{r \to 0} \frac{\mathcal{L}^n(B(x,r) \cap \{f < t\})}{\mathcal{L}^n(B(x,r))} = 0.$$

DEFINITION 1.29. *We say* $f : \mathbb{R}^n \to \mathbb{R}^m$ *is* **approximately continuous** *at* $x \in \mathbb{R}^n$ *if*

$$\text{ap} \lim_{y \to x} f(y) = f(x).$$

THEOREM 1.37 (Measurability and approximate continuity). *Suppose that* $f : \mathbb{R}^n \to \mathbb{R}^m$ *is* $\mathcal{L}^n$-*measurable.*

Then f *is approximately continuous* $\mathcal{L}^n$-*a.e.*

Remark. Thus a measurable function is "practically continuous at practically every point." The converse is also true; see Federer [F, Section 2.9.13]. □

Proof. 1. *Claim:* There exist disjoint, compact sets $\{K_i\}_{i=1}^{\infty} \subseteq \mathbb{R}^n$ such that

$$\mathcal{L}^n \left(\mathbb{R}^n - (\cup_{i=1}^{\infty} K_i) \right) = 0$$

and for each $i = 1, 2, \ldots$, $f|_{K_i}$ is continuous.

Proof of claim: For each positive integer m, set $B_m := B(m)$. By Lusin's Theorem, there exists a compact set $K_1 \subseteq B_1$ such that $\mathcal{L}^n(B_1 - K_1) \leq 1$ and $f|_{K_1}$ is continuous. Assuming now $K_1, \ldots, K_m$ have been constructed, there exists a compact set

$$K_{m+1} \subseteq B_{m+1} - \cup_{i=1}^{m} K_i$$

such that

$$\mathcal{L}^n \left(B_{m+1} - \cup_{i=1}^{m+1} K_i \right) \leq \frac{1}{m+1}$$

and $f|K_{m+1}$ is continuous.

2. For $\mathcal{L}^n$-a.e. $x \in K_i$,

$$\lim_{r \to 0} \frac{\mathcal{L}^n(B(x,r) - K_i)}{\mathcal{L}^n(B(x,r))} = 0. \qquad (\star)$$

Define $A := \{x \mid x \in K_i \text{ for some } i, \text{ and } (\star) \text{ holds}\}$; then $\mathcal{L}^n(\mathbb{R}^n - A) = 0$. Let $x \in A$, so that $x \in K_i$ and $(\star)$ holds for some fixed i. Fix $\epsilon > 0$. There exists $s > 0$ such that $y \in K_i$ and $|x - y| < s$ imply $|f(x) - f(y)| < \epsilon$.

Then if $0 < r < s, B(x,r) \cap \{y \mid |f(y) - f(x)| \geq \epsilon\} \subseteq B(x,r) - K_i$. In view of $(\star)$, we see that therefore

$$\text{ap} \lim_{y \to x} f(y) = f(x). \qquad \square$$

Remark. If $f \in L^1_{\text{loc}}(\mathbb{R}^n)$, the proof is much easier. Indeed, for each $\epsilon > 0$

$$\frac{\mathcal{L}^n(B(x,r) \cap \{|f - f(x)| > \epsilon\})}{\mathcal{L}^n(B(x,r))} \leq \frac{1}{\epsilon} \int_{B(x,r)} |f - f(x)| \, dy,$$

and the right-hand side goes to zero for $\mathcal{L}^n$-a.e. x. In particular *a Lebesgue point is a point of approximate continuity.* $\quad\square$

In Section 5.9 we will define and discuss the related notion of *approximate differentiability.*

1.8   Riesz Representation Theorem

In this book there will be two primary sources of measures to which we will apply the foregoing abstract theory. These are (a) Hausdorff measures, constructed in Chapter 2, and (b) Radon measures characterizing certain linear functionals. These arise as follows:

THEOREM 1.38 (Riesz Representation Theorem). *Let*

$$L : C_c(\mathbb{R}^n; \mathbb{R}^m) \to \mathbb{R}$$

be a linear functional satisfying

$$\sup\{L(f) \mid f \in C_c(\mathbb{R}^n; \mathbb{R}^m), |f| \leq 1, \text{spt}(f) \subseteq K\} < \infty \qquad (\star)$$

for each compact set $K \subset \mathbb{R}^n$. Then there exists a Radon measure μ on $\mathbb{R}^n$ and a μ-measurable function $\sigma : \mathbb{R}^n \to \mathbb{R}^m$ such that

$$|\sigma(x)| = 1 \quad \text{for } \mu\text{-a.e. } x,$$

and

$$L(f) = \int_{\mathbb{R}^n} f \cdot \sigma \, d\mu$$

for all $f \in C_c(\mathbb{R}^n; \mathbb{R}^m)$.

DEFINITION 1.30. *We call μ the **variation measure** associated with L. It is defined for each open set $V \subset \mathbb{R}^n$ by*

$$\mu(V) := \sup\{L(f) \mid f \in C_c(\mathbb{R}^n; \mathbb{R}^m), |f| \leq 1, \text{spt}(f) \subseteq V\}.$$

Proof. 1. Define μ on open sets V as above and then set

$$\mu(A) := \inf\{\mu(V) \mid A \subseteq V \text{ open}\}$$

for arbitrary $A \subseteq \mathbb{R}^n$.

2. *Claim #1*: μ is a measure.

Proof of claim: Let V, $\{V_i\}_{i=1}^{\infty}$ be open subsets of $\mathbb{R}^n$, with $V \subseteq \cup_{i=1}^{\infty} V_i$. Choose $g \in C_c(\mathbb{R}^n; \mathbb{R}^m)$ such that $|g| \leq 1$ and $\mathrm{spt}(g) \subseteq V$. Since $\mathrm{spt}(g)$ is compact, there exists an index k such that $\mathrm{spt}(g) \subseteq \cup_{j=1}^{k} V_j$.

Let $\{\zeta_j\}_{j=1}^{k}$ be a finite sequence of smooth nonnegative functions such that $\mathrm{spt}\,\zeta_j \subset V_j$ for $1 \leq j \leq k$ and $\sum_{j=1}^{k} \zeta_j = 1$ on $\mathrm{spt}\,g$. Then $g = \sum_{j=1}^{k} g\zeta_j$, and so

$$|L(g)| \leq \sum_{j=1}^{k} |L(g\zeta_j)| \leq \sum_{j=1}^{\infty} \mu(V_j).$$

Taking the supremum over g, we find $\mu(V) \leq \sum_{j=1}^{\infty} \mu(V_j)$.

Now let $\{A_j\}_{j=1}^{\infty}$ be arbitrary sets with $A \subseteq \cup_{j=1}^{\infty} A_j$. Fix $\epsilon > 0$. Choose open sets V_j such that $A_j \subseteq V_j$ and $\mu(A_j) + \frac{\epsilon}{2^j} \geq \mu(V_j)$. Then

$$\mu(A) \leq \mu\left(\bigcup_{j=1}^{\infty} V_j\right) \leq \sum_{j=1}^{\infty} \mu(V_j) \leq \sum_{j=1}^{\infty} \mu(A_j) + \epsilon.$$

3. *Claim #2*: μ is a Radon measure.

Proof of claim: Let U_1 and U_2 be open sets with $\mathrm{dist}(U_1, U_2) > 0$. Then $\mu(U_1 \cup U_2) = \mu(U_1) + \mu(U_2)$ by definition of μ. Hence if $A_1, A_2 \subseteq \mathbb{R}^n$ and $\mathrm{dist}(A_1, A_2) > 0$, then $\mu(A_1 \cup A_2) = \mu(A_1) + \mu(A_2)$. According to Caratheodory's criterion (Theorem 1.9), μ is a Borel measure.

Furthermore, by its definition, μ is Borel regular; indeed, given $A \subseteq \mathbb{R}^n$, there exist open sets V_k such that $A \subseteq V_k$ and $\mu(V_k) \leq \mu(A) + \frac{1}{k}$ for all k. Thus $\mu(A) = \mu\left(\cap_{k=1}^{\infty} V_k\right)$. Finally, the boundedness condition $(\star)$ implies $\mu(K) < \infty$ for all compact K.

4. Now, let $C_c^+(\mathbb{R}^n) := \{f \in C_c(\mathbb{R}^n) \mid f \geq 0\}$; and for $f \in C_c^+(\mathbb{R}^n)$, set

$$\lambda(f) := \sup\{|L(g)| \mid g \in C_c(\mathbb{R}^n; \mathbb{R}^m), |g| \leq f\}.$$

Observe that for all $f_1, f_2 \in C_c^+(\mathbb{R}^n)$, $f_1 \leq f_2$ implies $\lambda(f_1) \leq \lambda(f_2)$. Also $\lambda(cf) = c\lambda(f)$ for all $c \geq 0$ and $f \in C_c^+(\mathbb{R}^n)$.

5. *Claim #3:* For all $f_1, f_2 \in C_c^+(\mathbb{R}^n)$, $\lambda(f_1 + f_2) = \lambda(f_1) + \lambda(f_2)$.

Proof of claim: If $g_1, g_2 \in C_c(\mathbb{R}^n; \mathbb{R}^m)$ with $|g_1| \leq f_1$ and $|g_2| \leq f_2$, then $|g_1 + g_2| \leq f_1 + f_2$. We can furthermore assume $L(g_1), L(g_2) \geq 0$. Therefore,

$$|L(g_1)| + |L(g_2)| = L(g_1 + g_2) = |L(g_1 + g_2)| \leq \lambda(f_1 + f_2).$$

Taking suprema over g_1 and g_2 with $g_1, g_2 \in C_c(\mathbb{R}^n; \mathbb{R}^m)$ gives

$$\lambda(f_1) + \lambda(f_2) \leq \lambda(f_1 + f_2).$$

Now fix $g \in C_c(\mathbb{R}^n; \mathbb{R}^m)$, with $|g| \leq f_1 + f_2$. Set

$$g_i := \begin{cases} \frac{f_i g}{f_1 + f_2} & \text{if } f_1 + f_2 > 0 \\ 0 & \text{if } f_1 + f_2 = 0 \end{cases}$$

for $i = 1, 2$. Then $g_1, g_2 \in C_c(\mathbb{R}^n; \mathbb{R}^m)$ and $g = g_1 + g_2$. Moreover, $|g_i| \leq f_i, (i = 1, 2)$; and consequently

$$|L(g)| \leq |L(g_1)| + |L(g_2)| \leq \lambda(f_1) + \lambda(f_2).$$

Hence,

$$\lambda(f_1 + f_2) \leq \lambda(f_1) + \lambda(f_2).$$

6. *Claim #4:* $\lambda(f) = \int_{\mathbb{R}^n} f \, d\mu$ for all $f \in C_c^+(\mathbb{R}^n)$.

Proof of claim: Let $\epsilon > 0$. Choose $0 = t_0 < t_1 < \cdots < t_N$ such that $t_N := 2\|f\|_{L^\infty}, 0 < t_i - t_{i-1} < \epsilon$, and $\mu(f^{-1}\{t_i\}) = 0$ for $i = 1, \ldots, N$. Set $U_j = f^{-1}((t_{j-1}, t_j))$; then U_j is open and $\mu(U_j) < \infty$.

By Theorem 1.8, there exist compact sets K_j such that $K_j \subseteq U_j$ and $\mu(U_j - K_j) < \frac{\epsilon}{N}$ for $j = 1, 2, \ldots, N$. Furthermore there exist functions $g_j \in C_c(\mathbb{R}^n; \mathbb{R}^m)$ with $|g_j| \leq 1$, $\mathrm{spt}(g_j) \subseteq U_j$, and $|L(g_j)| \geq \mu(U_j) - \frac{\epsilon}{N}$. Note also that there exist functions $h_j \in C_c^+(\mathbb{R}^n)$ such that $\mathrm{spt}(h_j) \subseteq U_j$, $0 \leq h_j < 1$, and $h_j \equiv 1$ on the compact set $K_j \cup \mathrm{spt}(g_j)$. Then

$$\lambda(h_j) \geq |L(g_j)| \geq \mu(U_j) - \frac{\epsilon}{N}$$

and

$$\begin{aligned} \lambda(h_j) &= \sup \{|L(g)| \mid g \in C_c(\mathbb{R}^n; \mathbb{R}^m), |g| \leq h_j\} \\ &\leq \sup\{|L(g)| \mid g \in C_c(\mathbb{R}^n; \mathbb{R}^m), |g| \leq 1, \mathrm{spt}(g) \subseteq U_j\} \\ &= \mu(U_j); \end{aligned}$$

whence $\mu(U_j) - \frac{\epsilon}{N} \leq \lambda(h_j) \leq \mu(U_j)$.

Define
$$A := \left\{ x \mid f(x) \left(1 - \sum_{j=1}^{N} h_j(x) \right) > 0 \right\}.$$

Then A is open and

$$\mu(A) = \mu \left(\cup_{j=1}^{N}(U_j - \{h_j = 1\}) \right) \le \sum_{j=1}^{N} \mu(U_j - K_j) \le \epsilon.$$

Therefore

$$\lambda \left(f - f \sum_{j=1}^{N} h_j \right)$$

$$= \sup \left\{ |L(g)| \mid g \in C_c(\mathbb{R}^n; \mathbb{R}^m), |g| \le f - f \sum_{j=1}^{N} h_j \right\}$$

$$\le \sup \{ |L(g)| \mid g \in C_c(\mathbb{R}^n; \mathbb{R}^m), |g| \le \|f\|_{L^\infty} \chi_A \}$$

$$= \|f\|_{L^\infty} \sup \{ L(g) \mid g \in C_c(\mathbb{R}^n; \mathbb{R}^m), |g| \le \chi_A \}$$

$$= \|f\|_{L^\infty} \mu(A)$$

$$\le \epsilon \|f\|_{L^\infty}.$$

Hence

$$\lambda(f) = \lambda \left(f - f \sum_{j=1}^{N} h_j \right) + \lambda \left(f \sum_{j=1}^{N} h_j \right)$$

$$\le \epsilon \|f\|_{L^\infty} + \sum_{j=1}^{N} \lambda(f h_j) \le \epsilon \|f\|_{L^\infty} + \sum_{j=1}^{N} t_j \mu(U_j)$$

and

$$\lambda(f) \ge \sum_{j=1}^{N} \lambda(f h_j) \ge \sum_{j=1}^{N} t_{j-1} \left(\mu(U_j) - \frac{\epsilon}{N} \right) \ge \sum_{j=1}^{N} t_{j-1} \mu(U_j) - t_N \epsilon.$$

Finally, since

$$\sum_{j=1}^{N} t_{j-1} \mu(U_j) \le \int_{\mathbb{R}^n} f \, d\mu \le \sum_{j=1}^{N} t_j \mu(U_j),$$

we have

$$\left| \lambda(f) - \int f \, d\mu \right| \le \sum_{j=1}^{N} (t_j - t_{j-1}) \mu(U_j) + \epsilon \|f\|_{L^\infty} + \epsilon t_N$$

$$\le \epsilon \mu(\mathrm{spt}(f)) + 3\epsilon \|f\|_{L^\infty}.$$

7. *Claim #5*: There exists a μ-measurable function $\sigma : \mathbb{R}^n \to \mathbb{R}^m$ satisfying assertion (ii).

Proof of claim: Fix $e \in \mathbb{R}^m, |e| = 1$. Define $\lambda_e(f) := L(fe)$ for $f \in C_c(\mathbb{R}^n)$. Then λ_e is linear and

$$|\lambda_e(f)| = |L(fe)|$$
$$\leq \sup\{|L(g)| \mid g \in C_c(\mathbb{R}^n; \mathbb{R}^m), |g| \leq |f|\}$$
$$= \lambda(|f|) = \int_{\mathbb{R}^n} |f| \, d\mu;$$

thus we can extend λ_e to a bounded linear functional on $L^1(\mathbb{R}^n; \mu)$. Hence there exists $\sigma_e \in L^\infty(\mu)$ such that

$$\lambda_e(f) = \int_{\mathbb{R}^n} f\sigma_e \, d\mu$$

for $f \in C_c(\mathbb{R}^n)$).

Let $e_1, \ldots, e_m$ be the standard basis for $\mathbb{R}^m$ and define $\sigma := \sum_{j=1}^m \sigma_{e_j} e_j$. Then if $f \in C_c(\mathbb{R}^n; \mathbb{R}^m)$, we have

$$L(f) = \sum_{j=1}^m L((f \cdot e_j)e_j) = \sum_{j=1}^m \int (f \cdot e_j)\sigma_{e_j} \, d\mu = \int f \cdot \sigma \, d\mu.$$

8. *Claim #6*: $|\sigma| = 1$ μ-a.e.

Proof of claim: Let $U \subseteq \mathbb{R}^n$ be open, $\mu(U) < \infty$. By definition,

$$\mu(U) =$$
$$\sup\left\{\int f \cdot \sigma \, d\mu \mid f \in C_c(\mathbb{R}^n; \mathbb{R}^m), |f| \leq 1, \text{spt}(f) \subset U\right\}. \qquad (\star\star)$$

Now take $f_k \in C_c(\mathbb{R}^n; \mathbb{R}^m)$ such that $|f_k| \leq 1, \text{spt}(f_k) \subseteq U$, and $f_k \cdot \sigma \to |\sigma|$ μ-a.e.; such functions exist according to Theorem 1.15. Then

$$\int_U |\sigma| \, d\mu = \lim_{k\to\infty} \int f_k \cdot \sigma \, d\mu \leq \mu(U)$$

by $(\star\star)$.

On the other hand, if $f \in C_c(\mathbb{R}^n; \mathbb{R}^m)$ with $|f| \leq 1$ and spt $f \subseteq U$, then

$$\int f \cdot \sigma \, d\mu \leq \int_U |\sigma| \, d\mu.$$

Consequently (**) implies

$$\mu(U) \leq \int_U |\sigma| \, d\mu.$$

Thus $\mu(U) = \int_U |\sigma| \, d\mu$ for all open $U \subset \mathbb{R}^n$; hence $|\sigma| = 1$ μ-a.e. $\qquad\square$

An immediate and very useful application is the following characterization of nonnegative linear functionals.

THEOREM 1.39 (Nonnegative linear functionals). *Assume*

$$L : C_c^\infty(\mathbb{R}^n) \to \mathbb{R}$$

is linear and nonnegative; that is,

$$L(f) \geq 0 \quad \text{for all } f \in C_c^\infty(\mathbb{R}^n), f \geq 0. \tag{$\star$}$$

Then there exists a Radon measure μ on $\mathbb{R}^n$ such that

$$L(f) = \int_{\mathbb{R}^n} f \, d\mu$$

for all $f \in C_c^\infty(\mathbb{R}^n)$.

Proof. Choose any compact set $K \subseteq \mathbb{R}^n$, and select a smooth function ζ such that ζ has compact support, $\zeta \equiv 1$ on K, and $0 \leq \zeta \leq 1$.

For any $f \in C_c^\infty(\mathbb{R}^n)$ with spt $f \subseteq K$, set $g := \|f\|_{L^\infty} \zeta - f \geq 0$. Then $(\star)$ implies

$$0 \leq L(g) = \|f\|_{L^\infty} L(\zeta) - L(f),$$

and so

$$L(f) \leq C\|f\|_{L^\infty}$$

for $C := L(\zeta)$. Replacing f with $-f$, we deduce that

$$|L(f)| \leq C\|f\|_{L^\infty}.$$

The functional L thus extends to a linear mapping from $C_c(\mathbb{R}^n)$ into $\mathbb{R}$, satisfying the hypothesis of the Riesz Representation Theorem. Hence there exist μ, σ as above so that

$$L(f) = \int_{\mathbb{R}^n} f\sigma \, d\mu$$

for $f \in C_c^\infty(\mathbb{R}^n))$, with $\sigma = \pm 1$ μ-a.e. But then $(\star)$ implies $\sigma = 1$ μ-a.e. $\qquad\square$

1.9 Weak convergence

1.9.1 Weak convergence of measures

We introduce next a notion of weak convergence for measures.

THEOREM 1.40 (Weak convergence of measures). *Let* μ, μ_k $(k = 1, 2, \dots)$ *be Radon measures on* $\mathbb{R}^n$. *The following three statements are equivalent:*

(i) $\lim_{k \to \infty} \int_{\mathbb{R}^n} f \, d\mu_k = \int_{\mathbb{R}^n} f \, d\mu$ *for all* $f \in C_c(\mathbb{R}^n)$.

(ii) $\limsup_{k \to \infty} \mu_k(K) \leq \mu(K)$ *for each compact set* $K \subseteq \mathbb{R}^n$ *and* $\mu(U) \leq \liminf_{k \to \infty} \mu_k(U)$ *for each open set* $U \subseteq \mathbb{R}^n$.

(iii) $\lim_{k \to \infty} \mu_k(B) = \mu(B)$ *for each bounded Borel set* $B \subseteq \mathbb{R}^n$ *with* $\mu(\partial B) = 0$.

DEFINITION 1.31. *If* (i)–(iii) *hold, we say the measures* $\{\mu_k\}_{k=1}^{\infty}$ **converge weakly** *to the measure* μ, *written*

$$\mu_k \rightharpoonup \mu.$$

Proof. 1. Assume (i) holds and fix $\epsilon > 0$. Let $U \subset \mathbb{R}^n$ be open and choose a compact set $K \subseteq U$. Next, choose $f \in C_c(\mathbb{R}^n)$ such that spt $f \subset U$, $0 \leq f \leq 1$, $f \equiv 1$ on K. Then

$$\mu(K) \leq \int_{\mathbb{R}^n} f \, d\mu = \lim_{k \to \infty} \int_{\mathbb{R}^n} f \, d\mu_k \leq \liminf_{k \to \infty} \mu_k(U).$$

Thus

$$\mu(U) = \sup\{\mu(K) \mid K \text{ compact}, K \subseteq U\} \leq \liminf_{k \to \infty} \mu_k(U).$$

This proves the second part of (ii); the proof of the other part is similar.

2. Suppose now (ii) holds, $B \subseteq \mathbb{R}^n$ is a bounded Borel set, $\mu(\partial B) = 0$. Let B^0 denote the interior of B. Then

$$\mu(B) = \mu(B^0) \leq \liminf_{k \to \infty} \mu_k(B^0) \leq \limsup \mu_k(\bar{B}) \leq \mu(\bar{B}) = \mu(B).$$

3. Finally, assume (iii) holds. Fix $\epsilon > 0$, $f \in C_c^+(\mathbb{R}^n)$. Let $R > 0$ be such that $\operatorname{spt}(f) \subseteq B(0, R)$ and $\mu(\partial B(R)) = 0$. Choose $0 = t_0 < t_1, <$

$\cdots < t_N$ such that $t_N := 2\|f\|_{L^\infty}, 0 < t_i - t_{i-1} < \epsilon$, and $\mu(f^{-1}\{t_i\}) = 0$
for $i = 1, \ldots, N$. Set $B_i = f^{-1}(t_{i-1}, t_i)$; then $\mu(\partial B_i) = 0$ for $i = 2, \ldots$.
  Now

$$\sum_{i=2}^{N} t_{i-1} \mu_k(B_i) \leq \int_{\mathbb{R}^n} f \, d\mu_k \leq \sum_{i=2}^{N} t_i \mu_k(B_i) + t_1 \mu_k(B(R))$$

and

$$\sum_{i=2}^{N} t_{i-1} \mu(B_i) \leq \int_{\mathbb{R}^n} f \, d\mu \leq \sum_{i=2}^{N} t_i \mu(B_i) + t_1 \mu(B(R);$$

so (iii) implies

$$\limsup_{k \to \infty} \left| \int_{\mathbb{R}^n} f \, d\mu_k - \int_{\mathbb{R}^n} f \, d\mu \right| \leq 2\epsilon \mu(B(R)). \qquad \square$$

  The great advantage of weak convergence of measures is that compactness is had relatively easily.

THEOREM 1.41 (Weak compactness for measures). *Let $\{\mu_k\}_{k=1}^{\infty}$ be a sequence of Radon measures on $\mathbb{R}^n$ satisfying*

$$\sup_k \mu_k(K) < \infty \quad \text{for each compact set } K \subset \mathbb{R}^n.$$

Then there exists a subsequence $\{\mu_{k_j}\}_{j=1}^{\infty}$ and a Radon measure μ such that

$$\mu_{k_j} \rightharpoonup \mu.$$

Proof. 1. Assume first

$$\sup_k \mu_k(\mathbb{R}^n) < \infty. \qquad (\star)$$

  2. Let $\{f_k\}_{k=1}^{\infty}$ be a countable dense subset of $C_c(\mathbb{R}^n)$. As $(\star)$ implies $\{\int f_1 \, d\mu_j\}$ is bounded, we can find a subsequence $\{\mu_j^1\}_{j=1}^{\infty}$ and $a_1 \in \mathbb{R}$ such that

$$\int f_1 \, d\mu_j^1 \to a_1.$$

Continuing, we choose a subsequence $\{\mu_j^k\}_{j=1}^{\infty}$ of $\{\mu_j^{k-1}\}_{j=1}^{\infty}$ and $a_k \in \mathbb{R}$ such that

$$\int f_k \, d\mu_j^k \to a_k.$$

Set $\nu_j := \mu_j^j$; then

$$\int f_k \, d\nu_j \to a_k$$

for all $k \geq 1$.

Define $L(f_k) = a_k$, and note that L is linear, with $|L(f_k)| \leq \|f_k\|_{L^\infty} M$ by ($\star$), for $M := \sup_k \mu_k(\mathbb{R}^n)$. Thus L can be uniquely extended to a bounded linear functional $\bar{L}$ on $C_c(\mathbb{R}^n)$. Then according to the Riesz Representation Theorem, there exists a Radon measure μ on $\mathbb{R}^n$ such that

$$\bar{L}(f) = \int f \, d\mu$$

for all $f \in C_c(\mathbb{R}^n)$.

3. Choose any $f \in C_c(\mathbb{R}^n)$. The denseness of $\{f_k\}_{k=1}^\infty$ implies the existence of a subsequence $\{f_i\}_{i=1}^\infty$ such that $f_i \to f$ uniformly. Fix $\epsilon > 0$ and then choose i so large that

$$\|f - f_i\|_{L^\infty} < \frac{\epsilon}{4M}.$$

Next choose J so that for all $j > J$

$$\left| \int f_i \, d\nu_j - \int f_i \, d\mu \right| < \frac{\epsilon}{2}.$$

Then for $j > J$

$$\left| \int f \, d\nu_j - \int f \, d\mu \right| \leq \left| \int f - f_i \, d\nu_j \right| + \left| \int f - f_i \, d\mu \right|$$

$$+ \left| \int f_i \, d\nu_j - \int f_i \, d\mu \right|$$

$$\leq 2M \|f - f_i\|_{L^\infty} + \frac{\epsilon}{2} < \epsilon.$$

4. In the general case that ($\star$) fails to hold, but

$$\sup_k \mu_k(K) < \infty$$

for each compact set $K \subset \mathbb{R}^n$, we apply the reasoning above to the measures

$$\mu_k^l := \mu_k \, \llcorner \, B(l) \quad (k, l = 1, 2, \dots)$$

and use a diagonal argument. $\qquad \square$

1.9.2   Weak convergence of functions

Assume now that $U \subseteq \mathbb{R}^n$ is open, $1 \le p < \infty$.

DEFINITION 1.32. *A sequence $\{f_k\}_{k=1}^\infty \subset L^p(U)$ converges* **weakly** *to a function $f \in L^p(U)$, written*

$$f_k \rightharpoonup f \quad in \ L^p(U),$$

provided

$$\lim_{k \to \infty} \int_U f_k g \, dx = \int_U f g \, dx$$

for each $g \in L^q(U)$, where

$$\frac{1}{p} + \frac{1}{q} = 1 \quad (1 < q \le \infty).$$

THEOREM 1.42 (Weak compactness in L^p). *Suppose $1 < p < \infty$. Let $\{f_k\}_{k=1}^\infty$ be a sequence of functions in $L^p(U)$ satisfying*

$$\sup_k \|f_k\|_{L^p(U)} < \infty. \tag{$\star$}$$

Then there exists a subsequence $\{f_{k_j}\}_{j=1}^\infty$ and a function $f \in L^p(U)$ such that

$$f_{k_j} \rightharpoonup f \quad in \ L^p(U).$$

Remark. This assertion is in general false for $p = 1$, but see Section 1.9.3 below. $\qquad\qquad\qquad\qquad\qquad\qquad\qquad\qquad\qquad\qquad\quad \square$

Proof. 1. If $U \ne \mathbb{R}^n$, we extend each function f_k to all of $\mathbb{R}^n$ by setting it equal to zero on $\mathbb{R}^n - U$. This done, we may with no loss of generality assume $U = \mathbb{R}^n$. Furthermore, we may as well suppose

$$f_k \ge 0 \quad \mathcal{L}^n\text{-a.e};$$

for we could otherwise apply the following analysis to f_k^+ and f_k^-.

2. Define the Radon measures

$$\mu_k := \mathcal{L}^n \llcorner f_k \quad (k = 1, 2, \dots).$$

Then for each compact set $K \subset \mathbb{R}^n$,

$$\mu_k(K) = \int_K f_k \, dx \le \left(\int_K f_k^p \, dx \right)^{\frac{1}{p}} \mathcal{L}^n(K)^{1 - \frac{1}{p}},$$

and so
$$\sup_k \mu_k(K) < \infty.$$

Accordingly, we may apply Theorem 1.41, to find a Radon measure μ on $\mathbb{R}^n$ and a subsequence $\mu_{k_j} \rightharpoonup \mu$.

3. *Claim #1:* $\mu << \mathcal{L}^n$.

Proof of claim: Let $A \subset \mathbb{R}^n$ be bounded, $\mathcal{L}^n(A) = 0$. Fix $\epsilon > 0$ and choose an open, bounded set $V \supset A$ such that $\mathcal{L}^n(V) < \epsilon$. Then

$$\mu(V) \leq \liminf_{j \to \infty} \mu_{k_j}(V)$$

$$= \liminf_{j \to \infty} \int_V f_{k_j}\, dx$$

$$\leq \liminf_{j \to \infty} \left(\int_V f_{k_j}^p\, dx \right)^{\frac{1}{p}} \mathcal{L}^n(V)^{1-\frac{1}{p}}$$

$$\leq C\epsilon^{1-\frac{1}{p}}.$$

Thus $\mu(A) = 0$.

4. In view of Theorem 1.30, there exists a function $f \in L^1_{\text{loc}}$ satisfying

$$\mu(A) = \int_A f\, dx$$

for all Borel sets $A \subseteq \mathbb{R}^n$.

5. *Claim #2:* $f \in L^p(\mathbb{R}^n)$.

Proof of claim: Let $\phi \in C_c(\mathbb{R}^n)$. Then

$$\int_{\mathbb{R}^n} \phi f\, dx = \int_{\mathbb{R}^n} \phi\, d\mu = \lim_{j \to \infty} \int_{\mathbb{R}^n} \phi\, d\mu_{k_j}$$

$$= \lim_{j \to \infty} \int_{\mathbb{R}^n} \phi f_{k_j}\, dx \leq \sup_k \|f_k\|_{L^p} \|\phi\|_{L^q}$$

$$\leq C\|\phi\|_{L^q},$$

where $\frac{1}{p} + \frac{1}{q} = 1$, $1 < q < \infty$. Thus

$$\|f\|_{L^p} = \sup_{\substack{\phi \in C_c(\mathbb{R}^n) \\ \|\phi\|_{L^q} \leq 1}} \int_{\mathbb{R}^n} \phi f\, dx < \infty.$$

6. *Claim #3:* $f_{k_j} \rightharpoonup f$ in $L^p(\mathbb{R}^n)$.

Proof of claim: As noted above,

$$\int_{\mathbb{R}^n} f_{k_j}\phi\, dx \to \int_{\mathbb{R}^n} f\phi\, dx$$

for all $\phi \in C_c(\mathbb{R}^n)$. Given $g \in L^q(\mathbb{R}^n)$, we fix $\epsilon > 0$ and then choose $\phi \in C_c(\mathbb{R}^n)$ with

$$\|g - \phi\|_{L^q(\mathbb{R}^n)} < \epsilon.$$

Then

$$\int_{\mathbb{R}^n} (f_{k_j} - f)g\, dx = \int_{\mathbb{R}^n} (f_{k_j} - f)\phi\, dx + \int_{\mathbb{R}^n} (f_{k_j} - f)(g - \phi)\, dx.$$

The first term on the right goes to zero, and the last term is estimated by

$$\|f_{k_j} - f\|_{L^p}\|g - \phi\|_{L^q} \leq C\epsilon. \qquad \square$$

1.9.3   Weak convergence in L^1

The L^p weak compactness Theorem 1.42 fails for L^1, since its dual space L^∞ is not separable. We need more information to find weakly convergent sequences in L^1:

THEOREM 1.43 (Uniform integrability and weak convergence). *Assume U is bounded and let $\{f_k\}_{k=1}^\infty$ be a sequence of functions in $L^1(U)$ satisfying*

$$\sup_k \|f_k\|_{L^1(U)} < \infty. \qquad (\star)$$

Suppose also

$$\lim_{l\to\infty} \sup_k \int_{\{|f_k|\geq l\}} |f_k|\, dx = 0. \qquad (\star\star)$$

Then there exist a subsequence $\{f_{k_j}\}_{j=1}^\infty$ and $f \in L^1(U)$ such that

$$f_{k_j} \rightharpoonup f \quad in\ L^1(U).$$

Remark. We call condition $(\star\star)$ *uniform integrability.* $\qquad \square$

Proof. 1. *Claim #1:* For each $\epsilon > 0$, there exists $\delta > 0$ such that

$$\mathcal{L}^n(E) < \delta \quad \text{implies} \quad \sup_k \int_E |f_k|\, dx < \epsilon$$

for each $\mathcal{L}^n$-measurable set $E \subset U$.

Proof of claim: For each $j = 1, \ldots,$

$$\int_E |f_j| \, dx = \int_{E \cap \{|f_j| \geq l\}} |f_j| \, dx + \int_{E \cap \{|f_j| < l\}} |f_j| \, dx$$

$$\leq \sup_k \int_{\{|f_k| \geq l\}} |f_k| \, dx + l \mathcal{L}^n(E)$$

$$< \epsilon,$$

provided we employ $(\star\star)$ to fix l so large that

$$\sup_k \int_{\{|f_k| \geq l\}} |f_k| \, dx < \frac{\epsilon}{2}$$

and then let $\delta = \frac{\epsilon}{2l}$.

2. As in the proof of Theorem 1.42, we may assume $f_k \geq 0$ and define the Radon measures

$$\mu_k := \mathcal{L}^n \llcorner f_k \quad (k = 1, 2, \ldots).$$

Then

$$\sup_k \mu_k(U) < \infty.$$

We invoke Theorem 1.41, to find a Radon measure μ on $\mathbb{R}^n$ and a subsequence $\mu_{k_j} \rightharpoonup \mu$. We can use Claim #1 to prove that $\mu <\!< \mathcal{L}^n \llcorner U$; and consequently, according to Theorem 1.30, there exists a function $f \in L^1(U)$ satisfying

$$\mu(E) = \int_E f \, dx$$

for each Borel set $E \subseteq \mathbb{R}^n$.

3. *Claim #2:* $f_{k_j} \rightharpoonup f$ in $L^1(U)$.

Proof of claim: Select any function $g \in L^\infty(U)$; we must show that

$$\int_U f_{k_j} g \, dx \to \int_U f g \, dx.$$

Using mollifiers (see Theorem 4.1 later) we obtain a sequence $\{g_i\}_{i=1}^\infty$ of bounded, continuous functions that converge to g $\mathcal{L}^n$-a.e. Fix $\epsilon > 0$ and select the corresponding δ given by Claim #1. According the Egoroff's Theorem, there exists a measurable set $E \subset U$ such that

$$g_i \to g \quad \text{uniformly on } U - E, \ \mathcal{L}^n(E) \leq \delta.$$

Then

$$\left| \int_U (f_{k_j} - f) g \, dx \right| \le \int_U |f_{k_j} - f| |g - g_i| \, dx$$
$$+ \left| \int_U (f_{k_j} - f) g_i \, dx \right|$$
$$\le \int_E |f_{k_j} - f| |g - g_i| \, dx + \int_{U-E} |f_{k_j} - f| |g - g_i| \, dx$$
$$+ \left| \int_U (f_{k_j} - f) g_i \, dx \right|$$
$$\le C \int_E |f_{k_j}| + |f| \, dx + C \sup_{U-E} |g - g_i| + \left| \int_U (f_{k_j} - f) g_i \, dx \right|$$
$$\le C\epsilon + C \sup_{U-E} |g - g_i| + \left| \int_U (f_{k_j} - f) g_i \, dx \right|,$$

according to Claim #1. We fix i so large that the second term is smaller than ϵ, and then send $k_j \to \infty$, to deduce that

$$\limsup_{j \to \infty} \left| \int_U f_{k_j} g \, dx - \int_U f g \, dx \right| \le C\epsilon. \qquad \square$$

If a sequence bounded in L^1 fails to satisfy the uniform integrability condition $(\star\star)$ from Theorem 1.43, we can nevertheless still find a subsequence weakly convergent off a set of arbitrarily small $\mathcal{L}^n$ measure, a tiny "bite" taken from U:

THEOREM 1.44 (Biting Lemma). *Assume U is bounded and let $\{f_k\}_{k=1}^\infty$ be a sequence of functions in $L^1(U)$ satisfying*

$$\sup_k \|f_k\|_{L^1(U)} < \infty. \qquad (\star)$$

There exists a subsequence $\{f_{k_j}\}_{j=1}^\infty$ and a function $f \in L^1(U)$ such that for each $\delta > 0$ there exists an $\mathcal{L}^n$-measurable set $E \subset U$ with

$$\mathcal{L}^n(E) \le \delta$$

and

$$f_{k_j} \rightharpoonup f \quad \text{in } L^1(U - E).$$

Proof. 1. For $k = 1, \ldots$ and integers $l \ge 0$, define

$$\phi_k(l) := \int_{\{|f_k| \ge l\}} |f_k| \, dx.$$

Then the mapping $l \mapsto \phi_k(l)$ is nonincreasing for each k and the functions $\{\phi_k\}_{k=1}^\infty$ are uniformly bounded on $\mathbb{Z}^+$.

Using the standard diagonal argument, we can find a subsequence $k_j \to \infty$ such that the limits

$$\alpha(l) := \lim_{j \to \infty} \phi_{k_j}(l)$$

exist for all integers $l = 0, 1, \ldots$. Furthermore, $l \mapsto \alpha(l)$ is nonincreasing and consequently the limit

$$\alpha_\infty := \lim_{\substack{l \to \infty \\ l \text{ integer}}} \alpha(l)$$

exists.

2. *Case 1*: $\alpha_\infty = 0$. In this situation,

$$\lim_{l \to \infty} \sup_j \int_{\{|f_{k_j}| \geq l\}} |f_{k_j}| \, dx = 0,$$

and hence the L^1 weak convergence Theorem 1.43 applies. Consequently, passing if necessary to a further subsequence and reindexing, we have

$$f_{k_j} \rightharpoonup f \quad \text{in } L^1(U)$$

and so we can take $E = \emptyset$.

3. *Case 2*: $\alpha_\infty > 0$. We must construct a small set E off which a further subsequence converges weakly.

4. *Claim #1*: There exists a sequence $\{l_j\}_{j=1}^\infty$ of integers such that

$$l_j \to \infty, \quad \phi_{k_j}(l_j) \to \alpha_\infty.$$

Proof of claim: Define

$$l_j := \max\{l \in \mathbb{Z}^+ \mid \phi_{k_j}(l) \geq \alpha_\infty - \tfrac{1}{j}\};$$

the maximum exists since $\lim_{l \to \infty} \phi_k(l) = 0$ for each k. Also, if $\sup_j l_j$ is finite, then for $l' > \sup_j l_j$ we would have

$$\phi_{k_j}(l') < \alpha_\infty - \tfrac{1}{l'}$$

for all j. Then $\alpha(l') \leq \alpha_\infty - \tfrac{1}{l'}$, a contradiction since $l \mapsto \alpha(l)$ is nonincreasing. Hence, passing if necessary to a subsequence and reindexing, we may assume $l_j \to \infty$.

Now fix a positive integer m. Then for large enough j,

$$\alpha_\infty - \tfrac{1}{l_j} \leq \phi_{k_j}(l_j) \leq \phi_{k_j}(m)$$

and so

$$\alpha_\infty \leq \liminf_{j \to \infty} \phi_{k_j}(l_j) \leq \limsup_{j \to \infty} \phi_{k_j}(l_j) \leq \alpha(m).$$

Letting $m \to \infty$, we deduce that $\lim_{j \to \infty} \phi_{k_j}(l_j) = \alpha_\infty$.

5. *Claim #2*: We have

$$\lim_{m \to \infty} \sup_j \int_{\{m \leq |f_{k_j}| \leq l_j\}} |f_{k_j}| \, dx = 0.$$

Proof of claim: Select any $\epsilon > 0$ and then $m_0 \in \mathbb{Z}^+$ such that $\alpha(m_0) < \alpha_\infty + \epsilon$. Next, pick j_0 so that $j \geq j_0$ implies

$$\phi_{k_j}(m_0) \leq \alpha(m_0) + \epsilon, \quad \phi_{k_j}(l_j) \geq \alpha_\infty - \epsilon.$$

Then

$$\phi_{k_j}(m_0) - \phi_{k_j}(l_j) \leq \alpha(m_0) - \alpha_\infty + 2\epsilon \leq 3\epsilon.$$

Hence if $m_1 \geq \max\{m_0, \max_{0 \leq j \leq j_0} l_j\}$, we have

$$\sup_j \int_{\{m_1 \leq |f_{k_j}| \leq l_j\}} |f_{k_j}| \, dx = \sup_{j, \, l_j > m} \{\phi_j(m_1) - \phi_{k_j}(l_j)\} \leq 3\epsilon.$$

6. Given now $\delta > 0$, pass to a further subsequence if necessary to guarantee that

$$\sum_{j=1}^\infty \frac{1}{l_j} \leq \delta.$$

Define

$$E := \bigcup_{j=1}^\infty \{|f_{k_j}| \geq l_j\}.$$

Then

$$\mathcal{L}^n(E) \leq \sum_{j=1}^\infty \mathcal{L}^n(\{|f_{k_j}| \geq l_j\})$$

$$\leq \sum_{j=1}^\infty \frac{1}{l_j} \int_U |f_{k_j}| \, dx \leq C \sum_{j=1}^\infty \frac{1}{l_j} \leq C\delta,$$

and

$$\lim_{m \to \infty} \sup_j \int_{\{|f_{k_j}| \geq m\} - E} |f_{k_j}| \, dx$$

$$\leq \lim_{m \to \infty} \sup_j \int_{\{m \leq |f_{k_j}| \leq l_j\}} |f_{k_j}| \, dx = 0,$$

owing to Claim #2. We now apply Theorem 1.43 to extract a further subsequence weakly convergent in $L^1(U - E)$.

7. Repeating this construction for $\delta = 1, \frac{1}{2}, \ldots, \frac{1}{2^m}, \ldots$ and reindexing, we obtain the desired subsequence. $\square$

1.9.4 Measures of oscillation

Let μ be a finite Radon measure on $\mathbb{R}^{n+m}$.

DEFINITION 1.33. *The **projection** of μ onto $\mathbb{R}^n$ is the measure σ defined by*

$$\sigma(A) := \mu(A \times \mathbb{R}^m)$$

for $A \subseteq \mathbb{R}^n$.

THEOREM 1.45 (Slicing measures). *For σ-a.e. point $x \in \mathbb{R}^n$ there exists a Radon measure ν_x on $\mathbb{R}^m$ such that*

$$\nu_x(\mathbb{R}^m) = 1 \quad \sigma\text{-a.e.};$$

and for each bounded continuous function $f : \mathbb{R}^n \times \mathbb{R}^m \to \mathbb{R}$, the mapping $x \mapsto \int_{\mathbb{R}^m} f(x,y) \, d\nu_x(y)$ is σ-measurable and

$$\int_{\mathbb{R}^{n+m}} f \, d\mu = \int_{\mathbb{R}^n} \left(\int_{\mathbb{R}^m} f \, d\nu_x \right) d\sigma. \tag{$\star$}$$

Proof. 1. Let $\{f_k\}_{k=1}^{\infty}$ be a countable, dense subset of $C_c(\mathbb{R}^m)$. For each $k = 1, \ldots$, define the signed measure γ^k by

$$\gamma^k(A) := \int_{A \times \mathbb{R}^m} f_k(y) \, d\mu \quad (A \text{ Borel}, A \subseteq \mathbb{R}^n).$$

Then $\gamma^k \ll \sigma$. Hence Theorems 1.29 and 1.30 imply that for $k = 1, \ldots$ the limits

$$D_\sigma \gamma^k(x) := \lim_{r \to 0} \frac{\gamma^k(B(x,r))}{\sigma(B(x,r))} \tag{$\star\star$}$$

exist for σ-a.e. x; the mappings $x \mapsto D_\sigma \gamma^k$ are σ-measurable and bounded; and we can write

$$\int_{A \times \mathbb{R}^m} f_k(y)\, d\mu = \gamma^k(A) = \int_A D_\sigma \gamma^k \, d\sigma \qquad (\star\star\star)$$

for $k = 1, \ldots$ and Borel sets $A \subseteq \mathbb{R}^n$.

2. Since

$$\left| \frac{\gamma^k(B(x,r))}{\sigma(B(x,r))} - \frac{\gamma^l(B(x,r))}{\sigma(B(x,r))} \right| \le \max_{\mathbb{R}^m} |f_k - f_l|,$$

we have

$$|D_\sigma \gamma^k(x) - D_\sigma \gamma^l(x)| \le \max_{\mathbb{R}^m} |f_k - f_l|$$

at any point x where ($\star\star$) holds.

Given a function $f \in C_c(\mathbb{R}^m)$, select a subsequence $\{f_{k_j}\}_{j=1}^\infty \subset \{f_k\}_{k=1}^\infty$ such that $f_{k_j} \to f$ uniformly as $j \to \infty$. Then

$$\Lambda_x(f) := \lim_{j \to \infty} D_\sigma \gamma^{k_j}(x)$$

exists for each point x satisfying ($\star\star$) and is independent of the particular choice of the subsequence $f_{k_j} \to f$.

3. For such points x, the mapping $f \mapsto \Lambda_x(f)$ is linear, with

$$|\Lambda_x(f)| \le \max_{\mathbb{R}^n} |f|$$

for each $f \in C_c(\mathbb{R}^m)$. According to the Riesz Representation Theorem, there exists a Radon measure ν_x such that

$$\Lambda_x(f) = \int_{\mathbb{R}^m} f \, d\nu_x.$$

Passing to limits in ($\star\star\star$) as $k = k_j \to \infty$, we deduce that

$$\int_{A \times \mathbb{R}^m} f(y)\, d\mu = \int_A \left(\int_{\mathbb{R}^m} f \, d\nu_x \right) d\sigma.$$

An approximation now shows that

$$\int_{\mathbb{R}^{n+m}} g(x) f(y)\, d\mu = \int_{\mathbb{R}^n} g(x) \left(\int_{\mathbb{R}^m} f \, d\nu_x \right) d\sigma$$

for continuous functions f and g. Putting $f \equiv 1$ shows that $\nu_x(\mathbb{R}^m) = 1$. A further approximation gives the integration formula ($\star$). $\qquad \square$

A weakly convergent sequence $f_k \rightharpoonup f$ need not converge almost everywhere, even if we pass to a subsequence. It is possible for example that the functions f_k may oscillate more and more rapidly as $k \to \infty$. We can however introduce measures ν_x to "record" these oscillations near almost every point x.

THEOREM 1.46 (Weak limits and Young measures). *Suppose that U is bounded and the sequence $\{f_k\}_{k=1}^{\infty}$ is bounded in $L^{\infty}(U; \mathbb{R}^m)$.*

Then there exist a subsequence $\{f_{k_j}\}_{j=1}^{\infty} \subseteq \{f_k\}_{k=1}^{\infty}$ and, for $\mathcal{L}^n$-a.e. $x \in U$, a Radon measure ν_x on $\mathbb{R}^m$ such that

$$\nu_x(\mathbb{R}^m) = 1$$

and

$$F(f_{k_j}) \rightharpoonup \overline{F} := \int_{\mathbb{R}^m} F(y) \, d\nu_x$$

weakly in $L^2(U)$ for each continuous function F on $\mathbb{R}^m$.

DEFINITION 1.34. *We call $\{\nu_x\}_{x \in U}$ Young measures associated with the subsequence $\{f_{k_j}\}_{j=1}^{\infty}$.*

Remark. In fact $F(f_{k_j}) \rightharpoonup \overline{F}$ weakly in $L^p(U)$ for each $1 < p < \infty$. If $f_k \to f$ $\mathcal{L}^n$-a.e., then $\nu_x = \delta_{f(x)}$ almost everywhere. $\qquad \square$

Proof. 1. For each Borel set $A \subseteq \mathbb{R}^{n+m}$, define

$$\mu_k(A) := \int_U \chi_A(x, f_k(x)) \, dx.$$

According to Theorem 1.10,(iii), the integrand is $\mathcal{L}^n$-measurable. We extend μ_k to a Radon measure on $\mathbb{R}^{n+m}$ and observe that

$$\sup_k \mu_k(\mathbb{R}^{n+m}) < \infty.$$

Consequently, there exists a subsequence $\{\mu_{k_j}\}_{j=1}^{\infty}$ and a finite Radon measure μ such that

$$\mu_{k_j} \rightharpoonup \mu \quad \text{weakly on } \mathbb{R}^{n+m}.$$

2. *Claim:* The projection of μ onto $\mathbb{R}^n$ is $\sigma := \mathcal{L}^n \, \llcorner \, U$.

Proof of claim: Let $V \subset U$ be open. Then

$$\sigma(V) = \mu(V \times \mathbb{R}^m) \le \liminf_{j \to \infty} \mu_{k_j}(V \times \mathbb{R}^m) = \mathcal{L}^n(V).$$

Thus $\sigma \le \mathcal{L}^n \llcorner U$. Now let $K \subset U$ be compact. There exists a closed ball $B(R) \subset \mathbb{R}^m$ such that $\mathrm{spt}(\mu), \mathrm{spt}(\mu_{k_j}) \subset \bar{U} \times B(R)$ for $j = 1, \ldots$. Therefore

$$\sigma(K) = \mu(K \times \mathbb{R}^m) = \mu(K \times B(R))$$
$$\ge \limsup_{j \to \infty} \mu_{k_j}(K \times B(R)) = \mathcal{L}^n(K);$$

and consequently $\sigma \ge \mathcal{L}^n \llcorner U$.

3. According therefore to Theorem 1.45, we have

$$\int_{\mathbb{R}^{n+m}} f \, d\mu = \int_U \left(\int_{\mathbb{R}^m} f \, d\nu_x \right) dx$$

for each $f \in C(\mathbb{R}^{n+m})$. Let $f(x, y) = g(x)F(y)$, where $g \in C_c(U)$ and $F \in C(\mathbb{R}^m)$. Then

$$\lim_{j \to \infty} \int_U gF(f_{k_j}) \, dx = \lim_{j \to \infty} \int_{\mathbb{R}^{n+m}} f \, d\mu_{k_j}$$
$$= \int_{\mathbb{R}^{n+m}} f \, d\mu$$
$$= \int_U \left(\int_{\mathbb{R}^m} f \, d\nu_x \right) dx$$
$$= \int_U g\bar{F} \, dx.$$

This is valid for each $g \in C_c(U)$ and thus for each $g \in L^2(U)$. Consequently $F(f_{k_j}) \rightharpoonup \bar{F}$ weakly in $L^2(U)$. $\qquad \square$

1.10   References and notes

This long chapter mostly follows Federer [F], with additional material from Simon [S, Chapter 1] and Hardt [H].

Theorems 1.1–1.6 in Section 1.1 are from [F, Sections 2.1.3, 2.1.5].

The π-λ Theorem is from Durrett [D] and Stroock [Sk]. Lemma 1.1 is [F, Section 2.2.2] and Theorem 1.8 is [F, Section 2.2.5]. Theorem 1.9 (Caratheodory's criterion) and Theorem 1.11 are from [F, Section 2.3.2]; Theorem 1.12 is [F, Section 2.3.3].

Theorem 1.13 in Section 1.2 is a simplified version of [F, Sections 3.1.13, 3.1.14]. Lusin's and Egoroff's theorems are in [F, Sections 2.3.5, 2.3.7]. The general theory of integration may be found in [F, Section 2.4]; see in particular [F, Sections 2.4.6, 2.4.7, 2.4.9] for Fatou's Lemma, the Monotone Convergence Theorem, and the Dominated Convergence Theorem. Our treatment of product measure and Fubini's Theorem is taken directly from [F, Section 2.6]. The π-λ Theorem provides an alternative approach to assertion (iii) of Fubini's Theorem; see Durrett [D] or Stroock [Sk].

We relied heavily on Simon [S] for Vitali's Covering Theorem and Hardt [H] for Besicovitch's Covering Theorem. Consult also [F, Sections 2.9.12, 2.9.13]. Furedi and Loeb [F-L] discuss the best constant for Besicovitch's Theorem. The differentiation theory in Section 1.6 is based on [H], [S], and [F, Section 2.9]. Approximate limits and approximate continuity are in [F, Sections 2.9.12, 2.9.13].

We took the proof of the Riesz Representation Theorem from [S, Sections 4.1, 4.2] (cf. [F, Section 2.5]). A. Damlamian showed us the proof of Theorem 1.39 in Section 1.8. See Giusti [G, Appendix A] for Theorem 1.41 in Section 1.9. Theorem 1.43 is sometimes called the Dunford–Pettis Theorem. The Biting Lemma appears in Brooks–Chacon [B-C] and our proof follows Ball–Murat [B-M]. L. Tartar introduced Young measure methods for PDE theory; see for instance [T]. The proofs for Theorems 1.45 and 1.46 are from [E1].

There are many good introductory measure theory books. We recommend especially DiBenedetto [DB], Fitzpatrick–Royden [F-R], Folland [Fo] and Wheeden–Zygmund [W-Z]. Krantz–Parks [K-P] and Lin–Yang [L-Y] are two very good more advanced texts. Oxtoby [O] provides a fascinating discussion of measure theory versus Baire category for characterizing "negligible" sets.

Chapter 2

Hausdorff Measures

We introduce next certain "lower dimensional" measures on $\mathbb{R}^n$, which allow us to measure certain "very small" subsets of $\mathbb{R}^n$. These are the Hausdorff measures $\mathcal{H}^s$, defined in terms of the diameters of various efficient coverings. The idea is that A is an "s-dimensional subset" of $\mathbb{R}^n$ if $0 < \mathcal{H}^s(A) < \infty$, even if A is very complicated geometrically.

Section 2.1 provides the definitions and basic properties of Hausdorff measures. In Section 2.2 we prove n-dimensional Lebesgue and n-dimensional Hausdorff measure agree on $\mathbb{R}^n$. Density theorems for lower dimensional Hausdorff measures are established in Section 2.3. Section 2.4 records for later use some easy facts concerning the Hausdorff dimension of graphs and the sets where a summable function is large.

2.1 Definitions and elementary properties

DEFINITION 2.1.

(i) *Let* $A \subseteq \mathbb{R}^n, 0 \le s < \infty, 0 < \delta \le \infty$. *We write*

$$\mathcal{H}^s_\delta(A) :=$$
$$\inf \left\{ \sum_{j=1}^\infty \alpha(s) \left(\frac{\operatorname{diam} C_j}{2} \right)^s \mid A \subseteq \bigcup_{j=1}^\infty C_j, \operatorname{diam} C_j \le \delta \right\},$$

where
$$\alpha(s) := \frac{\pi^{\frac{s}{2}}}{\Gamma\left(\frac{s}{2} + 1\right)}.$$

(ii) *For A and s as above, define*
$$\mathcal{H}^s(A) := \lim_{\delta \to 0} \mathcal{H}^s_\delta(A) = \sup_{\delta > 0} \mathcal{H}^s_\delta(A).$$

We call $\mathcal{H}^s$ **s-dimensional Hausdorff measure** *on $\mathbb{R}^n$.*

Remarks.

(i) Our requiring $\delta \to 0$ forces the coverings to "follow the local geometry" of the set A.

(ii) In the definition, $\Gamma(s) := \int_0^\infty e^{-x} x^{s-1} \, dx$ $(0 < s < \infty)$ is the gamma function. Observe that

$$\mathcal{L}^n(B(x,r)) = \alpha(n) r^n$$

for balls $B(x,r) \subset \mathbb{R}^n$. We will see later in Chapter 3 that if $s = k$ is an integer, $\mathcal{H}^k$ agrees with ordinary "k-dimensional surface area" on nice sets; this is the reason we include the normalizing constant $\alpha(s)$ in the definition. $\qquad\square$

THEOREM 2.1 (Hausdorff measures are Borel). *For all $0 \le s < \infty$ $\mathcal{H}^s$ is a Borel regular measure in $\mathbb{R}^n$.*

Warning: $\mathcal{H}^s$ is *not* a Radon measure if $0 \le s < n$, since $\mathbb{R}^n$ is not σ-finite with respect to $\mathcal{H}^s$.

Proof. 1. *Claim #1: $\mathcal{H}_\delta^s$ is a measure.*
Proof of claim: Choose $\{A_k\}_{k=1}^\infty \subseteq \mathbb{R}^n$ and suppose $A_k \subseteq \cup_{j=1}^\infty C_j^k$, diam $C_j^k \le \delta$. Then $\{C_j^k\}_{j,k=1}^\infty$ covers $\cup_{k=1}^\infty A_k$. Thus

$$\mathcal{H}_\delta^s\left(\bigcup_{k=1}^\infty A_k\right) \le \sum_{k=1}^\infty \sum_{j=1}^\infty \alpha(s) \left(\frac{\operatorname{diam} C_j^k}{2}\right)^s.$$

Taking infima, we find

$$\mathcal{H}_\delta^s\left(\bigcup_{k=1}^\infty A_k\right) \le \sum_{k=1}^\infty \mathcal{H}_\delta^s(A_k).$$

2. *Claim #2: $\mathcal{H}^s$ is a measure.*
Proof of claim: Select $\{A_k\}_{k=1}^\infty \subseteq \mathbb{R}^n$. Then

$$\mathcal{H}_\delta^s\left(\bigcup_{k=1}^\infty A_k\right) \le \sum_{k=1}^\infty \mathcal{H}_\delta^s(A_k) \le \sum_{k=1}^\infty \mathcal{H}^s(A_k).$$

Let $\delta \to 0$.

3. *Claim #3*: $\mathcal{H}^s$ is a Borel measure.

Proof of claim: Choose $A, B \subseteq \mathbb{R}^n$ with $\text{dist}(A, B) > 0$. Select $0 < \delta < \frac{1}{4} \text{dist}(A, B)$. Suppose $A \cup B \subseteq \cup_{k=1}^\infty C_k$ and $\text{diam} \, C_k \leq \delta$.

Write $\mathcal{A} := \{C_j \mid C_j \cap A \neq \emptyset\}$, and let $\mathcal{B} := \{C_j \mid C_j \cap B \neq \emptyset\}$. Then $A \subseteq \cup_{C_j \in \mathcal{A}} C_j$ and, $B \subseteq \cup_{C_j \in \mathcal{B}} C_j$, $C_i \cap C_j = \emptyset$ if $C_i \in \mathcal{A}, C_j \in \mathcal{B}$. Hence

$$\sum_{j=1}^\infty \alpha(s) \left(\frac{\text{diam} \, C_j}{2}\right)^s \geq \sum_{C_j \in \mathcal{A}} \alpha(s) \left(\frac{\text{diam} \, C_j}{2}\right)^s$$

$$+ \sum_{C_j \in \mathcal{B}} \alpha(s) \left(\frac{\text{diam} \, C_j}{2}\right)^s$$

$$\geq \mathcal{H}_\delta^s(A) + \mathcal{H}_\delta^s(B).$$

Taking the infimum over all such sets $\{C_j\}_{j=1}^\infty$, we find $\mathcal{H}_\delta^s(A \cup B) \geq \mathcal{H}_\delta^s(A) + \mathcal{H}_\delta^s(B)$, provided $0 < 4\delta < \text{dist}(A, B)$. Letting $\delta \to 0$, we obtain $\mathcal{H}^s(A \cup B) \geq \mathcal{H}^s(A) + \mathcal{H}^s(B)$. Consequently,

$$\mathcal{H}^s(A \cup B) = \mathcal{H}^s(A) + \mathcal{H}^s(B)$$

for all $A, B \subseteq \mathbb{R}^n$ with $\text{dist}(A, B) > 0$. Hence Caratheodory's criterion implies $\mathcal{H}^s$ is a Borel measure.

4. *Claim #4*; $\mathcal{H}^s$ is a Borel regular measure.

Proof of claim: Note that $\text{diam} \, \bar{C} = \text{diam} \, C$ for all C; hence

$$\mathcal{H}_\delta^s(A) =$$

$$\inf \left\{ \sum_{j=1}^\infty \alpha(s) \left(\frac{\text{diam} \, C_j}{2}\right)^s \mid A \subseteq \bigcup_{j=1}^\infty C_j, \text{diam} \, C_j \leq \delta, C_j \text{ closed} \right\}.$$

Choose $A \subseteq \mathbb{R}^n$ such that $\mathcal{H}^s(A) < \infty$; then $\mathcal{H}_\delta^s(A) < \infty$ for all $\delta > 0$. For each $k \geq 1$, choose closed sets $\{C_j^k\}_{j=1}^\infty$ so that $\text{diam} \, C_j^k \leq \frac{1}{k}$, $A \subseteq \cup_{j=1}^\infty C_j^k$, and

$$\sum_{j=1}^\infty \alpha(s) \left(\frac{\text{diam} \, C_j^k}{2}\right)^s \leq \mathcal{H}_{\frac{1}{k}}^s(A) + \frac{1}{k}.$$

Let $A_k := \cup_{j=1}^{\infty} C_j^k$, $B := \cap_{k=1}^{\infty} A_k$; B is Borel. Also $A \subseteq A_k$ for each k, and so $A \subseteq B$, Furthermore,

$$\mathcal{H}_{\frac{1}{k}}^s(B) \leq \sum_{j=1}^{\infty} \alpha(s) \left(\frac{\operatorname{diam} C_j^k}{2} \right)^s \leq \mathcal{H}_{\frac{1}{k}}^s(A) + \frac{1}{k}.$$

Letting $k \to \infty$, we discover $\mathcal{H}^s(B) \leq \mathcal{H}^s(A)$. But $A \subseteq B$, and thus $\mathcal{H}^s(A) = \mathcal{H}^s(B)$. $\qquad\square$

THEOREM 2.2 (Properties of Hausdorff measure).

(i) $\mathcal{H}^0$ is counting measure.

(ii) $\mathcal{H}^1 = \mathcal{L}^1$ on $\mathbb{R}^1$.

(iii) $\mathcal{H}^s \equiv 0$ on $\mathbb{R}^n$ for all $s > n$.

(iv) $\mathcal{H}^s(\lambda A) = \lambda^s \mathcal{H}^s(A)$ for all $\lambda > 0, A \subseteq \mathbb{R}^n$.

(v) $\mathcal{H}^s(L(A)) = \mathcal{H}^s(A)$ for each affine isometry $L : \mathbb{R}^n \to \mathbb{R}^n, A \subseteq \mathbb{R}^n$.

Proof. 1. Statements (iv) and (v) are easy.

2. First observe $\alpha(0) = 1$. Thus obviously $\mathcal{H}^0(\{a\}) = 1$ for all $a \in \mathbb{R}^n$, and (i) follows.

3. Choose $A \subseteq \mathbb{R}^1$ and $\delta > 0$. Then

$$\mathcal{L}^1(A) = \inf \left\{ \sum_{j=1}^{\infty} \operatorname{diam} C_j \mid A \subseteq \bigcup_{j=1}^{\infty} C_j \right\}$$

$$\leq \inf \left\{ \sum_{j=1}^{\infty} \operatorname{diam} C_j \mid A \subseteq \bigcup_{j=1}^{\infty} C_j, \operatorname{diam} C_j \leq \delta \right\}$$

$$= \mathcal{H}_\delta^1(A),$$

since $\Gamma(\frac{3}{2}) = \frac{\sqrt{\pi}}{2}$ and thus $\alpha(1) = 2$. Hence $\mathcal{L}^1(A) \leq \mathcal{H}^1(A)$.

On the other hand, set $I_k := [k\delta, (k+1)\delta]$ for $k \in \mathbb{Z}$. Then $\operatorname{diam}(C_j \cap I_k) \leq \delta$ and

$$\sum_{k=-\infty}^{\infty} \operatorname{diam}(C_j \cap I_k) \leq \operatorname{diam} C_j.$$

Hence

$$\mathcal{L}^1(A) = \inf \left\{ \sum_{j=1}^{\infty} \operatorname{diam} C_j \mid A \subseteq \bigcup_{j=1}^{\infty} C_j \right\}$$

$$\geq \inf \left\{ \sum_{j=1}^{\infty} \sum_{k=-\infty}^{\infty} \operatorname{diam}(C_j \cap I_k) \mid A \subseteq \bigcup_{j=1}^{\infty} C_j \right\}$$

$$\geq H_\delta^1(A).$$

Thus $\mathcal{L}^1 = H_\delta^1$ for all $\delta > 0$, and so $\mathcal{L}^1 = \mathcal{H}^1$ on $\mathbb{R}^1$.

4. Fix an integer $m \geq 1$. The unit cube Q in $\mathbb{R}^n$ can be decomposed into m^n cubes with side $\frac{1}{m}$ and diameter $\frac{\sqrt{n}}{m}$. Therefore

$$H_{\frac{\sqrt{n}}{m}}^s(Q) \leq \sum_{i=1}^{m^n} \alpha(s) \left(\frac{\sqrt{n}}{m} \right)^s = \alpha(s) n^{\frac{s}{2}} m^{n-s};$$

and the last term goes to zero as $m \to \infty$, if $s > n$. Hence $\mathcal{H}^s(Q) = 0$, and so $\mathcal{H}^s(\mathbb{R}^n) = 0$. $\qquad\square$

A convenient way to verify that $\mathcal{H}^s$ vanishes on a set is the following:

LEMMA 2.1. *Suppose $A \subset \mathbb{R}^n$ and $\mathcal{H}_\delta^s(A) = 0$ for some $0 < \delta < \infty$. Then $\mathcal{H}^s(A) = 0$.*

Proof. The conclusion is obvious for $s = 0$, and so we may assume $s > 0$.

Fix $\epsilon > 0$. There then exist sets $\{C_j\}_{j=1}^{\infty}$ such that $A \subseteq \cup_{j=1}^{\infty} C_j$, and

$$\sum_{j=1}^{\infty} \alpha(s) \left(\frac{\operatorname{diam} C_j}{2} \right)^s \leq \epsilon.$$

In particular for each i,

$$\operatorname{diam} C_i \leq 2 \left(\frac{\epsilon}{\alpha(s)} \right)^{\frac{1}{s}} =: \delta(\epsilon).$$

Hence

$$H_{\delta(\epsilon)}^s(A) \leq \epsilon.$$

Since $\delta(\epsilon) \to 0$ as $\epsilon \to 0$, we see that $\mathcal{H}^s(A) = 0$. $\qquad\square$

Next we define the Hausdorff dimension of a subset of $\mathbb{R}^n$.

LEMMA 2.2. *Let* $A \subset \mathbb{R}^n$ *and* $0 \le s < t < \infty$.

(i) *If* $\mathcal{H}^s(A) < \infty$, *then* $\mathcal{H}^t(A) = 0$.

(ii) *If* $\mathcal{H}^t(A) > 0$, *then* $\mathcal{H}^s(A) = +\infty$.

Proof. Let $\mathcal{H}^s(A) < \infty$ and $\delta > 0$. Then there exist sets $\{C_j\}_{j=1}^\infty$ such that $\operatorname{diam} C_j \le \delta$, $A \subseteq \cup_{i=1}^\infty C_j$ and

$$\sum_{j=1}^\infty \alpha(s) \left(\frac{\operatorname{diam} C_j}{2} \right)^s \le \mathcal{H}^s_\delta(A) + 1 \le \mathcal{H}^s(A) + 1.$$

Consequently,

$$\begin{aligned}
\mathcal{H}^t_\delta(A) &\le \sum_{j=1}^\infty \alpha(t) \left(\frac{\operatorname{diam} C_j}{2} \right)^t \\
&= \frac{\alpha(t)}{\alpha(s)} 2^{s-t} \sum_{j=1}^\infty \alpha(s) \left(\frac{\operatorname{diam} C_j}{2} \right)^s (\operatorname{diam} C_j)^{t-s} \\
&\le \frac{\alpha(t)}{\alpha(s)} 2^{s-t} \delta^{t-s} (\mathcal{H}^s(A) + 1).
\end{aligned}$$

We send $\delta \to 0$ to conclude $\mathcal{H}^t(A) = 0$. This proves assertion (i). Assertion (ii) follows at once from (i). $\qquad\square$

DEFINITION 2.2. *The **Hausdorff dimension** of a set* $A \subseteq \mathbb{R}^n$ *is*

$$H_{\dim}(A) := \inf\{0 \le s < \infty \mid \mathcal{H}^s(A) = 0\}.$$

Remark. Observe $H_{\dim}(A) \le n$. Let $s = H_{\dim}(A)$. Then $\mathcal{H}^t(A) = 0$ for all $t > s$ and $\mathcal{H}^t(A) = +\infty$ for all $t < s$; $\mathcal{H}^s(A)$ may be any number between 0 and ∞, inclusive.

Furthermore, $H_{\dim}(A)$ need not be an integer. Even if $H_{\dim}(A) = k$ is an integer and $0 < \mathcal{H}^k(A) < \infty$, A need not be a "k-dimensional surface" in any sense; see Falconer [Fa1], [Fa2] or Federer [F] for examples of extremely complicated Cantor-like subsets A of $\mathbb{R}^n$, with $0 < \mathcal{H}^k(A) < \infty$. $\qquad\square$

2.2 Isodiametric inequality, $\mathcal{H}^n = \mathcal{L}^n$

Our goal in this section is to prove $\mathcal{H}^n = \mathcal{L}^n$ on $\mathbb{R}^n$. This is not obvious, since $\mathcal{L}^n$ is defined as the n-fold product of one-dimensional Lebesgue measure $\mathcal{L}^1$ and therefore

$$\mathcal{L}^n(A) = \inf\{\sum_{i=1}^{\infty} \mathcal{L}^n(Q_i) \mid Q_i \text{ cubes}, A \subseteq \cup_{i=1}^{\infty} Q_i\}.$$

On the other hand, $\mathcal{H}^n(A)$ is computed in terms of arbitrary coverings of small diameter.

LEMMA 2.3. *Let $f : \mathbb{R}^n \to [0, \infty]$ be $\mathcal{L}^n$-measurable. Then the region "under the graph of f"*

$$A := \{(x, y) \mid x \in \mathbb{R}^n, y \in \mathbb{R}, 0 \le y \le f(x)\},$$

is $\mathcal{L}^{n+1}$-measurable.

Proof. Set

$$g(x, y) := f(x) - y$$

for $x \in \mathbb{R}^n$ and $y \in \mathbb{R}$. Then g is $\mathcal{L}^{n+1}$-measurable and thus

$$A = \{(x, y) \mid y \ge 0\} \cap \{(x, y) \mid g(x, y) \ge 0\}$$

is $\mathcal{L}^{n+1}$-measurable. $\qquad\qquad\square$

NOTATION Fix $a, b \in \mathbb{R}^n$, with $|a| = 1$. We define

$$L_b^a := \{b + ta \mid t \in \mathbb{R}\},$$

the line through b in the direction a, and

$$P_a := \{x \in \mathbb{R}^n \mid x \cdot a = 0\},$$

the plane through the origin perpendicular to a.

DEFINITION 2.3. *Choose $a \in \mathbb{R}^n$ with $|a| = 1$, and let $A \subset \mathbb{R}^n$. We define the **Steiner symmetrization** of A with respect to the plane P_a to be the set*

$$S_a(A) := \bigcup_{\substack{b \in P_a \\ A \cap L_b^a \ne \emptyset}} \left\{ b + ta \mid |t| \le \frac{1}{2} \mathcal{H}^1(A \cap L_b^a) \right\}.$$

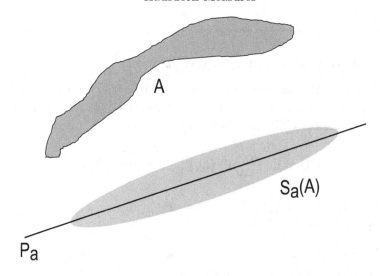

THEOREM 2.3 (Properties of Steiner symmetrization).

(i) $\operatorname{diam} S_a(A) \leq \operatorname{diam} A$.

(ii) *If A is $\mathcal{L}^n$-measurable, then so is $S_a(A)$; and*

$$\mathcal{L}^n(S_a(A)) = \mathcal{L}^n(A).$$

Proof. 1. Statement (i) is trivial if $\operatorname{diam} A = \infty$; assume therefore $\operatorname{diam} A < \infty$. We may also suppose A is closed. Fix $\epsilon > 0$ and select $x, y \in S_a(A)$ such that

$$\operatorname{diam} S_a(A) \leq |x - y| + \epsilon.$$

Write $b := x - (x \cdot a)a$ and $c = y - (y \cdot a)a$; then $b, c \in P_a$. Set

$$r := \inf\{t \mid b + ta \in A\},$$
$$s := \sup\{t \mid b + ta \in A\},$$
$$u := \inf\{t \mid c + ta \in A\},$$
$$v := \sup\{t \mid c + ta \in A\}.$$

Without loss of generality, we may assume $v - r \geq s - u$. Then

$$v - r \geq \frac{1}{2}(v - r) + \frac{1}{2}(s - u)$$

$$= \frac{1}{2}(s - r) + \frac{1}{2}(v - u)$$

$$\geq \frac{1}{2}\mathcal{H}^1\left(A \cap L_b^a\right) + \frac{1}{2}\mathcal{H}^1\left(A \cap L_c^a\right).$$

Now $|x \cdot a| \leq \frac{1}{2}\mathcal{H}^1(A \cap L_b^a)$ and $|y \cdot a| \leq \frac{1}{2}\mathcal{H}^1(A \cap L_c^a)$. Consequently

$$v - r \geq |x \cdot a| + |y \cdot a| \geq |x \cdot a - y \cdot a|.$$

Therefore

$$(\text{diam } S_a(A) - \epsilon)^2 \leq |x - y|^2$$
$$= |b - c|^2 + |x \cdot a - y \cdot a|^2$$
$$\leq |b - c|^2 + (v - r)^2$$
$$= |(b + ra) - (c + va)|^2$$
$$\leq (\text{diam } A)^2,$$

since A is closed and so $b + ra, c + va \in A$. It follows that diam $S_a(A) - \epsilon \leq$ diam A. This establishes (i).

2. As $\mathcal{L}^n$ is rotation invariant, we may assume $a = e_n = (0, \ldots, 0, 1)$. Then $P_a = P_{e_n} = \mathbb{R}^{n-1}$. Since $\mathcal{L}^1 = \mathcal{H}^1$ on $\mathbb{R}^1$, Fubini's Theorem implies the map $f \colon \mathbb{R}^{n-1} \to \mathbb{R}$ defined by $f(b) = \mathcal{H}^1(A \cap L_b^a)$ is $\mathcal{L}^{n-1}$-measurable and $\mathcal{L}^n(A) = \int_{\mathbb{R}^{n-1}} f(b)\, db$. Hence

$$S_a(A) := \left\{(b, y) \mid \frac{-f(b)}{2} \leq y \leq \frac{f(b)}{2}\right\} - \{(b, 0) | L_b^a \cap A = \emptyset\}$$

is $\mathcal{L}^n$-measurable by Lemma 1, and

$$\mathcal{L}^n(S_a(A)) = \int_{\mathbb{R}^{n-1}} f(b)\, db = \mathcal{L}^n(A). \qquad \square$$

Remark. In proving $\mathcal{H}^n = \mathcal{L}^n$ below, observe we only use statement (ii) above in the special case that a is a standard coordinate vector. Since $\mathcal{H}^n$ is obviously rotation invariant, we therefore in fact prove $\mathcal{L}^n$ is rotation invariant. $\qquad \square$

THEOREM 2.4 (Isodiametric inequality). *For all sets $A \subseteq \mathbb{R}^n$,*

$$\mathcal{L}^n(A) \leq \alpha(n)\left(\frac{\text{diam } A}{2}\right)^n.$$

Remark. This is interesting since it is not necessarily the case that A is contained in a ball whose diameter is $\operatorname{diam} A$.   □

Proof. 1. If $\operatorname{diam} A = \infty$, this is trivial; let us therefore suppose $\operatorname{diam} A < \infty$. Let $\{e_1, \ldots, e_n\}$ be the standard basis for $\mathbb{R}^n$. Define $A_1 := S_{e_1}(A)$, $A_2 := S_{e_2}(A_1), \ldots, A_n := S_{e_n}(A_{n-1})$. Write $A^* = A_n$.

2. *Claim #1:* A^* is symmetric with respect to the origin.

Proof of claim: Clearly A_1 is symmetric with respect to P_{e_1}. Let $1 \le k < n$ and suppose A_k is symmetric with respect to $P_{e_1}, \ldots P_{e_k}$. Clearly $A_{k+1} = S_{e_{k+1}}(A_k)$ is symmetric with respect to $P_{e_{k+1}}$ Fix $1 \le j \le k$ and let $S_j : \mathbb{R}^n \to \mathbb{R}^n$ be reflection through P_{e_j}. Let $b \in P_{e_{k+1}}$. Since $S_j(A_k) = A_k$,

$$\mathcal{H}^1(A_k \cap L_b^{e_{k+1}}) = \mathcal{H}^1(A_k \cap L_{S_j b}^{e_{k+1}});$$

consequently

$$\{t \mid b + te_{k+1} \in A_{k+1}\} = \{t \mid S_j b + te_{k+1} \in A_{k+1}\}.$$

Thus $S_j(A_{k+1}) = A_{k+1}$; that is, A_{k+1} is symmetric with respect to P_{e_j}. Thus $A^* = A_n$ is symmetric with respect to $P_{e_1}, \ldots, P_{e_n}$ and so with respect to the origin.

3. *Claim #2:* $\mathcal{L}^n(A^*) \le \alpha(n) \left(\frac{\operatorname{diam} A^*}{2}\right)^n$.

Proof of claim: Choose $x \in A^*$. Then $-x \in A^*$ by Claim #1, and so $\operatorname{diam} A^* \ge 2|x|$. Thus $A^* \subseteq B(0, \frac{\operatorname{diam} A^*}{2})$ and consequently

$$\mathcal{L}^n(A^*) \le \mathcal{L}^n \left(B\left(0, \frac{\operatorname{diam} A^*}{2}\right)\right) = \alpha(n) \left(\frac{\operatorname{diam} A^*}{2}\right)^n.$$

4. *Claim #3:* $\mathcal{L}^n(A) \le \alpha(n) \left(\frac{\operatorname{diam} A}{2}\right)^n$.

Proof of claim: $\bar{A}$ is $\mathcal{L}^n$-measurable, and thus Lemma 2 implies

$$\mathcal{L}^n((\bar{A})^*) = \mathcal{L}^n(\bar{A}), \ \operatorname{diam}(\bar{A})^* \le \operatorname{diam} \bar{A}.$$

Hence Claim # 2 lets us compute

$$\mathcal{L}^n(A) \le \mathcal{L}^n(\bar{A}) = \mathcal{L}^n((\bar{A})^*) \le \alpha(n) \left(\frac{\operatorname{diam}(\bar{A})^*}{2}\right)^n$$

$$\le \alpha(n) \left(\frac{\operatorname{diam} \bar{A}}{2}\right)^n = \alpha(n) \left(\frac{\operatorname{diam} A}{2}\right)^n. \quad □$$

THEOREM 2.5 (n-dimensional Hausdorff and Lebesgue measures). *We have*

$$\mathcal{H}^n = \mathcal{L}^n \quad on \ \mathbb{R}^n.$$

Proof. 1. *Claim #1:* $\mathcal{L}^n(A) \leq \mathcal{H}^n(A)$ for all $A \subseteq \mathbb{R}^n$.

Proof of claim: Fix $\delta > 0$. Choose sets $\{C_j\}_{j=1}^\infty$ such that $A \subseteq \cup_{j=1}^\infty C_j$ and diam $C_j \leq \delta$. Then by the isodiametric inequality,

$$\mathcal{L}^n(A) \leq \sum_{j=1}^\infty \mathcal{L}^n(C_j) \leq \sum_{j=1}^\infty \alpha(n) \left(\frac{\operatorname{diam} C_j}{2} \right)^n.$$

Taking infima, we find $\mathcal{L}^n(A) \leq H_\delta^n(A)$, and thus $\mathcal{L}^n(A) \leq \mathcal{H}^n(A)$.

2. Now, from the definition of $\mathcal{L}^n$ as $\mathcal{L}^1 \times \cdots \times \mathcal{L}^1$, we see that for all $A \subseteq \mathbb{R}^n$ and $\delta > 0$,

$$\mathcal{L}^n(A) = \inf \left\{ \sum_{i=1}^\infty \mathcal{L}^n(Q_i) \mid Q_i \text{ cubes}, A \subseteq \bigcup_{i=1}^\infty Q_i, \operatorname{diam} Q_i \leq \delta \right\}.$$

Here and afterwards we consider only cubes parallel to the coordinate axes in $\mathbb{R}^n$.

3. *Claim #2:* $\mathcal{H}^n$ is absolutely continuous with respect to $\mathcal{L}^n$.

Proof of claim: Set $C_n := \alpha(n)(\frac{\sqrt{n}}{2})^n$. Then for each cube $Q \subset \mathbb{R}^n$,

$$\alpha(n) \left(\frac{\operatorname{diam} Q}{2} \right)^n = C_n \mathcal{L}^n(Q).$$

Thus

$$H_\delta^n(A)$$
$$\leq \inf \left\{ \sum_{i=1}^\infty \alpha(n) \left(\frac{\operatorname{diam} Q_i}{2} \right)^n \mid A \subseteq \bigcup_{i=1}^\infty Q_i, \operatorname{diam} Q_i \leq \delta \right\}$$
$$= C_n \mathcal{L}^n(A),$$

where in the second line the Q_i are cubes. Let $\delta \to 0$.

4. *Claim #3:* $\mathcal{H}^n(A) \leq \mathcal{L}^n(A)$ for all $A \subseteq \mathbb{R}^n$.

Proof of claim: Fix $\delta > 0, \epsilon > 0$. We can select cubes $\{Q_i\}_{i=1}^\infty$, such that $A \subseteq \cup_{i=1}^\infty Q_i$, diam $Q_i < \delta$, and

$$\sum_{i=1}^\infty \mathcal{L}^n(Q_i) \leq \mathcal{L}^n(A) + \epsilon.$$

According to Theorem 1.26, for each i there exist disjoint closed balls $\{B_k^i\}_{k=1}^\infty$ contained in Q_i^o (= interior of Q_i) such that

$$\text{diam } B_k^i \le \delta, \mathcal{L}^n\left(Q_i - \bigcup_{k=1}^\infty B_k^i\right) = \mathcal{L}^n\left(Q_i^o - \bigcup_{k=1}^\infty B_k^i\right) = 0.$$

By Claim #2, $\mathcal{H}^n(Q_i - \cup_{k=1}^\infty B_k^i) = 0$. Thus

$$\mathcal{H}_\delta^n(A) \le \sum_{i=1}^\infty \mathcal{H}_\delta^n(Q_i) = \sum_{i=1}^\infty \mathcal{H}_\delta^n\left(\bigcup_{k=1}^\infty B_k^i\right) \le \sum_{i=1}^\infty \sum_{k=1}^\infty \mathcal{H}_\delta^n(B_k^i)$$

$$\le \sum_{i=1}^\infty \sum_{k=1}^\infty \alpha(n) \left(\frac{\text{diam } B_k^i}{2}\right)^n = \sum_{i=1}^\infty \sum_{k=1}^\infty \mathcal{L}^n(B_k^i)$$

$$= \sum_{i=1}^\infty \mathcal{L}^n\left(\bigcup_{k=1}^\infty B_k^i\right) = \sum_{i=1}^\infty \mathcal{L}^n(Q_i) \le \mathcal{L}^n(A) + \epsilon.$$

Let $\delta, \epsilon \to 0$. $\square$

2.3   Densities

We proved in Section 1.7 that

$$\lim_{r \to 0} \frac{\mathcal{L}^n(B(x,r) \cap E)}{\alpha(n)r^n} = \begin{cases} 1 & \text{for } \mathcal{L}^n\text{-a.e. } x \in E \\ 0 & \text{for } \mathcal{L}^n\text{-a.e. } x \in \mathbb{R}^n - E, \end{cases}$$

provided $E \subseteq \mathbb{R}^n$ is $\mathcal{L}^n$-measurable. This section develops some analogous statements for lower dimensional Hausdorff measures. We assume throughout

$$0 < s < n.$$

THEOREM 2.6 (Density at points not in E). *Assume $E \subset \mathbb{R}^n$, E is $\mathcal{H}^s$-measurable, and $\mathcal{H}^s(E) < \infty$. Then*

$$\lim_{r \to 0} \frac{\mathcal{H}^s(B(x,r) \cap E)}{\alpha(s)r^s} = 0$$

for $\mathcal{H}^s$-a.e. $x \in \mathbb{R}^n - E$.

Proof. Fix $t > 0$ and define

$$A_t := \left\{ x \in \mathbb{R}^n - E \mid \limsup_{r \to 0} \frac{\mathcal{H}^s(B(x,r) \cap E)}{\alpha(s)r^s} > t \right\}.$$

Now $\mathcal{H}^s \llcorner E$ is a Radon measure, and so given $\epsilon > 0$, there exists a compact set $K \subseteq E$ such that

$$\mathcal{H}^s(E - K) \le \epsilon. \tag{$\star$}$$

Set $U := \mathbb{R}^n - K$; then U is open and $A_t \subseteq U$. Fix $\delta > 0$ and consider the family of balls

$$\mathcal{F} := \left\{ B(x,r) \mid B(x,r) \subseteq U, 0 < r < \delta, \frac{\mathcal{H}^s(B(x,r) \cap E)}{\alpha(s)r^s} > t \right\}.$$

By the Vitali Covering Theorem, there exists a countable disjoint family of balls $\{B_i\}_{i=1}^\infty$ in $\mathcal{F}$ such that

$$A_t \subseteq \bigcup_{i=1}^\infty \hat{B}_i.$$

Write $B_i = B(x_i, r_i)$. Then

$$H_{10\delta}^s(A_t) \le \sum_{i=1}^\infty \alpha(s)(5r_i)^s \le \frac{5^s}{t} \sum_{i=1}^\infty \mathcal{H}^s(B_i \cap E)$$

$$\le \frac{5^s}{t} \mathcal{H}^s(U \cap E) = \frac{5^s}{t} \mathcal{H}^s(E - K) \le \frac{5^s}{t} \epsilon,$$

by ($\star$). Let $\delta \to 0$ to find $\mathcal{H}^s(A_t) \le 5^s t^{-1} \epsilon$. Therefore $\mathcal{H}^s(A_t) = 0$ for each $t > 0$, and the theorem follows. $\qquad\square$

THEOREM 2.7 (Density bounds for points in E). *Assume $E \subset \mathbb{R}^n$, E is $\mathcal{H}^s$-measurable, and $\mathcal{H}^s(E) < \infty$. Then*

$$\frac{1}{2^s} \le \limsup_{r \to 0} \frac{\mathcal{H}^s(B(x,r) \cap E)}{\alpha(s)r^s} \le 1$$

for $\mathcal{H}^s$-a.e. $x \in E$.

Remark. It is possible to have

$$\limsup_{r \to 0} \frac{\mathcal{H}^s(B(x,r) \cap E)}{\alpha(s)r^s} < 1$$

and

$$\liminf_{r \to 0} \frac{\mathcal{H}^s(B(x,r) \cap E)}{\alpha(s)r^s} = 0$$

for $\mathcal{H}^s$-a.e. $x \in E$, even if $0 < \mathcal{H}^s(E) < \infty$.   □

Proof. 1. *Claim #1:* $\limsup_{r \to 0} \frac{\mathcal{H}^s(B(x,r) \cap E)}{\alpha(s)r^s} \le 1$ for $\mathcal{H}^s$-a.e. $x \in E$.
Proof of claim: Fix $\epsilon > 0, t > 1$ and define

$$B_t := \left\{ x \in E \mid \limsup_{r \to 0} \frac{\mathcal{H}^s(B(x,r) \cap E)}{\alpha(s)r^s} > t \right\}.$$

Since $\mathcal{H}^s \llcorner E$ is a Radon measure according to Theorem 1.7, there exists an open set U containing B_t with

$$\mathcal{H}^s(U \cap E) \le \mathcal{H}^s(B_t) + \epsilon. \tag{$\star$}$$

Define

$$\mathcal{F} := \left\{ B(x,r) \mid B(x,r) \subset U, 0 < r < \delta, \frac{\mathcal{H}^s(B(x,r) \cap E)}{\alpha(s)r^s} > t \right\}.$$

According to Theorem 1.25, there exists a countable disjoint family of balls $\{B_i\}_{i=1}^{\infty}$ in $\mathcal{F}$ such that

$$B_t \subseteq \bigcup_{i=1}^{m} B_i \cup \bigcup_{i=m+1}^{\infty} \hat{B}_i$$

for each $m = 1, 2, \ldots$. Write $B_i = B(x_i, r_i)$. Then

$$H_{10\delta}^s(B_t) \le \sum_{i=1}^{m} \alpha(s)r_i^s + \sum_{i=m+1}^{\infty} \alpha(s)(5r_i)^s$$

$$\le \frac{1}{t} \sum_{i=1}^{m} \mathcal{H}^s(B_i \cap E) + \frac{5^s}{t} \sum_{i=m+1}^{\infty} \mathcal{H}^s(B_i \cap E)$$

$$\le \frac{1}{t} \mathcal{H}^s(U \cap E) + \frac{5^s}{t} \mathcal{H}^s \left(\bigcup_{i=m+1}^{\infty} B_i \cap E \right).$$

This estimate is valid for $m = 1, \dots$; and thus our sending m to infinity yields the estimate

$$H^s_{10\delta}(B_t) \leq t^{-1} \mathcal{H}^s(U \cap E) \leq t^{-1}(\mathcal{H}^s(B_t) + \epsilon)$$

by $(\star)$. Let $\delta \to 0$ and then $\epsilon \to 0$:

$$\mathcal{H}^s(B_t) \leq t^{-1} \mathcal{H}^s(B_t).$$

Since $\mathcal{H}^s(B_t) \leq \mathcal{H}^s(E) < \infty$, this implies $\mathcal{H}^s(B_t) = 0$ for each $t > 1$.

2. *Claim #2:* We have $\limsup_{r \to 0} \frac{\mathcal{H}^s_\infty(B(x,r) \cap E)}{\alpha(s)r^s} \geq \frac{1}{2^s}$ for $\mathcal{H}^s$-a.e. $x \in E$.

Proof of claim: For $\delta > 0, 1 > \tau > 0$, denote by $E(\delta, \tau)$ the set of points $x \in E$ such that

$$\mathcal{H}^s_\delta(C \cap E) \leq \tau \alpha(s) \left(\frac{\operatorname{diam} C}{2} \right)^s$$

whenever $C \subseteq \mathbb{R}^n$, $x \in C$, $\operatorname{diam} C \leq \delta$. Then if $\{C_i\}_{i=1}^\infty$ are subsets of $\mathbb{R}^n$ with $\operatorname{diam} C_i \leq \delta$, $E(\delta, \tau) \subseteq \cup_{i=1}^\infty C_i$, and $C_i \cap E(\delta, \tau) \neq \emptyset$, we have

$$\mathcal{H}^s_\delta(E(\delta, \tau)) \leq \sum_{i=1}^\infty \mathcal{H}^s_\delta(C_i \cap E(\delta, \tau))$$

$$\leq \sum_{i=1}^\infty \mathcal{H}^s_\delta(C_i \cap E)$$

$$\leq \tau \sum_{i=1}^\infty \alpha(s) \left(\frac{\operatorname{diam} C_i}{2} \right)^s.$$

Hence $\mathcal{H}^s_\delta(E(\delta, \tau)) \leq \tau \mathcal{H}^s_\delta(E(\delta, \tau))$. Consequently $\mathcal{H}^s_\delta(E(\delta, \tau)) = 0$, since $0 < \tau < 1$ and $\mathcal{H}^s_\delta(E(\delta, \tau)) \leq \mathcal{H}^s_\delta(E) \leq \mathcal{H}^s(E) < \infty$. In particular,

$$\mathcal{H}^s(E(\delta, 1 - \delta)) = 0. \qquad (\star)$$

Now if $x \in E$ and

$$\limsup_{r \to 0} \frac{\mathcal{H}^s_\infty(B(x,r) \cap E)}{\alpha(s)r^s} < \frac{1}{2^s},$$

there exists $\delta > 0$ such that

$$\frac{\mathcal{H}^s_\infty(B(x,r) \cap E)}{\alpha(s)r^s} \leq \frac{1 - \delta}{2^s} \qquad (\star\star)$$

for all $0 < r \leq \delta$. Thus if $x \in C$ and $\operatorname{diam} C \leq \delta$, we have

$$
\begin{aligned}
\mathcal{H}_\infty^s(C \cap E) &= \mathcal{H}_\infty^s(C \cap E) \\
&\leq \mathcal{H}_\infty^s(B(x, \operatorname{diam} C) \cap E) \\
&\leq (1 - \delta)\alpha(s) \left(\frac{\operatorname{diam} C}{2} \right)^s
\end{aligned}
$$

by ($\star\star$); consequently $x \in E(\delta, 1 - \delta)$. But then

$$
\left\{ x \in E \mid \limsup_{r \to 0} \frac{\mathcal{H}_\infty^s(B(x, r) \cap E)}{\alpha(s) r^s} < \frac{1}{2^s} \right\} \subseteq \bigcup_{k=1}^\infty E\left(\frac{1}{k}, 1 - \frac{1}{k} \right),
$$

and so ($\star$) finishes the proof of Claim #2.

3. Since $\mathcal{H}^s(B(x, r) \cap E) \geq \mathcal{H}_\infty^s(B(x, r) \cap E)$, Claim #2 at once implies the lower estimate in the statement of the theorem. □

2.4   Functions and Hausdorff measure

In this section we record for later use some simple properties relating the behavior of functions and Hausdorff measure.

2.4.1   Hausdorff measure and Lipschitz mappings

DEFINITION 2.4.

(i) *A function $f : \mathbb{R}^n \to \mathbb{R}^m$ is called **Lipschitz continuous** if there exists a constant C such that*

$$
|f(x) - f(y)| \leq C|x - y| \quad \text{for all } x, y \in \mathbb{R}^n. \tag{$\star$}
$$

(ii) *The smallest constant C such that ($\star$) holds for all x, y is the* **Lipschitz constant** *for f, denoted*

$$
\operatorname{Lip}(f) := \sup \left\{ \frac{|f(x) - f(y)|}{|x - y|} \mid x, y \in \mathbb{R}^n, x \neq y \right\}.
$$

We will sometimes refer to a Lipschitz continuous function as a "Lipschitz function".

THEOREM 2.8 (Hausdorff measure under Lipschitz maps).

(i) *Let* $f : \mathbb{R}^n \to \mathbb{R}^m$ *be Lipschitz continuous,* $A \subseteq \mathbb{R}^n$ *and* $0 \leq s < \infty$. *Then*

$$\mathcal{H}^s(f(A)) \leq (\mathrm{Lip}(f))^s \mathcal{H}^s(A).$$

(ii) *Suppose* $n > k$ *and let* $P : \mathbb{R}^n \to \mathbb{R}^k$ *denote the projection. Assume* $A \subseteq \mathbb{R}^n$ *and* $0 \leq s < \infty$. *Then*

$$\mathcal{H}^s(P(A)) \leq \mathcal{H}^s(A).$$

Proof. 1. Fix $\delta > 0$ and choose sets $\{C_i\}_{i=1}^\infty \subseteq \mathbb{R}^n$ such that $\mathrm{diam}\, C_i \leq \delta$, $A \subseteq \cup_{i=1}^\infty C_i$. Then $\mathrm{diam}\, f(C_i) \leq \mathrm{Lip}(f)\, \mathrm{diam}\, C_i \leq \mathrm{Lip}(f)\delta$ and $f(A) \subseteq \cup_{i=1}^\infty f(C_i)$. Thus

$$\mathcal{H}^s_{\mathrm{Lip}(f)\delta}(f(A)) \leq \sum_{i=1}^\infty \alpha(s) \left(\frac{\mathrm{diam}\, f(C_i)}{2} \right)^s$$

$$\leq (\mathrm{Lip}(f))^s \sum_{i=1}^\infty \alpha(s) \left(\frac{\mathrm{diam}\, C_i}{2} \right)^s.$$

Taking infima over all such sets $\{C_i\}_{i=1}^\infty$, we find

$$\mathcal{H}^s_{\mathrm{Lip}(f)\delta}(f(A)) \leq (\mathrm{Lip}(f))^s \mathcal{H}^s_\delta(A).$$

Send $\delta \to 0$ to finish the proof of (i).

2. Assertion (ii) follows at once, since $\mathrm{Lip}(P) = 1$. $\qquad\square$

2.4.2 Graphs of Lipschitz functions

DEFINITION 2.5. *For* $f : \mathbb{R}^n \to \mathbb{R}^m$ *and* $A \subseteq \mathbb{R}^n$, *write*

$$G(f; A) := \{(x, f(x)) \mid x \in A\} \subset \mathbb{R}^n \times \mathbb{R}^m = \mathbb{R}^{n+m};$$

$G(f; A)$ *is the **graph** of* f *over* A.

THEOREM 2.9 (Hausdorff dimension of graphs). *Assume that* $f : \mathbb{R}^n \to \mathbb{R}^m$ *and* $\mathcal{L}^n(A) > 0$.

(i) *Then* $H_{\dim}(G(f; A)) \geq n$.

(ii) *If* f *is Lipschitz continuous,* $H_{\dim}(G(f; A)) = n$.

Remark. We thus see the graph of a Lipschitz continuous function f has the expected Hausdorff dimension. We will later discover from the area formula in Section 3.3 that $\mathcal{H}^n(G(f; A))$ can be computed according to the usual rules of calculus. ☐

Proof. 1. Let $P : \mathbb{R}^{n+m} \to \mathbb{R}^n$ be the standard projection. Then $\mathcal{H}^n(G(f; A)) \geq \mathcal{H}^n(A) > 0$ and thus $H_{\dim}(G(f; A)) \geq n$.

2. Let Q denote any cube in $\mathbb{R}^n$ of side length 1. Subdivide Q into k^n subcubes of side length $\frac{1}{k}$. Call these subcubes $Q_1, \ldots, Q_{k^n}$. Note $\operatorname{diam} Q_i = \frac{\sqrt{n}}{k}$. Define

$$a_j^i := \min_{x \in Q_j} f^i(x), \quad b_j^i := \max_{x \in Q_j} f^i(x)$$

for $i = 1, \ldots, m; \ j = 1, \ldots, k^n$. Since f is Lipschitz continuous,

$$|b_j^i - a_j^i| \leq \operatorname{Lip}(f) \operatorname{diam} Q_j = \operatorname{Lip}(f) \frac{\sqrt{n}}{k}.$$

Next, let $C_i := Q_j \times \prod_{i=1}^m (a_j^i, b_j^i)$. Then

$$\{(x, f(x)) | x \in Q_j \cap A\} \subseteq C_j$$

and $\operatorname{diam} C_j < \frac{C}{k}$. Since $G(f; A \cap Q) \subseteq \cup_{j=1}^{k^n} C_j$, we have

$$\mathcal{H}_{\frac{C}{k}}^n(G(f; A \cap Q)) \leq \sum_{j=1}^{k^n} \alpha(n) \left(\frac{\operatorname{diam} C_j}{2}\right)^n$$

$$\leq k^n \alpha(n) \left(\frac{C}{2k}\right)^n = \alpha(n) \left(\frac{C}{2}\right)^n.$$

Then, letting $k \to \infty$, we find $\mathcal{H}^n(G(f; A \cap Q)) < \infty$, and so $H_{\dim}(G(f; A \cap Q)) \leq n$. This estimate is valid for each cube Q in $\mathbb{R}^n$ of side length 1, and consequently $H_{\dim}(G(f; A)) \leq n$. ☐

2.4.3 Integrals over balls

If a function is locally summable, we can estimate the Hausdorff measure of the set where it is locally large.

THEOREM 2.10 (Hausdorff measure and integrals over balls). *Let $f \in L^1_{\text{loc}}(\mathbb{R}^n)$, suppose $0 \leq s < n$, and define*

$$\Lambda_s := \left\{ x \in \mathbb{R}^n \mid \limsup_{r \to 0} \frac{1}{r^s} \int_{B(x,r)} |f| \, dy > 0 \right\}.$$

Then

$$\mathcal{H}^s(\Lambda_s) = 0.$$

Proof. 1. We may as well assume $f \in L^1(\mathbb{R}^n)$. According to the Lebesgue–Besicovitch Differentiation Theorem,

$$\lim_{r \to 0} \fint_{B(x,r)} |f| \, dy = |f(x)|,$$

and thus

$$\lim_{r \to 0} \frac{1}{r^s} \int_{B(x,\tau)} |f| \, dy = 0$$

for $\mathcal{L}^n$-a.e. x, since $0 \leq s < n$. Hence

$$\mathcal{L}^n(\Lambda_s) = 0.$$

2. Now fix $\epsilon > 0, \delta > 0, \sigma > 0$. As f is $\mathcal{L}^n$-summable, there exists $\eta > 0$ such that $\mathcal{L}^n(U) \leq \eta$ implies $\int_U |f| \, dx < \sigma$.

Define

$$\Lambda_s^e := \left\{ x \in \mathbb{R}^n \mid \lim_{r \to 0} \frac{1}{r^s} \int_{B(x,r)} |f| \, dy > \epsilon \right\};$$

then

$$\mathcal{L}^n(\Lambda_s^\epsilon) = 0.$$

There thus exists an open subset U with $U \supset \Lambda_s^\epsilon$, $\mathcal{L}^n(U) < \eta$. Define

$$\mathcal{F} :=$$

$$\left\{ B(x,r) \mid x \in \Lambda_s^\epsilon, 0 < r < \delta, B(x,r) \subseteq U, \int_{B(x,r)} |f| \, dy > \epsilon r^s \right\}.$$

By the Vitali Covering Theorem, there exist disjoint balls $\{B_i\}_{i=1}^{\infty}$ in $\mathcal{F}$ such that

$$\Lambda_s^\epsilon \subseteq \bigcup_{i=1}^{\infty} \hat{B}_i.$$

Hence, writing r_i for the radius of B_i, we compute

$$\mathcal{H}^s_{10\delta}(\Lambda^\epsilon_s) \le \sum_{i=1}^{\infty} \alpha(s)(5r_i)^s$$

$$\le \frac{\alpha(s)5^s}{\epsilon} \sum_{i=1}^{\infty} \int_{B_i} |f| \, dy$$

$$\le \frac{\alpha(s)5^s}{\epsilon} \int_U |f| \, dy$$

$$\le \frac{\alpha(s)5^s}{\epsilon} \sigma.$$

Send $\delta \to 0$, and then $\sigma \to 0$, to discover

$$\mathcal{H}^s(\Lambda^\epsilon_s) = 0.$$

This holds for all $\epsilon > 0$ and hence $\mathcal{H}^s(\Lambda_s) = 0$.   □

2.5   References and notes

Again, our primary source is Federer [F], especially [F, Section 2.10]. Steiner symmetrization may be found in [F, Sections 2.10.30, 2.10.31]. We closely follow Hardt [H] for the proof of the isodiametric inequality, but incorporated a simplification due to L–F Tam, who noted that we need to symmetrize only in coordinate directions. R. Hardt told us about Tam's observation.

The proof that $\mathcal{H}^n = \mathcal{L}^n$ is from Hardt [H] and Simon [S, Sections 2.3–2.6]. We consulted [S, Section 3] for the density theorems in Section 2.3.

Falconer [Fa1, Fa2] and Morgan [Mo] provide nice introductions to Hausdorff measure. A good advanced text is Mattila [Ma].

Chapter 3

Area and Coarea Formulas

In this chapter we study Lipschitz continuous mappings

$$f : \mathbb{R}^n \to \mathbb{R}^m$$

and derive corresponding change of variables formulas. There are two essentially different cases depending on the relative size of n and m.

If $m \geq n$, the *area formula* asserts that the n-dimensional measure of $f(A)$, counting multiplicity, can be calculated by integrating the appropriate Jacobian of f over A.

If $m \leq n$, the *coarea formula* states that the integral of the $n - m$ dimensional measure of the level sets of f can be computed by integrating the Jacobin. This assertion is a far-reaching generalization of Fubini's Theorem. (The word "coarea" is pronounced, and often spelled, "co-area.")

We begin in Section 3.1 with a detailed study of the differentiability properties of Lipschitz continuous functions and prove Rademacher's Theorem. In Section 3.2 we discuss linear maps from $\mathbb{R}^n$ to $\mathbb{R}^m$ and introduce Jacobians. The area formula is proved in Section 3.3, the coarea formula in Section 3.4.

3.1 Lipschitz functions, Rademacher's Theorem

3.1.1 Lipschitz continuous functions

We recall and extend slightly some terminology from Section 2.4.

DEFINITION 3.1.

(i) *Let $A \subseteq \mathbb{R}^n$. A function $f : A \to \mathbb{R}^m$ is called* **Lipschitz continuous** *provided*

$$|f(x) - f(y)| \leq C|x - y| \qquad (\star)$$

for some constant C and all $x, y \in A$.

(ii) *The smallest constant C such that $(\star)$ holds for all x, y is denoted*

$$\operatorname{Lip}(f) := \sup\left\{\frac{|f(x)-f(y)|}{|x-y|} \mid x,y \in A, x \neq y\right\}.$$

Thus

$$|f(x) - f(y)| \leq \operatorname{Lip}(f)|x-y| \quad (x,y \in A).$$

(iii) *A function $f : A \to \mathbb{R}^m$ is called **locally Lipschitz continuous** if for each compact $K \subseteq A$, there exists a constant C_K such that*

$$|f(x) - f(y)| \leq C_K|x-y|$$

for all $x,y \in K$.

THEOREM 3.1 (Extension of Lipschitz mappings). *Assume $A \subset \mathbb{R}^n$, and let $f : A \to \mathbb{R}^m$ be Lipschitz continuous.*

Then there exists a Lipschitz continuous function $\bar{f} : \mathbb{R}^n \to \mathbb{R}^m$ such that

(i) $\bar{f} = f$ *on A,*

(ii) $\operatorname{Lip}(\bar{f}) \leq \sqrt{m}\,\operatorname{Lip}(f)$.

Proof. 1. First assume $f : A \to \mathbb{R}$. Define

$$\bar{f}(x) := \inf_{a \in A}\left\{f(a) + \operatorname{Lip}(f)|x-a|\right\} \quad (x \in \mathbb{R}^n).$$

If $b \in A$, then we have $\bar{f}(b) = f(b)$. This follows since for all $a \in A$,

$$f(a) + \operatorname{Lip}(f)|b-a| \geq f(b);$$

whereas obviously $\bar{f}(b) \leq f(b)$. If $x,y \in \mathbb{R}^n$, then

$$\bar{f}(x) \leq \inf_{a \in A}\left\{f(a) + \operatorname{Lip}(f)(|y-a| + |x-y|)\right\}$$
$$= \bar{f}(y) + \operatorname{Lip}(f)|x-y|.$$

Likewise

$$\bar{f}(y) \leq \bar{f}(x) + \operatorname{Lip}(f)|x-y|.$$

2. In the general case that $f : A \to \mathbb{R}^m$, $f = (f^1, \ldots, f^m)$, we define $\bar{f} := (\bar{f}^1, \ldots, \bar{f}^m)$. Then

$$|\bar{f}(x) - \bar{f}(y)|^2 = \sum_{i=1}^{m}|\bar{f}^i(x) - \bar{f}^i(y)|^2 \leq m(\operatorname{Lip}(f))^2|x-y|^2. \quad \square$$

Remark. Kirszbraun's Theorem ([F, Section 2.10.43]) asserts that there in fact exists an extension $\bar{f}$ with $\operatorname{Lip}(\bar{f}) = \operatorname{Lip}(f)$. $\quad \square$

3.1.2   Rademacher's Theorem

We next prove Rademacher's remarkable theorem that a Lipschitz continuous function is differentiable $\mathcal{L}^n$-a.e. This is surprising since the inequality

$$|f(x) - f(y)| \leq \operatorname{Lip}(f)|x - y|$$

apparently says nothing about the possibility of locally approximating f by a linear map. (In Section 6.4 we prove Aleksandrov's Theorem, stating that a convex function is twice differentiable-a.e.)

DEFINITION 3.2. *The function* $f : \mathbb{R}^n \to \mathbb{R}^m$ *is* ***differentiable*** *at* $x \in \mathbb{R}^n$ *if there exists a linear mapping*

$$L : \mathbb{R}^n \to \mathbb{R}^m$$

such that
$$\lim_{y \to x} \frac{|f(y) - f(x) - L(y - x)|}{|x - y|} = 0,$$

or, equivalently,

$$f(y) = f(x) + L(y - x) + o(|y - x|) \quad \text{as } y \to x.$$

NOTATION   If such a linear mapping L exists, it is clearly unique, and we write

$$Df(x)$$

for L We call $Df(x)$ the **derivative** of f at x.

THEOREM 3.2 (Rademacher's Theorem). *Assume that* $f : \mathbb{R}^n \to \mathbb{R}^m$ *is a locally Lipschitz continuous function.*

Then f is differentiable $\mathcal{L}^n$-a.e.

Proof. 1. We may assume $m = 1$. Since differentiability is a local property, we may as well also suppose f is Lipschitz continuous.

Fix any $v \in \mathbb{R}^n$ with $|v| = 1$, and define

$$D_v f(x) := \lim_{t \to 0} \frac{f(x + tv) - f(x)}{t} \quad (x \in \mathbb{R}^n),$$

provided this limit exists.

2. *Claim #1*: $D_v f(x)$ exists for $\mathcal{L}^n$-a.e. x.

Proof of claim: Since f is continuous,

$$\overline{D}_v f(x) := \limsup_{t \to 0} \frac{f(x+tv) - f(x)}{t}$$

$$= \lim_{k \to \infty} \sup_{\substack{0 < |t| < \frac{1}{k} \\ t \text{ rational}}} \frac{f(x+tv) - f(x)}{t}$$

is Borel measurable, as is

$$\underline{D}_v f(x) := \liminf_{t \to 0} \frac{f(x+tv) - f(x)}{t}.$$

Thus

$$A_v = \{x \in \mathbb{R}^n \mid D_v f(x) \text{ does not exist}\}$$
$$= \{x \in \mathbb{R}^n \mid \underline{D}_v f(x) < \overline{D}_v f(x)\}$$

is Borel measurable.

For each $x, v \in \mathbb{R}^n$ with $|v| = 1$, define $\phi : \mathbb{R} \to \mathbb{R}$ by

$$\phi(t) := f(x+tv) \quad (t \in \mathbb{R}).$$

Then ϕ is Lipschitz continuous, thus absolutely continuous, and thus differentiable $\mathcal{L}^1$-a.e. Hence

$$\mathcal{H}^1(A_v \cap L) = 0$$

for each line L parallel to v. Fubini's Theorem then implies

$$\mathcal{L}^n(A_v) = 0.$$

3. As a consequence of Claim #1, we see that

$$\operatorname{grad} f(x) := (f_{x_1}(x), \dots, f_{x_n}(x))$$

exists for $\mathcal{L}^n$-a.e. point x.

Claim #2:

$$D_v f(x) = v \cdot \operatorname{grad} f(x) \quad \text{for } \mathcal{L}^n\text{-a.e. } x.$$

Proof of claim: Write $v = (v_1, \ldots, v_n)$. Let $\zeta \in C_c^\infty(\mathbb{R}^n)$. Then

$$\int_{\mathbb{R}^n} \left[\frac{f(x + tv) - f(x)}{t} \right] \zeta(x)\, dx$$

$$= -\int_{\mathbb{R}^n} f(x) \left[\frac{\zeta(x) - \zeta(x - tv)}{t} \right] dx.$$

Put $t = \frac{1}{k}$ in the above equality and observe

$$\left| \frac{f\left(x + \frac{1}{k}v\right) - f(x)}{\frac{1}{k}} \right| \le \mathrm{Lip}(f)|v| = \mathrm{Lip}(f).$$

Thus the Dominated Convergence Theorem implies

$$\int_{\mathbb{R}^n} D_v f(x) \zeta(x)\, dx = -\int_{\mathbb{R}^n} f(x) D_v \zeta(x)\, dx$$

$$= -\sum_{i=1}^{n} v_i \int_{\mathbb{R}^n} f(x) \zeta_{x_i}(x)\, dx$$

$$= \sum_{i=1}^{n} v_i \int_{\mathbb{R}^n} f_{x_i}(x) \zeta(x)\, dx$$

$$= \int_{\mathbb{R}^n} (v \cdot \mathrm{grad}\, f(x)) \zeta(x)\, dx,$$

where we used Fubini's Theorem and the absolute continuity of f on lines. The above equality holding for each $\zeta \in C_c(\mathbb{R}^n)$ implies $D_v f = v \cdot \mathrm{grad}\, f$ $\mathcal{L}^n$-a.e.

4. Now choose $\{v_k\}_{k=1}^\infty$ to be a countable, dense subset of $\partial B(1)$. Set for $k = 1, 2, \ldots$

$$A_k := $$
$$\{x \in \mathbb{R}^n \mid D_{v_k} f(x), \mathrm{grad}\, f(x) \text{ exist}, D_{v_k} f(x) = v_k \cdot \mathrm{grad}\, f(x)\},$$

and define

$$A := \bigcap_{k=1}^{\infty} A_k.$$

Observe

$$\mathcal{L}^n(\mathbb{R}^n - A) = 0.$$

5. *Claim #3:* f is differentiable at each point $x \in A$.

Proof of claim: Fix any $x \in A$. Choose $v \in \partial B(1), t \in \mathbb{R}, t \neq 0$, and write

$$Q(x, v, t) := \frac{f(x + tv) - f(x)}{t} - v \cdot \operatorname{grad} f(x).$$

Then if $v' \in \partial B(1)$, we have

$$
\begin{aligned}
|Q(x, v, t) &- Q(x, v', t)| \\
&\leq \left| \frac{f(x + tv) - f(x + tv')}{t} \right| + |(v - v') \cdot \operatorname{grad} f(x)| \\
&\leq \operatorname{Lip}(f)|v - v'| + |\operatorname{grad} f(x)||v - v'| \\
&\leq (\sqrt{n} + 1) \operatorname{Lip}(f)|v - v'|. \qquad\qquad (\star)
\end{aligned}
$$

Now fix $\epsilon > 0$, and choose N so large that if $v \in \partial B(1)$, then

$$|v - v_k| \leq \frac{\epsilon}{2(\sqrt{n} + 1) \operatorname{Lip,}(f)} \qquad\qquad (\star\star)$$

for some $k \in \{1, \ldots, N\}$.

Now

$$\lim_{t \to 0} Q(x, v_k, t) = 0 \quad (k = 1, \ldots, N),$$

and thus there exists $\delta > 0$ so that

$$|Q(x, v_k, t)| < \frac{\epsilon}{2} \quad \text{for all } 0 < |t| < \delta, k = 1, \ldots, N. \qquad (\star\star\star)$$

Consequently, for each $v \in \partial B(1)$, there exists $k \in \{1, \ldots, N\}$ such that

$$|Q(x, v, t)| \leq |Q(x, v_k, t)| + |Q(x, v, t) - Q(x, v_k, t)| < \epsilon$$

if $0 < |t| < \delta$, according to $(\star) - (\star\star\star)$. Note the same $\delta > 0$ works for all $v \in \partial B(1)$.

Now choose any $y \in \mathbb{R}^n, y \neq x$. Write $v := \frac{y-x}{|y-x|}$; so that $y = x + tv$ for $t := |x - y|$. Then

$$
\begin{aligned}
f(y) - f(x) - \operatorname{grad} f(x) \cdot (y - x) &= f(x + tv) - f(x) - tv \cdot \operatorname{grad} f(x) \\
&= o(t) \\
&= o(|x - y|) \quad \text{as } y \to x.
\end{aligned}
$$

Hence f is differentiable at x, with

$$Df(x) = \operatorname{grad} f(x). \qquad\qquad \square$$

Remark. See Theorem 6.6 for another proof of Rademacher's Theorem and Theorem 6.5 for a generalization.   □

We next record some technical facts for use later.

THEOREM 3.3 (Differentiability on level sets).

(i) *Let* $f : \mathbb{R}^n \to \mathbb{R}^m$ *be locally Lipschitz continuous, and*

$$Z := \{x \in \mathbb{R}^n \mid f(x) = 0\}.$$

Then $Df(x) = 0$ *for* $\mathcal{L}^n$-*a.e.* $x \in Z$.

(ii) *Let* $f, g : \mathbb{R}^n \to \mathbb{R}^n$ *be locally Lipschitz continuous, and*

$$Y := \{x \in \mathbb{R}^n \mid g(f(x)) = x\}.$$

Then

$$Dg(f(x))Df(x) = I \quad for \ \mathcal{L}^n\text{-}a.e. \ x \in Y.$$

Proof. 1. We may assume $m = 1$ in assertion (i).

Choose $x \in Z$ so that $Df(x)$ exists, and

$$\lim_{r \to 0} \frac{\mathcal{L}^n(Z \cap B(x, r))}{\mathcal{L}^n(B(x, r))} = 1; \tag{$\star$}$$

$\mathcal{L}^n$-a.e. point $x \in Z$ will do. Then

$$f(y) = Df(x) \cdot (y - x) + o(|y - x|) \quad \text{as } y \to x. \tag{$\star\star$}$$

Assume $a := Df(x) \neq 0$, and set

$$S := \left\{ v \in \partial B(1) \mid a \cdot v \geq \frac{1}{2}|a| \right\}.$$

For each $v \in S$ and $t > 0$, put $y = x + tv$ in ($\star\star$):

$$f(x + tv) = a \cdot tv + o(|tv|) \geq \frac{t|a|}{2} + o(t) \quad \text{as } t \to 0.$$

Hence there exists $t_0 > 0$ such that

$$f(x + tv) > 0 \quad \text{for } 0 < t < t_0, v \in S,$$

a contradiction to ($\star$). This proves assertion (i).

2. To prove assertion (ii), first define

$$A := \{x \mid Df(x) \text{ exists}\}, \quad B := \{x \mid Dg(x) \text{ exists}\}.$$

Let

$$X := Y \cap A \cap f^{-1}(B).$$

Then

$$Y - X \subseteq (\mathbb{R}^n - A) \cup g(\mathbb{R}^n - B). \qquad (\star\star\star)$$

This follows since

$$x \in Y - f^{-1}(B)$$

implies

$$f(x) \in \mathbb{R}^n - B,$$

and so

$$x = g(f(x)) \in g(\mathbb{R}^n - B).$$

According to $(\star\star\star)$ and Rademacher's Theorem,

$$\mathcal{L}^n(Y - X) = 0.$$

Now if $x \in X$, then $Dg(f(x))$ and $Df(x)$ exist; and consequently

$$Dg(f(x))Df(x) = D(g \circ f)(x)$$

exists. Since $(g \circ f)(x) - x = 0$ on Y, assertion (i) implies

$$D(g \circ f) = I \quad \mathcal{L}^n\text{-a.e. on } Y. \qquad \square$$

3.2   Linear maps and Jacobians

We next review some linear algebra. Our goal thereafter will be to define the Jacobian of a map $f : \mathbb{R}^n \to \mathbb{R}^m$.

3.2.1   Linear mappings

DEFINITION 3.3.

(i) *A linear map $O : \mathbb{R}^n \to \mathbb{R}^m$ is **orthogonal** if*

$$(Ox) \cdot (Oy) = x \cdot y$$

for all $x, y \in \mathbb{R}^n$.

(ii) *A linear map $S : \mathbb{R}^n \to \mathbb{R}^n$ is **symmetric** if*

$$x \cdot (Sy) = (Sx) \cdot y$$

for all $x, y \in \mathbb{R}^n$.

(iii) *A linear map $D : \mathbb{R}^n \to \mathbb{R}^n$ is **diagonal** if there exist $d_1, \ldots, d_n \in \mathbb{R}$ such that*

$$Dx = (d_1 x_1, \ldots, d_n x_n)$$

for all $x \in \mathbb{R}^n$.

(iv) *Let $A : \mathbb{R}^n \to \mathbb{R}^m$ be linear. The **adjoint** of A is the linear map $A^* : \mathbb{R}^m \to \mathbb{R}^n$ defined by*

$$x \cdot (A^* y) = (Ax) \cdot y$$

for all $x \in \mathbb{R}^n, y \in \mathbb{R}^m$.

First we record some standard facts from linear algebra.

THEOREM 3.4 (Linear algebra).

(i) $A^{**} = A$.

(ii) $(A \circ B)^* = B^* \circ A^*$.

(iii) $O^* = O^{-1}$ *if $O : \mathbb{R}^n \to \mathbb{R}^n$ is orthogonal.*

(iv) $S^* = S$ *if $S : \mathbb{R}^n \to \mathbb{R}^n$ is symmetric.*

(v) *If $S : \mathbb{R}^n \to \mathbb{R}^n$ is symmetric, there exists an orthogonal map $O : \mathbb{R}^n \to \mathbb{R}^n$ and a diagonal map $D : \mathbb{R}^n \to \mathbb{R}^n$ such that*

$$S = O \circ D \circ O^{-1}.$$

(vi) *If $O : \mathbb{R}^n \to \mathbb{R}^m$ is orthogonal, then $n \le m$ and*

$$O^* \circ O = I \quad \text{on } \mathbb{R}^n,$$
$$O \circ O^* = I \quad \text{on } \mathbb{R}^m.$$

THEOREM 3.5 (Polar decomposition).
Let $L : \mathbb{R}^n \to \mathbb{R}^m$ be a linear mapping.

(i) *If $n \leq m$, there exists a symmetric map $S : \mathbb{R}^n \to \mathbb{R}^n$ and an orthogonal map $O : \mathbb{R}^n \to \mathbb{R}^m$ such that*

$$L = O \circ S.$$

(ii) *If $n \geq m$, there exists a symmetric map $S : \mathbb{R}^m \to \mathbb{R}^m$ and an orthogonal map $O : \mathbb{R}^m \to \mathbb{R}^n$ such that*

$$L = S \circ O^*.$$

Proof. 1. First suppose $n \leq m$. Define $C := L^* \circ L$; then $C : \mathbb{R}^n \to \mathbb{R}^n$. Now

$$(Cx) \cdot y = (L^* \circ Lx) \cdot y = Lx \cdot Ly = x \cdot (L^* \circ L)y = x \cdot Cy$$

and also

$$(Cx) \cdot x = Lx \cdot Lx \geq 0.$$

Thus C is symmetric, nonnegative definite. Hence there exist $\mu_1, \ldots, \mu_n \geq 0$ and an orthogonal basis $\{x_k\}_{k=1}^n$ of $\mathbb{R}^n$ such that

$$Cx_k = \mu_k x_k \quad (k = l, \ldots, n).$$

Write $\mu_k := \lambda_k^2, \lambda_k \geq 0 \ (k = 1, \ldots, n)$.

2. *Claim*: There exists an orthonormal set $\{z_k\}_{k=1}^n$ in $\mathbb{R}^m$ such that

$$Lx_k = \lambda_k z_k \quad (k = 1, \ldots, n).$$

Proof of claim: If $\lambda_k \neq 0$, define

$$z_k := \frac{1}{\lambda_k} Lx_k.$$

Then if $\lambda_k, \lambda_l \neq 0$,

$$z_k \cdot z_l = \frac{1}{\lambda_k \lambda_l} Lx_k \cdot Lx_l = \frac{1}{\lambda_k \lambda_l}(Cx_k) \cdot x_l = \frac{\lambda_k^2}{\lambda_k \lambda_l} x_k \cdot x_l = \frac{\lambda_k}{\lambda_l} \delta_{kl}.$$

Thus the set $\{z_k \mid \lambda_k \neq 0\}$ is orthogonal. If $\lambda_k = 0$, define z_k to be any unit vector such that $\{z_k\}_{k=1}^n$ is orthonormal.

3. Now define $S : \mathbb{R}^n \to \mathbb{R}^n$ by

$$Sx_k = \lambda_k x_k \quad (k = 1, \ldots, n)$$

and $O : \mathbb{R}^n \to \mathbb{R}^m$ by

$$Ox_k = z_k \quad (k = 1, \ldots, n).$$

Then $O \circ Sx_k = \lambda_k Ox_k = \lambda_k z_k = Lx_k$, and so

$$L = O \circ S.$$

The mapping S is clearly symmetric, and O is orthogonal since

$$Ox_k \cdot Ox_l = z_k \cdot z_l = \delta_{kl}.$$

4. Assertion (ii) follows from our applying (i) to $L : \mathbb{R}^m \to \mathbb{R}^n$. □

DEFINITION 3.4. *Assume* $L : \mathbb{R}^n \to \mathbb{R}^m$ *is linear.*

(i) *If* $n \leq m$, *we write* $L = O \circ S$ *as above, and we define the* **Jacobian** *of* L *to be*

$$[\![L]\!] = |\det S|.$$

(ii) *If* $n \geq m$, *we write* $L = S \circ O^*$ *as above, and we define the* **Jacobian** *of* L *to be*

$$[\![L]\!] = |\det S|.$$

Remark. It follows from Theorem 3.6 below that the definition of $[\![L]\!]$ is independent of the particular choices of O and S. Observe also that

$$[\![L]\!] = [\![L^*]\!].$$ □

THEOREM 3.6 (Jacobians and adjoints).

(i) *If* $n \leq m$,
$$[\![L]\!]^2 = \det(L^* \circ L).$$

(ii) *If* $n \geq m$,
$$[\![L]\!]^2 = \det(L \circ L^*).$$

Proof. Assume $n \leq m$ and write

$$L = O \circ S, \quad L^* = S \circ O^*;$$

then

$$L^* \circ L = S \circ O^* \circ O \circ S = S^2,$$

since O is orthogonal, and thus $O^* \circ O = I$. Hence

$$\det(L^* \circ L) = (\det S)^2 = [\![L]\!]^2.$$

The proof of (ii) is similar.   □

Theorem 3.6 provides us with a useful method for computing $[\![L]\!]$, which we augment with the Binet–Cauchy formula below.

DEFINITION 3.5.

(i) *If $n \leq m$, we define*

$$\Lambda(m,n) = \{\lambda : \{1,\ldots,n\} \to \{1,\ldots,m\} \mid \lambda \text{ is increasing}\}.$$

(ii) *For each $\lambda \in \Lambda(m,n)$, we define $P_\lambda : \mathbb{R}^m \to \mathbb{R}^n$ by*

$$P_\lambda(x_1,\ldots,x_m) := (x_{\lambda(1)},\ldots,x_{\lambda(n)}).$$

(iii) *For each $\lambda \in \Lambda(m,n)$, define the n-dimensional subspace*

$$S_\lambda := \mathrm{span}\{e_{\lambda(1)},\ldots,e_{\lambda(n)}\} \subseteq \mathbb{R}^m.$$

Then P_λ is the projection of $\mathbb{R}^m$ onto S_λ.

THEOREM 3.7 (Binet–Cauchy formula). *Assume that $n \leq m$ and $L : \mathbb{R}^n \to \mathbb{R}^m$ is linear. Then*

$$[\![L]\!]^2 = \sum_{\lambda \in \Lambda(m,n)} (\det(P_\lambda \circ L))^2.$$

Remark. Thus to calculate $[\![L]\!]^2$, we compute the sums of the squares of the determinants of each $(n \times n)$-submatrix of the $(m \times n)$-matrix representing L (with respect to the standard bases of $\mathbb{R}^n$ and $\mathbb{R}^m$).

In view of Lemma 3.1 below, this is a higher dimensional generalization of the Pythagorean Theorem.   □

Proof. 1. Identifying linear maps with their matrices with respect to the standard bases of $\mathbb{R}^n$ and $\mathbb{R}^m$, we write

$$L = ((l_{ij}))_{m \times n}, A = L^* \circ L = ((a_{ij}))_{n \times n};$$

so that

$$a_{ij} = \sum_{k=1}^{m} l_{ki} l_{kj} \quad (i,j = 1,\ldots,n).$$

2. Then

$$[\![L]\!]^2 = \det A = \sum_{\sigma \in \Sigma} \mathrm{sgn}(\sigma) \prod_{i=1}^{n} a_{i,\sigma(i)},$$

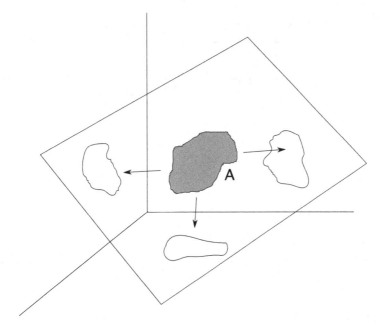

Σ denoting the set of all permutations of $\{1, \ldots, n\}$. Thus

$$[\![L]\!]^2 = \sum_{\sigma \in \Sigma} \mathrm{sgn}(\sigma) \prod_{i=1}^{n} \sum_{k=1}^{m} l_{ki} l_{k\sigma(i)}$$

$$= \sum_{\sigma \in \Sigma} \mathrm{sgn}(\sigma) \sum_{\phi \in \Phi} \prod_{i=1}^{n} l_{\phi(i)i} l_{\phi(i)\sigma(i)},$$

Φ denoting the set of all one-to-one mappings of $\{1, \ldots, n\}$ into $\{1, \ldots, m\}$.

3. For each $\phi \in \Phi$, we can uniquely write $\phi = \lambda \circ \theta$, where $\theta \in \Sigma$ and $\lambda \in \Lambda(m, n)$. Consequently,

$$[\![L]\!]^2 = \sum_{\sigma \in \Sigma} \mathrm{sgn}(\sigma) \sum_{\lambda \in \Lambda(m,n)} \sum_{\theta \in \Sigma} \prod_{i=1}^{n} l_{\lambda \circ \theta(i),i} l_{\lambda \circ \theta(i),\sigma(i)}$$

$$= \sum_{\sigma \in \Sigma} \mathrm{sgn}(\sigma) \sum_{\lambda \in \Lambda(m,n)} \sum_{\theta \in \Sigma} \prod_{i=1}^{n} l_{\lambda(i),\theta^{-1}(i)} l_{\lambda(i),\sigma \circ \theta^{-1}(i)}$$

$$= \sum_{\lambda \in \Lambda(m,n)} \sum_{\theta \in \Sigma} \sum_{\sigma \in \Sigma} \mathrm{sgn}(\sigma) \prod_{i=1}^{n} l_{\lambda(i),\theta(i)} l_{\lambda(i),\sigma \circ \theta(i)}$$

$$= \sum_{\lambda \in \Lambda(m,n)} \sum_{\rho \in \Sigma} \sum_{\theta \in \Sigma} \mathrm{sgn}(\theta)\,\mathrm{sgn}(\rho) \prod_{i=1}^{n} l_{\lambda(i),\theta(i)} l_{\lambda(i),\rho(i)}$$

$$= \sum_{\lambda \in \Lambda(m,n)} \left(\sum_{\theta \in \Sigma} \mathrm{sgn}(\theta) \prod_{i=1}^{n} l_{\lambda(i),\theta(i)} \right)^2$$

$$= \sum_{\lambda \in \Lambda(m,n)} (\det(P_\lambda \circ L))^2,$$

where we set $\rho = \sigma \circ \theta$. $\qquad\qquad\square$

3.2.2 Jacobians

Now let $f : \mathbb{R}^n \to \mathbb{R}^m$ be Lipschitz continuous. By Rademacher's Theorem, f is differentiable $\mathcal{L}^n$-a.e., and therefore $Df(x)$ exists, and can be regarded as a linear mapping from $\mathbb{R}^n$ into $\mathbb{R}^m$, for $\mathcal{L}^n$-a.e. $x \in \mathbb{R}^n$.

NOTATION If $f : \mathbb{R}^n \to \mathbb{R}^m$, $f = (f^1, \dots, f^m)$, we write the gradient matrix

$$Df = \begin{pmatrix} f^1_{x_1} & \cdots & f^1_{x_n} \\ \vdots & \ddots & \vdots \\ f^m_{x_1} & \cdots & f^m_{x_n} \end{pmatrix}_{m \times n}$$

at each point where Df exists.

DEFINITION 3.6. *For $\mathcal{L}^n$ a.e point x, we define the **Jacobian** of f to be*

$$Jf(x) := [\![\, Df(x) \,]\!].$$

3.3   The area formula

Through this section, we assume

$$n \le m.$$

3.3.1   Preliminaries

LEMMA 3.1. *Suppose $L : \mathbb{R}^n \to \mathbb{R}^m$ is linear, $n \le m$. Then*

$$\mathcal{H}^n(L(A)) = [\![\, L \,]\!]\, \mathcal{L}^n(A)$$

for all $A \subseteq \mathbb{R}^n$.

Proof. 1. Write $L = O \circ S$ as in Section 3.2; $[\![L]\!] = |\det S|$.

2. If $[\![L]\!] = 0$, then $\dim S(\mathbb{R}^n) \leq n - 1$ and so $\dim L(\mathbb{R}^n) \leq n - 1$. Consequently, $\mathcal{H}^n(L(\mathbb{R}^n)) = 0$.

If $[\![L]\!] > 0$, then

$$\frac{\mathcal{H}^n(L(B(x,r)))}{\mathcal{L}^n(B(x,r))} = \frac{\mathcal{L}^n(O^* \circ L(B(x,r)))}{\mathcal{L}^n(B(x,r))} = \frac{\mathcal{L}^n(O^* \circ O \circ S(B(x,r)))}{\mathcal{L}^n(B(x,r))}$$

$$= \frac{\mathcal{L}^n(S(B(x,r)))}{\mathcal{L}^n(B(x,r))} = \frac{\mathcal{L}^n(S(B(1)))}{\alpha(n)}$$

$$= |\det S| = [\![L]\!].$$

3. Define $\nu(A) := \mathcal{H}^n(L(A))$ for all $A \subseteq \mathbb{R}^n$. Then ν is a Radon measure, $\nu \ll \mathcal{L}^n$, and

$$D_{\mathcal{L}^n}\nu(x) = \lim_{r \to 0} \frac{\nu(B(x,r))}{\mathcal{L}^n(B(x,r))} = [\![L]\!].$$

Thus for all Borel sets $B \subseteq \mathbb{R}^n$, Theorem 1.30 implies

$$\mathcal{H}^n(L(B)) = [\![L]\!]\,\mathcal{L}^n(B).$$

Since ν and $\mathcal{L}^n$ are Radon measures, the same formula holds for all sets $A \subseteq \mathbb{R}^n$. $\qquad\square$

Henceforth we assume $f : \mathbb{R}^n \to \mathbb{R}^m$ is Lipschitz continuous.

LEMMA 3.2. *Let $A \subseteq \mathbb{R}^n$ be $\mathcal{L}^n$-measurable. Then*

(i) *$f(A)$ is $\mathcal{H}^n$-measurable,*

(ii) *the mapping $y \mapsto \mathcal{H}^0(A \cap f^{-1}\{y\})$ is $\mathcal{H}^n$-measurable on $\mathbb{R}^m$, and*

(iii)

$$\int_{\mathbb{R}^m} \mathcal{H}^0(A \cap f^{-1}\{y\})\, d\mathcal{H}^n \leq (\mathrm{Lip}(f))^n \mathcal{L}^n(A).$$

DEFINITION 3.7. *The mapping $y \mapsto \mathcal{H}^0(A \cap f^{-1}\{y\})$ is the **multiplicity function**.*

Proof. 1. We may assume with no loss of generality that A is bounded.

By Theorem 1.8, there exist compact sets $K_i \subseteq A$ such that

$$\mathcal{L}^n(K_i) \geq \mathcal{L}^n(A) - \frac{1}{i} \quad (i = 1, \dots).$$

As $\mathcal{L}^n(A) < \infty$ and A is $\mathcal{L}^n$-measurable, $\mathcal{L}^n(A - K_i) < \frac{1}{i}$. Since f is continuous, $f(K_i)$ is compact and thus $\mathcal{H}^n$-measurable. Hence $f(\cup_{i=1}^\infty K_i) = \cup_{i=1}^\infty f(K_i)$ is $\mathcal{H}^n$-measurable. Furthermore.

$$\mathcal{H}^n\left(f(A) - f\left(\bigcup_{i=1}^\infty K_i\right)\right) \leq \mathcal{H}^n\left(f\left(A - \bigcup_{i=1}^\infty K_i\right)\right)$$

$$\leq (\mathrm{Lip}(f))^n \mathcal{L}^n\left(A - \bigcup_{i=1}^\infty K_i\right) = 0.$$

Thus $f(A)$ is $\mathcal{H}^n$-measurable, and this proves (i).

2. Let

$$B_k := \left\{Q \mid Q = (a_1, b_1] \times \cdots \times (a_n, b_n], a_i = \frac{c_i}{k}, b_i = \frac{c_i + 1}{k}, c_i \in \mathbb{Z}\right\},$$

and note that

$$\mathbb{R}^n = \bigcup_{Q \in B_k} Q.$$

Now

$$g_k := \sum_{Q \in B_k} \chi_{f(A \cap Q)}$$

is $\mathcal{H}^n$-measurable by (i), and

$$g_k(y) = \text{number of cubes } Q \in B_k \text{ such that } f^{-1}\{y\} \cap (A \cap Q) \neq \emptyset.$$

Thus as $k \to \infty$,

$$g_k(y) \to \mathcal{H}^0(A \cap f^{-1}\{y\})$$

for each $y \in \mathbb{R}^m$; and so $y \mapsto \mathcal{H}^0(A \cap f^{-1}\{y\})$ is $\mathcal{H}^n$-measurable.

3. By the Monotone Convergence Theorem,

$$\int_{\mathbb{R}^m} \mathcal{H}^0(A \cap f^{-1}\{y\})\, d\mathcal{H}^n = \lim_{k \to \infty} \int_{\mathbb{R}^m} g_k\, d\mathcal{H}^n$$

$$= \lim_{k \to \infty} \sum_{Q \in B_k} \mathcal{H}^n(f(A \cap Q))$$

$$\leq \limsup_{k \to \infty} \sum_{Q \in B_k} (\mathrm{Lip}(f))^n \mathcal{L}^n(A \cap Q)$$

$$= (\mathrm{Lip}(f))^n \mathcal{L}^n(A). \qquad \square$$

LEMMA 3.3. *Let $t > 1$ and*

$$B := \{x \mid Df(x) \text{ exists}, Jf(x) > 0\}.$$

Then there is a countable collection $\{E_k\}_{k=1}^\infty$ of Borel subsets of $\mathbb{R}^n$ such that

(i) $B = \cup_{k=1}^\infty E_k$;

(ii) $f|_{E_k}$ *is one-to-one $(k = 1, 2, \ldots)$; and*

(iii) *for each $k = 1, 2, \ldots$, there exists a symmetric automorphism $T_k : \mathbb{R}^n \to \mathbb{R}^n$ such that*

$$\mathrm{Lip}((f|_{E_k}) \circ T_k^{-1}) \leq t, \ \mathrm{Lip}(T_k \circ (f|_{E_k})^{-1}) \leq t,$$
$$t^{-n}|\det T_k| \leq Jf|_{E_k} \leq t^n|\det T_k|.$$

Proof. 1. Fix $\epsilon > 0$ so that

$$t^{-1} + \epsilon < 1 < t - \epsilon.$$

Let C be a countable dense subset of B and let S be a countable dense subset of symmetric automorphisms of $\mathbb{R}^n$.

2. For each $c \in C, T \in S$, and $i = 1, 2, \ldots$, define $E(c, T, i)$ to be the set of all $b \in B \cap B(c, \frac{1}{i})$ satisfying

$$\left(t^{-1} + \epsilon\right) |Tv| \leq |Df(b)v| \leq (t - \epsilon)|Tv| \qquad (\star)$$

for all $v \in \mathbb{R}^n$ and

$$|f(a) - f(b) - Df(b) \cdot (a - b)| \leq \epsilon|T(a - b)| \qquad (\star\star)$$

for all $a \in B(b, \frac{2}{i})$. Note that $E(c, T, i)$ is a Borel set since Df is Borel measurable. From $(\star)$ and $(\star\star)$ follows the estimate

$$t^{-1}|T(a - b)| \leq |f(a) - f(b)| \leq t|T(a - b)| \qquad (\star\star\star)$$

for $b \in E(c,T,i), a \in B(b,\frac{2}{i})$.

3. *Claim:* If $b \in E(c,T,i)$, then

$$\left(t^{-1} + \epsilon\right)^n |\det T| \le Jf(b) \le (t - \epsilon)^n |\det T|.$$

Proof of claim: Write $Df(b) = L = O \circ S$, as above;

$$Jf(b) = [\![Df(b)]\!] = |\det S|.$$

According to $(\star)$,

$$\left(t^{-1} + \epsilon\right) |Tv| \le |(O \circ S)v| = |Sv| \le (t - \epsilon)|Tv|$$

for $v \in \mathbb{R}^n$, and so

$$\left(t^{-1} + \epsilon\right) |v| \le |(S \circ T^{-1}) v| \le (t - \epsilon)|v| \quad (v \in \mathbb{R}^n).$$

Thus

$$(S \circ T^{-1})(B(1)) \subseteq B(t - \epsilon);$$

whence

$$|\det(S \circ T^{-1})|\alpha(n) \le \mathcal{L}^n(B(t - \epsilon)) = \alpha(n)(t - \epsilon)^n,$$

and hence

$$|\det S| \le (t - \epsilon)^n |\det T|.$$

The proof of the other inequality is similar.

4. Relabel the countable collection $\{E(c,T,i)|c \in C, T \in S, i = 1, 2, \dots\}$ as $\{E_k\}_{k=1}^\infty$. Select any $b \in B$, write $Df(b) = O \circ S$ as above, and choose $T \in S$ such that

$$\mathrm{Lip}(T \circ S^{-1}) \le \left(t^{-1} + \epsilon\right)^{-1}, \ \mathrm{Lip}(S \circ T^{-1}) \le t - \epsilon.$$

Now select $i \in \{1, 2, \dots\}$ and $c \in C$ so that $|b - c| < \frac{1}{i}$,

$$|f(a) - f(b) - Df(b) \cdot (a - b)| \le \frac{\epsilon}{\mathrm{Lip}(T^{-1})}|a - b| \le \epsilon|T(a - b)|$$

for all $a \in B(b, \frac{2}{i})$. Then $b \in E(c,T,i)$. As this conclusion holds for all $b \in B$, statement (i) is proved.

5. Next choose any set E_k, which is of the form $E(c,T,i)$ for some $c \in C, T \in S, i = 1, 2, \dots$ Let $T_k = T$. According to $(\star\star\star)$,

$$t^{-1}|T_k(a - b)| \le |f(a) - f(b)| \le t|T_k(a - b)|$$

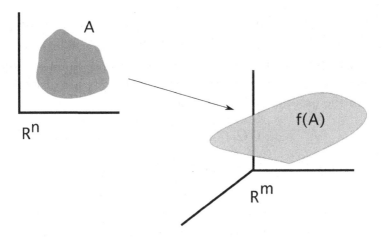

for all $b \in E_k, a \in B(b, \frac{2}{i})$. As $E_k \subseteq B(c, \frac{1}{i}) \subseteq B(b, \frac{2}{i})$, we thus have

$$t^{-1}|T_k(a - b)| \leq |f(a) - f(b)| \leq t|T_k(a - b)| \qquad (\star\star\star\star)$$

for all $a, b \in E_k$; hence $f|_{E_k}$ is one-to-one. Finally, notice the above implies

$$\mathrm{Lip}((f|_{E_k} \circ T_k^{-1}) \leq t, \; \mathrm{Lip}(T_k \circ (f|_{E_k})^{-1}) \leq t,$$

whereas the claim provides the estimate

$$t^{-n}|\det T_k| \leq Jf|_{E_k} \leq t^n|\det T_k|.$$

Assertion (iii) is proved. $\qquad\qquad\qquad\qquad\qquad\qquad\qquad\qquad\qquad\qquad$ □

3.3.2 Proof of the area formula

THEOREM 3.8 (Area formula). *Let $f : \mathbb{R}^n \to \mathbb{R}^m$ be Lipschitz continuous, $n \leq m$. There for each $\mathcal{L}^n$-measurable subset $A \subset \mathbb{R}^n$,*

$$\int_A Jf \, dx = \int_{\mathbb{R}^m} \mathcal{H}^0(A \cap f^{-1}\{y\}) \, d\mathcal{H}^n(y).$$

Remark. The area formula tells us that the $\mathcal{H}^n$-measure of the image $f(A) \subset \mathbb{R}^n$, counting multiplicity, can be computed by integrating the Jacobian Jf over A.

We also see that $f^{-1}\{y\}$ is at most countable for $\mathcal{H}^n$-a.e. $y \in \mathbb{R}^m$.

$\qquad\qquad\qquad\qquad\qquad\qquad\qquad\qquad\qquad\qquad\qquad\qquad\qquad\qquad\qquad\qquad$ □

Proof. 1. In view of Rademacher's Theorem, we may as well assume $Df(x)$ and $Jf(x)$ exist for all $x \in A$. We may also suppose $\mathcal{L}^n(A) < \infty$.

2. *Case 1:* $A \subseteq \{Jf > 0\}$. Fix $t > 1$ and choose Borel sets $\{E_k\}_{k=1}^{\infty}$ as in Lemma 3.3. We may assume the sets $\{E_k\}_{k=1}^{\infty}$ are disjoint. Define B_k as in the proof of Lemma 3.2. Set

$$F_j^i := E_j \cap Q_i \cap A \quad (Q_i \in B_k, j = 1, 2, \dots).$$

Then the sets F_j^i are disjoint and $A = \cup_{i,j=1}^{\infty} F_j^i$.

3. *Claim #1:*

$$\lim_{k \to \infty} \sum_{i,j=1}^{\infty} \mathcal{H}^n(f(F_j^i)) = \int_{\mathbb{R}^m} \mathcal{H}^0(A \cap f^{-1}\{y\}) d\mathcal{H}^n.$$

Proof of claim: Let

$$g_k := \sum_{i,j=1}^{\infty} \chi_{f(F_j^i)};$$

so that $g_k(y)$ is the number of the sets $\{F_j^i\}$ such that $F_j^i \cap f^{-1}\{y\} \neq \emptyset$. Then $g_k(y) \to \mathcal{H}^0(A \cap f^{-1}\{y\})$ as $k \to \infty$. Apply the Monotone Convergence Theorem.

4. Note

$$\mathcal{H}^n(f(F_j^i)) = \mathcal{H}^n(f|_{E_j} \circ T_j^{-1} \circ T_j(F_j^i)) \leq t^n \mathcal{L}^n(T_j(F_j^i))$$

and

$$\mathcal{L}^n(T_j(F_j^i)) = \mathcal{H}^n(T_j \circ (f|_{E_j})^{-1} \circ f(F_j^i)) \leq t^n \mathcal{H}^n(f(F_j^i))$$

by Lemma 3.3. Thus

$$\begin{aligned}
t^{-2n}\mathcal{H}^n(f(F_j^i)) &\leq t^{-n}\mathcal{L}^n(T_j(F_j^i)) \\
&= t^{-n}|\det T_j|\mathcal{L}^n(F_j^i) \\
&\leq \int_{F_i^j} Jf \, dx \\
&\leq t^n|\det T_j|\mathcal{L}^n(F_j^i) \\
&= t^n\mathcal{L}^n(T_j(F_j^i)) \\
&\leq t^{2n}\mathcal{H}^n(f(F_j^i)),
\end{aligned}$$

where we repeatedly used Lemmas 3.1 and 3.3. Now sum on i and j:

$$t^{-2n} \sum_{i,j=1}^{\infty} \mathcal{H}^n(f(F_j^i)) \leq \int_A Jf \, dx \leq t^{2n} \sum_{i,j=1}^{\infty} \mathcal{H}^n(f(F_j^i)).$$

Now let $k \to \infty$ and recall Claim #1:

$$t^{-2n} \int_{\mathbb{R}^m} \mathcal{H}^0(A \cap f^{-1}\{y\}) \, d\mathcal{H}^n \leq \int_A Jf \, dx$$

$$\leq t^{2n} \int_{\mathbb{R}^m} \mathcal{H}^0(A \cap f^{-1}\{y\}) \, d\mathcal{H}^n.$$

Finally, send $t \to 1^+$.

5. *Case 2.* $A \subseteq \{Jf = 0\}$. Fix $0 < \epsilon \leq 1$. We factor

$$f = p \circ g,$$

where $g : \mathbb{R}^n \to \mathbb{R}^m \times \mathbb{R}^n$ is the mapping

$$g(x) := (f(x), \epsilon x),$$

and $p : \mathbb{R}^m \times \mathbb{R}^n \to \mathbb{R}^m$ is the projection

$$p(y, z) = y.$$

6. *Claim #2:* There exists a constant C such that

$$0 < Jg(x) \leq C\epsilon$$

for $x \in A$.

Proof of claim: Write $g = (f^1, \ldots, f^m, \epsilon x_1, \ldots, \epsilon x_n)$; then

$$Dg(x) = \begin{pmatrix} Df(x) \\ \epsilon I \end{pmatrix}_{(n+m) \times n}.$$

Since $Jg(x)^2$ equals the sum of the squares of the $(n \times n)$-subdeterminants of $Dg(x)$, according to the Binet–Cauchy formula, we have $Jg(x)^2 \geq \epsilon^{2n} > 0$. Furthermore, since $|Df| \leq \mathrm{Lip}(f) < \infty$, we may also employ the Binet–Cauchy formula to compute

$$Jg(x)^2 = Jf(x)^2 + \left\{ \begin{array}{c} \text{sum of squares of terms, each} \\ \text{involving at least one } \epsilon \end{array} \right\} \leq C\epsilon^2.$$

for each $x \in A$.

7. Since $p : \mathbb{R}^m \times \mathbb{R}^n \to \mathbb{R}^m$ is a projection, we can compute, using Case 1 above,

$$
\begin{aligned}
\mathcal{H}^n(f(A)) &\leq \mathcal{H}^n(g(A)) \\
&\leq \int_{\mathbb{R}^{n+m}} \mathcal{H}^0(A \cap g^{-1}\{y, z\}) d\mathcal{H}^n(y, z) \\
&= \int_A Jg(x)\, dx \\
&\leq \epsilon C \mathcal{L}^n(A).
\end{aligned}
$$

Let $\epsilon \to 0$ to conclude $\mathcal{H}^n(f(A)) = 0$, and thus

$$
\int_{\mathbb{R}^n} \mathcal{H}^0(A \cap f^{-1}\{y\})\, d\mathcal{H}^n = 0,
$$

since spt $\mathcal{H}^0(A \cap f^{-1}\{y\}) \subseteq f(A)$. But then

$$
\int_{\mathbb{R}^n} \mathcal{H}^0(A \cap f^{-1}\{y\})\, d\mathcal{H}^n = 0 = \int_A Jf\, dx.
$$

8. In the general case, we write $A = A_1 \cup A_2$ with $A_1 \subseteq \{Jf > 0\}$, $A_2 \subseteq \{Jf = 0\}$, and apply Cases 1 and 2 above. $\qquad\square$

3.3.3   Change of variables formula

THEOREM 3.9 (Changing variables). *Let $f : \mathbb{R}^n \to \mathbb{R}^m$ be Lipschitz continuous, $n \leq m$. Then for each $\mathcal{L}^n$-summable function $g : \mathbb{R}^n \to \mathbb{R}$,*

$$
\int_{\mathbb{R}^n} g(x) Jf(x)\, dx = \int_{\mathbb{R}^m} \left[\sum_{x \in f^{-1}\{y\}} g(x) \right] d\mathcal{H}^n(y).
$$

Proof. 1. *Case 1.* $g \geq 0$. According to Theorem 1.12, we can write

$$
g = \sum_{i=1}^{\infty} \frac{1}{i} \chi_{A_i}
$$

for appropriate $\mathcal{L}^n$-measurable sets $\{A_i\}_{i=1}^\infty$. Then the Monotone Convergence Theorem implies

$$\int_{\mathbb{R}^n} gJf\,dx = \sum_{i=j}^\infty \frac{1}{i}\int_{\mathbb{R}^n} \chi_{A_i} Jf\,dx$$

$$= \sum_{i=1}^\infty \frac{1}{i}\int_{A_i} Jf\,dx$$

$$= \sum_{i=1}^\infty \frac{1}{i}\int_{\mathbb{R}^m} \mathcal{H}^0(A_i \cap f^{-1}\{y\})\,d\mathcal{H}^n(y)$$

$$= \int_{\mathbb{R}^m} \sum_{i=1}^\infty \frac{1}{i} \sum_{x \in f^{-1}\{y\}} \chi_{A_i}(x)\,d\mathcal{H}^n(y)$$

$$= \int_{\mathbb{R}^m} \sum_{x \in f^{-1}\{y\}} \sum_{i=1}^\infty \frac{1}{i}\chi_{A_i}(x)\,d\mathcal{H}^n(y)$$

$$= \int_{\mathbb{R}^m} \left[\sum_{x \in f^{-1}\{y\}} g(x) \right] d\mathcal{H}^n(y).$$

2. *Case 2.* g is any $\mathcal{L}^n$-summable function. Write $g = g^+ - g^-$ and apply Case 1. $\qquad\qquad\square$

3.3.4 Applications

A. Length of a curve. $(n = 1, m \geq 1)$. Assume $f : \mathbb{R} \to \mathbb{R}^m$ is Lipschitz continuous and one-to-one. Write

$$f = (f^1, \ldots, f^m), \quad Df = (\dot{f}^1, \ldots, \dot{f}^m) \quad (\dot{} = \tfrac{d}{dt});$$

so that

$$Jf = |Df| = |\dot{f}|.$$

For $-\infty < a < b < \infty$, define the curve $C := f([a,b]) \subseteq \mathbb{R}^m$. Then

$$\mathcal{H}^1(C) = \text{length of } C = \int_a^b |\dot{f}|\,dt.$$

$\square$

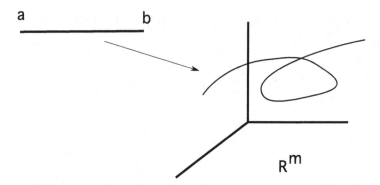

B. Surface area of a graph ($n \geq 1, m = n+1$). Assume $g : \mathbb{R}^n \to \mathbb{R}$ is Lipschitz continuous and define $f : \mathbb{R}^n \to \mathbb{R}^{n+1}$ by

$$f(x) := (x, g(x)).$$

Then

$$Df = \begin{pmatrix} 1 & \cdots & 0 \\ \vdots & \ddots & \vdots \\ 0 & \cdots & 1 \\ g_{x_1} & \cdots & g_{x_n} \end{pmatrix}_{(n+1) \times n} ;$$

consequently,

$$(Jf)^2 = \text{sum of squares of } n \times n \text{ subdeterminants} = 1 + |Dg|^2.$$

For each open set $U \subseteq \mathbb{R}^n$, define the graph of g over U,

$$G = G(g; U) := \{(x, g(x)) \mid x \in U\} \subset \mathbb{R}^{n+1}.$$

Then

$$\mathcal{H}^n(G) = \text{surface area of } G = \int_U (1 + |Dg|^2)^{\frac{1}{2}} \, dx.$$

$\square$

C. Surface area of a parametric hypersurface ($n \geq 1, m = n+1$). Suppose $f : \mathbb{R}^n \to \mathbb{R}^{n+1}$ is Lipschitz continuous and one-to-one. Write

$$f = (f^1, \ldots, f^{n+1}),$$

$$Df = \begin{pmatrix} f^1_{x_1} & \cdots & f^1_{x_n} \\ \vdots & \ddots & \vdots \\ f^{n+1}_{x_1} & \cdots & f^{n+1}_{x_n} \end{pmatrix}_{(n+1) \times n} ;$$

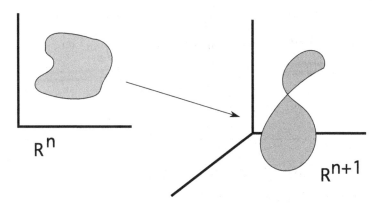

so that

$$(Jf)^2 = \text{sum of squares of } n \times n \text{ subdeterminants}$$
$$= \sum_{k=1}^{n+1} \left[\frac{\partial(f^1, \ldots, f^{k-1}, f^{k+1}, \ldots, f^{n+1})}{\partial(x_1, \ldots, x_n)} \right]^2.$$

For each open set $U \subseteq \mathbb{R}^n$, write

$$S := f(U) \subseteq \mathbb{R}^{n+1}.$$

Then

$$\mathcal{H}^n(S) = n\text{-dimensional surface area of } S$$
$$= \int_U \left(\sum_{k=1}^{n+1} \left[\frac{\partial(f^1, \ldots, f^{k-1}, f^{k+1}, \ldots, f^{n+1})}{\partial(x_1, \ldots, x_n)} \right]^2 \right)^{\frac{1}{2}} dx.$$

$\square$

D. Submanifolds. Let $M \subseteq \mathbb{R}^m$ be a Lipschitz continuous, n-dimensional embedded submanifold. Suppose that $U \subseteq \mathbb{R}^n$ and $f : U \to M$ is a chart for M. Let $A \subseteq f(U)$, where A is Borel, and set $B := f^{-1}(A)$.

Define

$$g_{ij} := f_{x_i} \cdot f_{x_j} \quad (i, j = 1, \ldots, n).$$

Then

$$(Df)^* \circ Df = ((g_{ij}))$$

and so

$$Jf = g^{\frac{1}{2}} \quad \text{for } g := \det((g_{ij})).$$

Therefore

$$\mathcal{H}^n(A) = \text{volume of } A \text{ in } M = \int_B g^{\frac{1}{2}} \, dx.$$

$\square$

3.4   The coarea formula

Throughout this section we assume

$$n \geq m.$$

3.4.1   Preliminaries

LEMMA 3.4. *Suppose* $L : \mathbb{R}^n \to \mathbb{R}^m$ *is linear and* $A \subseteq \mathbb{R}^n$ *is* $\mathcal{L}^n$-*measurable. Then*

(i) *the mapping* $y \mapsto \mathcal{H}^{n-m}(A \cap L^{-1}\{y\})$ *is* $\mathcal{L}^m$-*measurable, and*

(ii)

$$\int_{\mathbb{R}^m} \mathcal{H}^{n-m}(A \cap L^{-1}\{y\}) \, dy = [\![L]\!] \, \mathcal{L}^n(A).$$

Proof. 1. *Case* 1. $\dim L(\mathbb{R}^n) < m$.

Then for $\mathcal{L}^m$-a.e. $y \in \mathbb{R}^m$, we have $A \cap L^{-1}\{y\} = \emptyset$ and consequently $\mathcal{H}^{n-m}(A \cap L^{-1}\{y\}) = 0$. Also, if we write $L = S \circ O^*$ as in the Polar Decomposition Theorem 3.5, we have $L(\mathbb{R}^n) = S(\mathbb{R}^m)$. Thus $\dim S(\mathbb{R}^m) < m$ and hence $[\![L]\!] = |\det S| = 0$.

2. *Case* 2. $L = P = $ orthogonal projection of $\mathbb{R}^n$ onto $\mathbb{R}^m$.

Then for each $y \in \mathbb{R}^m$, $P^{-1}\{y\}$ is an $(n-m)$ -dimensional affine subspace of $\mathbb{R}^n$, a translate of $P^{-1}\{0\}$. By Fubini's Theorem,

$$y \mapsto \mathcal{H}^{n-m}(A \cap P^{-1}\{y\}) \quad \text{is } \mathcal{L}^m \text{ measurable}$$

and

$$\int_{\mathbb{R}^m} \mathcal{H}^{n-m}(A \cap P^{-1}\{y\}) \, dy = \mathcal{L}^n(A). \qquad (\star)$$

3. *Case* 3. $L : \mathbb{R}^n \to \mathbb{R}^m, \dim L(\mathbb{R}^n) = m$.

Using the Polar Decomposition Theorem, we can write

$$L = S \circ O^*$$

where

$$S : \mathbb{R}^m \to \mathbb{R}^m \text{ is symmetric,}$$
$$O : \mathbb{R}^m \to \mathbb{R}^n \text{ is orthogonal,}$$

and

$$[\![L]\!] = |\det S| > 0.$$

4. *Claim:* We can write $O^* = P \circ Q$, where P is the orthogonal projection of $\mathbb{R}^n$ onto $\mathbb{R}^m$ and $Q : \mathbb{R}^n \to \mathbb{R}^n$ is orthogonal.

Proof of claim : Let Q be any orthogonal map of $\mathbb{R}^n$ onto $\mathbb{R}^n$ such that

$$Q^*(x_1, \ldots, x_m, 0, \ldots, 0) = O(x_1, \ldots, x_m)$$

for all $x \in \mathbb{R}^m$. Note

$$P^*(x_1, \ldots, x_m) = (x_1, \ldots, x_m, 0, \ldots, 0) \in \mathbb{R}^n$$

for all $x \in \mathbb{R}^m$. Thus $O = Q^* \circ P^*$ and hence $O^* = P \circ Q$.

5. $L^{-1}\{0\}$ is an $(n-m)$ -dimensional subspace of $\mathbb{R}^n$ and $L^{-1}\{y\}$ is a translate of $L^{-1}\{0\}$ for each $y \in \mathbb{R}^m$. Thus by Fubini's Theorem, $y \to \mathcal{H}^{n-m}(A \cap L^{-1}\{y\})$ is $\mathcal{L}^m$-measurable, and we may calculate

$$
\begin{aligned}
\mathcal{L}^n(A) &= \mathcal{L}^n(Q(A)) \\
&= \int_{\mathbb{R}^m} \mathcal{H}^{n-m}(Q(A) \cap P^{-1}\{y\}) \, dy \quad \text{by } (\star) \\
&= \int_{\mathbb{R}^m} \mathcal{H}^{n-m}(A \cap (Q^{-1} \circ P^{-1}\{y\})) \, dy.
\end{aligned}
$$

Now set $z = Sy$, to compute using Theorem 3.9 that

$$|\det S| \mathcal{L}^n(A) = \int_{\mathbb{R}^m} \mathcal{H}^{n-m}(A \cap (Q^{-1} \circ P^{-1} \circ S^{-1}\{z\})) \, dz.$$

But $L = S \circ O^* = S \circ P \circ Q$, and so

$$[\![L]\!] \mathcal{L}^n(A) = \int_{\mathbb{R}^m} \mathcal{H}^{n-m}(A \cap L^{-1}\{z\}) \, dz. \qquad \square$$

Henceforth we assume $f : \mathbb{R}^n \to \mathbb{R}^m$ is Lipschitz continuous.

LEMMA 3.5. *Let $A \subseteq \mathbb{R}^n$ be $\mathcal{L}^n$-measurable, $n \geq m$. Then*

(i) $A \cap f^{-1}\{y\}$ *is $\mathcal{H}^{n-m}$-measurable for $\mathcal{L}^m$-a.e. y,*

(ii) *the mapping $y \mapsto \mathcal{H}^{n-m}(A \cap f^{-1}\{y\})$ is $\mathcal{L}^m$-measurable, and*

(iii)

$$\int_{\mathbb{R}^m} \mathcal{H}^{n-m}(A \cap f^{-1}\{y\}) \, dy \leq \frac{\alpha(n-m)\alpha(m)}{\alpha(n)} (\operatorname{Lip} f)^m \mathcal{L}^n(A).$$

Proof. 1. For each $j = 1, 2, \ldots$, there exist closed balls $\{B_i^j\}_{i=1}^\infty$ such that

$$A \subseteq \bigcup_{i=1}^\infty B_i^j, \quad \operatorname{diam} B_i^j \leq \frac{1}{j}, \quad \sum_{i=1}^\infty \mathcal{L}^n(B_i^j) \leq \mathcal{L}^n(A) + \frac{1}{j}.$$

Define

$$g_i^j := \alpha(n-m) \left(\frac{\operatorname{diam} B_i^j}{2} \right)^{n-m} \chi_{f(B_i^j)};$$

g_i^j is $\mathcal{L}^m$-measurable. Note also for all $y \in \mathbb{R}^m$,

$$\mathcal{H}_{\frac{1}{j}}^{n-m}(A \cap f^{-1}\{y\}) \leq \sum_{i=1}^\infty g_i^j(y).$$

Thus, using Fatou's Lemma and the isodiametric inequality (Section 2.2), we compute

$$\int_{\mathbb{R}^m}^* \mathcal{H}^{n-m}(A \cap f^{-1}\{y\}) \, dy$$

$$= \int_{\mathbb{R}^m}^* \lim_{j \to \infty} \mathcal{H}_{\frac{1}{j}}^{n-m}(A \cap f^{-1}\{y\}) \, dy$$

$$\leq \int_{\mathbb{R}^m} \liminf_{j \to \infty} \sum_{i=1}^\infty g_i^j \, dy$$

$$\leq \liminf_{j \to \infty} \sum_{i=1}^\infty \int_{\mathbb{R}^m} g_i^j \, dy$$

$$= \liminf_{j \to \infty} \sum_{i=1}^\infty \alpha(n-m) \left(\frac{\operatorname{diam} B_i^j}{2} \right)^{n-m} \mathcal{L}^m(f(B_i^j))$$

$$\leq \liminf_{j\to\infty} \sum_{i=1}^{\infty} \alpha(n-m) \left(\frac{\text{diam } B_i^j}{2} \right)^{n-m}$$

$$\alpha(m) \left(\frac{\text{diam } f(B_i^j)}{2} \right)^{m}$$

$$\leq \frac{\alpha(n-m)\alpha(m)}{\alpha(n)} (\text{Lip } f)^m \liminf_{j\to\infty} \sum_{i=1}^{\infty} \mathcal{L}^n(B_i^j)$$

$$\leq \frac{\alpha(n-m)\alpha(m)}{\alpha(n)} (\text{Lip } f)^m \mathcal{L}^n(A).$$

Thus

$$\int_{\mathbb{R}^m}^{*} \mathcal{H}^{n-m}(A \cap f^{-1}\{y\})\, dy \leq \frac{\alpha(n-m)\alpha(m)}{\alpha(n)} (\text{Lip } f)^m \mathcal{L}^n(A). \quad (\star)$$

This will prove (iii) once we establish (ii).

2. *Case* 1: A compact.

Fix $t \geq 0$, and for each positive integer i, let U_i denote the points $y \in \mathbb{R}^m$ for which there exist finitely many open sets $S_1, \ldots, S_l$ such that

$$\begin{cases} A \cap f^{-1}\{y\} \subseteq \cup_{j=1}^{l} S_j, \\ \text{diam } S_j \leq \frac{1}{i} \quad (j = 1, \ldots, l), \\ \sum_{j=1}^{l} \alpha(n-m) \left(\frac{\text{diam } S_j}{2} \right)^{n-m} \leq t + \frac{1}{i}. \end{cases}$$

3. *Claim* #1: U_i is open.

Proof of claim: Assume $y \in U_i, A \cap f^{-1}\{y\} \subseteq \cup_{j=1}^{l} S_j$, as above. Then, since f is continuous and A is compact,

$$A \cap f^{-1}\{z\} \subseteq \bigcup_{j=1}^{l} S_j$$

for all z sufficiently close to y.

4. *Claim* #2.

$$\{y \mid \mathcal{H}^{n-m}(A \cap f^{-1}\{y\}) \leq t\} = \bigcap_{i=1}^{\infty} U_i$$

and hence the set on the left is Borel.

Proof of claim: If $\mathcal{H}^{n-m}(A \cap f^{-1}\{y\}) \leq t$, then for each $\delta > 0$,

$$\mathcal{H}^{n-m}_\delta(A \cap f^{-1}\{y\}) \leq t.$$

Given i, choose $\delta \in (0, \frac{1}{i})$. Then there exist sets $\{S_j\}_{j=1}^\infty$ such that

$$\begin{cases} A \cap f^{-1}\{y\} \subseteq \cup_{j=1}^\infty S_j, \\[1mm] \operatorname{diam} S_j \leq \delta < \frac{1}{i}, \\[1mm] \sum_{j=1}^\infty \alpha(n-m) \left(\frac{\operatorname{diam} S_j}{2} \right)^{n-m} < t + \frac{1}{i}. \end{cases}$$

We may assume the S_j are open. Since $A \cap f^{-1}\{y\}$ is compact, a finite subcollection $\{S_1, \ldots, S_l\}$ covers $A \cap f^{-1}\{y\}$; and hence $y \in U_i$. Thus

$$\{y \mid \mathcal{H}^{n-m}(A \cap f^{-1}\{y\}) \leq t\} \subseteq \bigcap_{i=1}^\infty U_i.$$

On the other hand, if $y \in \cap_{i=1}^\infty U_i$, then for each i,

$$\mathcal{H}^{n-m}_{\frac{1}{i}}(A \cap f^{-1}\{y\}) \leq t + \frac{1}{i};$$

and so

$$\mathcal{H}^{n-m}(A \cap f^{-1}\{y\}) \leq t.$$

Thus

$$\bigcap_{i=1}^\infty U_i \subseteq \{y \mid \mathcal{H}^{n-m}(A \cap f^{-1}\{y\}) \leq t\}.$$

5. According to Claim #2, for compact A the mapping

$$y \to \mathcal{H}^{n-m}(A \cap f^{-1}\{y\})$$

is a Borel function.

6. *Case* 2: A is open. There exist compact sets $K_1 \subset K_2 \subset \cdots \subset A$ such that

$$A = \bigcup_{i=1}^\infty K_i.$$

Hence for each $y \in \mathbb{R}^m$,

$$\mathcal{H}^{n-m}(A \cap f^{-1}\{y\}) = \lim_{i \to \infty} \mathcal{H}^{n-m}(K_i \cap f^{-1}\{y\});$$

and therefore the mapping

$$y \mapsto \mathcal{H}^{n-m}(A \cap f^{-1}\{y\})$$

is Borel measurable.

7. *Case 3*: $\mathcal{L}^n(A) < \infty$. There exist open sets $V_1 \supset V_2 \supset \cdots \supset A$ such that

$$\lim_{i \to \infty} \mathcal{L}^n(V_i - A) = 0, \quad \mathcal{L}^n(V_1) < \infty.$$

Now

$$\mathcal{H}^{n-m}(V_i \cap f^{-1}\{y\})$$
$$\leq \mathcal{H}^{n-m}(A \cap f^{-1}\{y\}) + \mathcal{H}^{n-m}((V_i - A) \cap f^{-1}\{y\});$$

and thus by $(\star)$,

$$\limsup_{i \to \infty} \int_{\mathbb{R}^m}^* |\mathcal{H}^{n-m}(V_i \cap f^{-1}\{y\}) - \mathcal{H}^{n-m}(A \cap f^{-1}\{y\})| \, dy$$

$$\leq \limsup_{i \to \infty} \int_{\mathbb{R}^n}^* \mathcal{H}^{n-m}((V_i - A) \cap f^{-1}\{y\} \, dy$$

$$\leq \limsup_{i \to \infty} \frac{\alpha(n-m)\alpha(m)}{\alpha(n)} (\mathrm{Lip}\, f)^m \mathcal{L}^n(V_i - A) = 0.$$

Consequently,

$$\mathcal{H}^{n-m}(V_i \cap f^{-1}\{y\}) \to \mathcal{H}^{n-m}(A \cap f^{-1}\{y\})$$

$\mathcal{L}^m$-a.e.. According then to Case 2, it follows that

$$y \mapsto \mathcal{H}^{n-m}(A \cap f^{-1}\{y\})$$

is $\mathcal{L}^m$-measurable. In addition, we see $\mathcal{H}^{n-m}((V_i - A) \cap f^{-1}\{y\}) \to 0$ $\mathcal{L}^m$-a.e. and so $A \cap f^{-1}\{y\}$ is $\mathcal{H}^{n-m}$-measurable for $\mathcal{L}^m$-a.e. y.

8. *Case 4.* $\mathcal{L}^n(A) = \infty$. Write A as a union of an increasing sequence of bounded $\mathcal{L}^n$-measurable sets and apply Case 3 to prove $A \cap f^{-1}\{y\}$ is $\mathcal{H}^{n-m}$-measurable for $\mathcal{L}^m$-a.e. y, and

$$y \to \mathcal{H}^{n-m}(A \cap f^{-1}\{y\})$$

is $\mathcal{L}^m$-measurable. This proves (i) and (ii), and (iii) follows from $(\star)$.

$$\square$$

Remark. A proof similar to that of (iii) shows

$$\int_{\mathbb{R}^m}^* \mathcal{H}^k(A \cap f^{-1}\{y\}) \, d\mathcal{H}^l \leq \frac{\alpha(k)\alpha(l)}{\alpha(k+l)} (\operatorname{Lip} f)^l \mathcal{H}^{k+1}(A)$$

for each $A \subseteq \mathbb{R}^n$; see Federer [F, Sections 2.10.25 and 2.10.26].   □

LEMMA 3.6. *Let* $t > 1$, *assume* $h : \mathbb{R}^n \to \mathbb{R}^n$ *is Lipschitz continuous, and set*

$$B = \{x \mid Dh(x) \text{ exists}, Jh(x) > 0\}.$$

Then there exists a countable collection $\{D_k\}_{k=1}^{\infty}$ *of Borel subsets of* $\mathbb{R}^n$ *such that*

(i) $\mathcal{L}^n(B - \cup_{k=1}^{\infty} D_k) = 0$;

(ii) $h|_{D_k}$ *is one-to-one for* $k = 1, 2, \ldots$; *and*

(iii) *for each* $k = 1, 2, \ldots$, *there exists a symmetric automorphism* $S_k : \mathbb{R}^n \to \mathbb{R}^n$ *such that*

$$\operatorname{Lip}(S_k^{-1} \circ (h|_{D_k})) \leq t, \; \operatorname{Lip}((h|_{D_k})^{-1} \circ S_k) \leq t,$$
$$t^{-n}|\det S_k| \leq Jh|_{D_k} \leq t^n |\det S_k|.$$

Proof. 1. Apply Lemma 3.3 with h in place of f, to find Borel sets $\{E_k\}_{k=1}^{\infty}$ and symmetric automorphisms $T_k : \mathbb{R}^n \to \mathbb{R}^n$ such that

(a) $B = \cup_{k=1}^{\infty} E_k$,

(b) $h|_{E_k}$ is one-to–one,

(c) For $k = 1, 2, \ldots$

$$\operatorname{Lip}((h|_{E_k}) \circ T_k^{-1}) \leq t, \; \operatorname{Lip}(T_k \circ (h|_{E_k})^{-1}) \leq t$$
$$t^{-n}|\det T_k| \leq Jh|_{E_k} \leq t^n |\det T_k|.$$

According to (c), $(h|_{E_k})^{-1}$ is Lipschitz continuous and thus by Theorem 3.1, there exists a Lipschitz continuous mapping $h_k : \mathbb{R}^n \to \mathbb{R}^n$ such that $h_k = (h|_{E_k})^{-1}$ on $h(E_k)$.

2. *Claim #1:* $Jh_k > 0$ $\mathcal{L}^n$-a.e. on $h(E_k)$.

Proof of claim: Since $h_k \circ h(x) = x$ for $x \in E_k$, Theorem 3.3 implies

$$Dh_k(h(x)) \circ Dh(x) = 1 \quad \mathcal{L}^n\text{-a.e. on } E_k,$$

and so

$$Jh_k(h(x))Jh(x) = 1 \quad \mathcal{L}^n\text{-a.e. on } E_k.$$

In view of (c), this implies $Jh_k(h(x)) > 0$ for $\mathcal{L}^n$-a.e. $x \in E_k$, and the claim follows since h is Lipschitz continuous.

3. Now apply Lemma 3.3. There exist Borel sets $\{F_j^k\}_{j=1}^\infty$ and symmetric automorphisms $\{R_j^k\}_{j=1}^\infty$ such that

(d) $\mathcal{L}^n\left(h(E_k) - \cup_{j=1}^\infty F_j^k\right) = 0$;

(e) $h_k|_{F_j^k}$ is one-to-one;

(f) For $k = 1, 2, \ldots$

$$\mathrm{Lip}((h_k|_{F_j^k}) \circ (R_j^k)^{-1}) \leq t, \ \mathrm{Lip}(R_j^k \circ (h_k|_{F_j^k})^{-1}) \leq t$$
$$t^{-n}|\det R_j^k| \leq Jh_k|_{F_j^k} \leq t^n|\det R_j^k|.$$

Set

$$D_j^k := E_k \cap h^{-1}(F_j^k), \ S_j^k := (R_j^k)^{-1} \quad (k = 1, 2, \ldots).$$

4. *Claim #2:* $\mathcal{L}^n\left(B - \cup_{k,j=1}^\infty D_j^k\right) = 0.$

Proof of claim: Note that

$$h_k\left(h(E_k) - \cup_{j=1}^\infty F_j^k\right) = h^{-1}\left(h(E_k) - \cup_{j=1}^\infty F_j^k\right) = E_k - \cup_{j=1}^\infty D_j^k.$$

Thus, according to (d),

$$\mathcal{L}^n\left(E_k - \cup_{j=1}^\infty D_j^k\right) = 0 \quad (k = 1, \ldots).$$

Now recall (a).

5. Clearly (b) implies $h|_{D_j^k}$ is one-to-one.

6. *Claim #3:* For $k, j = 1, 2, \ldots$, we have

$$\mathrm{Lip}((S_j^k)^{-1} \circ (h|_{D_j^k})) \leq t, \quad \mathrm{Lip}((h|_{D_j^k})^{-1} \circ S_j^k) \leq t$$

$$t^{-n}|\det S_j^k| \le Jh|_{D_j^k} \le t^n|\det S_j^k|.$$

Proof of claim:

$$\text{Lip}((S_j^k)^{-1} \circ (h|_{D_j^k})) = \text{Lip}(R_j^k \circ (h|_{D_j^k}))$$

$$\le \text{Lip}(R_j^k \circ (h_k|_{F_j^k})^{-1}) \le t$$

by (f); similarly,

$$\text{Lip}((h|_{D_j^k})^{-1} \circ S_j^k = \text{Lip}((h|_{D_j^k})^{-1} \circ (R_j^k)^{-1})$$

$$\le \text{Lip}((h_k|_{F_j^k}) \circ (R_j^k)^{-1}) \le t.$$

Furthermore, as noted above,

$$Jh_k(h(x))Jh(x) = 1 \quad \mathcal{L}^n\text{-a.e. on } D_j^k.$$

Thus (f) implies

$$t^{-n}|\det S_j^k| = t^{-n}|\det R_j^k|^{-1}$$

$$\le Jh|_{D_j^k} \le t^n|\det R_j^k|^{-1} = t^n|\det S_j^k|. \quad \square$$

3.4.2   Proof of the coarea formula

THEOREM 3.10 (Coarea formula). *Let $f : \mathbb{R}^n \to \mathbb{R}^m$ be Lipschitz continuous, $n \ge m$. Then for each $\mathcal{L}^n$ -measurable set $A \subseteq \mathbb{R}^n$,*

$$\int_A Jf \, dx = \int_{\mathbb{R}^m} \mathcal{H}^{n-m}(A \cap f^{-1}\{y\}) \, dy.$$

Observe that the coarea formula is a kind of "curvilinear" generalization of Fubini's Theorem.

Remark. Applying the coarea formula to $A = \{Jf = 0\}$, we discover

$$\mathcal{H}^{n-m}(\{Jf = 0\} \cap f^{-1}\{y\}) = 0 \qquad (\star)$$

for $\mathcal{L}^m$-a.e. $y \in \mathbb{R}^m$. This is a weak variant of the **Morse–Sard Theorem**, which asserts

$$\{Jf = 0\} \cap f^{-1}\{y\} = \emptyset$$

for $\mathcal{L}^m$-a.e. y, provided $f \in C^k(\mathbb{R}^n; \mathbb{R}^m)$ for

$$k = 1 + n - m.$$

Note however $(\star)$ only requires that f be Lipschitz continuous. $\square$

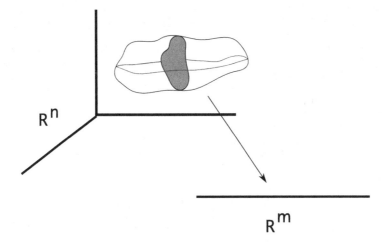

Proof. 1. In view of Lemma 3.5, we may assume that $Df(x)$, and thus $Jf(x)$, exist for all $x \in A$ and that $\mathcal{L}^n(A) < \infty$.

2. *Case 1.* $A \subseteq \{Jf > 0\}$. For each $\lambda \in \Lambda(n, n-m)$, write

$$f = q \circ h_\lambda,$$

where $h_\lambda : \mathbb{R}^n \to \mathbb{R}^m \times \mathbb{R}^{n-m}$ and $q : \mathbb{R}^m \times \mathbb{R}^{n-m} \to \mathbb{R}^m$ are the functions

$$h_\lambda(x) := (f(x), P_\lambda(x)) \quad (x \in \mathbb{R}^n)$$
$$q(y, z) := y \quad (y \in \mathbb{R}^m, z \in \mathbb{R}^{n-m}),$$

and P_λ is the projection defined in Section 3.2. Set

$$A_\lambda := \{x \in A \mid \det Dh_\lambda \neq 0\}$$
$$= \{x \in A \mid P_\lambda|_{[Df(x)]^{-1}(0)} \text{ is injective}\}.$$

Now $A = \cup_{\lambda \in \Lambda(n, n-m)} A_\lambda$; therefore we may as well for simplicity assume $A = A_\lambda$ for some $\lambda \in \Lambda(n, n-m)$.

3. Fix $t > 1$ and apply Lemma 3.6 to $h = h_\lambda$ to obtain disjoint Borel sets $\{D_k\}_{k=1}^\infty$ and symmetric automorphisms $\{S_k\}_{k=1}^\infty$ satisfying assertions (i)–(iii) in Lemma 3. Set $G_k := A \cap D_k$.

4. *Claim #1:* $t^{-n} \llbracket q \circ S_k \rrbracket \leq Jf|_{G_k} \leq t^n \llbracket q \circ S_k \rrbracket$.

Proof of claim: Since $f = q \circ h$, we have $\mathcal{L}^n$-a.e.

$$Df = q \circ Dh$$

$$= q \circ S_k \circ S_k^{-1} \circ Dh$$
$$= q \circ S_k \circ D(S_k^{-1} \circ h)$$
$$= q \circ S_k \circ C,$$

where $C := D(S_k^{-1} \circ h)$.

By Lemma 3,

$$t^{-1} \leq \mathrm{Lip}(S_k^{-1} \circ h) = \mathrm{Lip}(C) \leq t \quad \text{on } G_k. \qquad (\star)$$

Now write

$$Df = S \circ O^*, \quad q \circ S_k = T \circ P^*$$

for symmetric $S, T : \mathbb{R}^m \to \mathbb{R}^m$ and orthogonal $O, P : \mathbb{R}^m \to \mathbb{R}^n$.

We have then

$$S \circ O^* = T \circ P^* \circ C. \qquad (\star\star)$$

Consequently,

$$S = T \circ P^* \circ C \circ O.$$

As $G_k \subseteq A \subseteq \{Jf > 0\}, \det S \neq 0$ and so $\det T \neq 0$.

Therefore if $v \in \mathbb{R}^m$,

$$|T^{-1} \circ Sv| = |P^* \circ C \circ Ov|$$
$$\leq |C \circ Ov|$$
$$\leq t|Ov| \quad \text{by } (\star)$$
$$= t|v|.$$

Therefore

$$(T^{-1} \circ S)(B(1)) \subseteq B(t),$$

and so

$$Jf = |\det S| \leq t^n |\det T| = t^n \, [\![q \circ S_k]\!].$$

Similarly, if $v \in \mathbb{R}^m$, we have from $(\star)$ and $(\star\star)$ that

$$|S^{-1} \circ Tv| = |O^* \circ C^{-1} \circ Pv|$$
$$\leq |C^{-1} \circ Pv|$$
$$\leq t|Pv|$$
$$= t|v|.$$

Thus

$$[\![qoS_k]\!] = |\det T| \leq t^n |\det S| = t^n Jf.$$

5. Now calculate:

$$t^{-3n+m} \int_{\mathbb{R}^m} \mathcal{H}^{n-m}(G_k \cap f^{-1}\{y\}) \, dy$$

$$= t^{-3n+m} \int_{\mathbb{R}^m} \mathcal{H}^{n-m}(h^{-1}(h(G_k) \cap q^{-1}\{y\})) \, dy$$

$$\leq t^{-2n} \int_{\mathbb{R}^m} \mathcal{H}^{n-m}(S_k^{-1}(h(G_k) \cap q^{-1}\{y\})) \, dy$$

$$= t^{-2n} \int_{\mathbb{R}^m} \mathcal{H}^{n-m}(S_k^{-1} \circ h(G_k) \cap (q \circ S_k)^{-1}\{y\}) \, dy$$

$$= t^{-2n} \llbracket q \circ S_k \rrbracket \mathcal{L}^n(S_k^{-1} \circ h(G_k)) \quad \text{(by Lemma 3.4)}$$

$$\leq t^{-n} \llbracket q \circ S_k \rrbracket \mathcal{L}^n(G_k)$$

$$\leq \int_{G_k} Jf \, dx$$

$$\leq t^n \llbracket q \circ S_k \rrbracket \mathcal{L}^n(G_k)$$

$$\leq t^{2n} \llbracket q \circ S_k \rrbracket \mathcal{L}^n(S_k^{-1} \circ h(G_k))$$

$$= t^{2n} \int_{\mathbb{R}^m} \mathcal{H}^{n-m}(S_k^{-1} \circ h(G_k) \cap (q \circ S_k)^{-1}\{y\}) \, dy$$

$$\leq t^{3n-m} \int_{\mathbb{R}^m} \mathcal{H}^{n-m}(h^{-1}(h(G_k) \cap q^{-1}\{y\})) \, dy$$

$$= t^{3n-m} \int_{\mathbb{R}^m} \mathcal{H}^{n-m}(G_k \cap f^{-1}\{y\}) \, dy.$$

Since

$$\mathcal{L}^n \left(A - \cup_{k=1}^{\infty} G_k \right) = 0,$$

we can sum on k, use Lemma 3.5, and let $t \to 1^+$ to conclude

$$\int_{\mathbb{R}^m} \mathcal{H}^{n-m}(A \cap f^{-1}\{y\}) \, dy = \int_A Jf \, dx.$$

6. *Case 2.* $A \subseteq \{Jf = 0\}$. Fix $0 < \epsilon \leq 1$ and define

$$g(x,y) := f(x) + \epsilon y, \quad p(x,y) := y$$

for $x \in \mathbb{R}^n$, $y \in \mathbb{R}^m$. Then

$$Dg = (Df, \epsilon I)_{m \times (n+m)},$$

and

$$\epsilon^m \leq Jg = \llbracket Dg \rrbracket = \llbracket Dg^* \rrbracket \leq C\epsilon.$$

7. Observe

$$\int_{\mathbb{R}^m} \mathcal{H}^{n-m}(A \cap f^{-1}\{y\}) \, dy$$

$$= \int_{\mathbb{R}^m} \mathcal{H}^{n-m}(A \cap f^{-1}\{y - \epsilon w\}) \, dy \quad \text{for all } w \in \mathbb{R}^m$$

$$= \frac{1}{\alpha(m)} \int_{B(1)} \int_{\mathbb{R}^m} \mathcal{H}^{n-m}(A \cap f^{-1}\{y - \epsilon w\}) \, dy dw.$$

8. *Claim #2:* Fix $y \in \mathbb{R}^m, w \in \mathbb{R}^m$, and set $B := A \times B(1) \subset \mathbb{R}^{n+m}$. Then

$$B \cap g^{-1}\{y\} \cap p^{-1}\{w\}$$

$$= \begin{cases} \emptyset & \text{if } w \notin B(1) \\ (A \cap f^{-1}\{y - \epsilon w\}) \times \{w\} & \text{if } w \in B(1). \end{cases}$$

Proof of claim: We have $(x, z) \in B \cap g^{-1}\{y\} \cap p^{-1}\{w\}$ if and only if

$$x \in A, z \in B(1), f(x) + \epsilon z = y, z = w;$$

if any only if

$$x \in A, z = w \in B(1), f(x) = y - \epsilon w;$$

if and only if

$$w \in B(1), (x, z) \in (A \cap f^{-1}\{y - \epsilon w\}) \times \{w\}.$$

9. Now use Claim #2 to continue the calculation from Step 7:

$$\int_{\mathbb{R}^m} \mathcal{H}^{n-m}(A \cap f^{-1}\{y\}) \, dy$$

$$= \frac{1}{\alpha(m)} \int_{\mathbb{R}^m} \int_{\mathbb{R}^m} \mathcal{H}^{n-m}(B \cap g^{-1}\{y\} \cap p^{-1}\{w\}) \, dw dy$$

$$\leq \frac{\alpha(n-m)}{\alpha(n)} \int_{\mathbb{R}^m} \mathcal{H}^n(B \cap g^{-1}\{y\}) \, dy$$

$$= \frac{\alpha(n-m)}{\alpha(n)} \int_B Jg \, dx dz$$

$$\leq \frac{\alpha(n-m)\alpha(m)}{\alpha(n)} \mathcal{L}^n(A) \sup_B Jg$$

$$\leq C\epsilon.$$

The third line above follows from the Remark on page 132. Let $\epsilon \to 0$, to obtain

$$\int_{\mathbb{R}^m} \mathcal{H}^{n-m}(A \cap f^{-1}\{y\})dy = 0 = \int_A Jf\,dx.$$

10. In the general case we write $A = A_1 \cup A_2$ where $A_1 \subseteq \{Jf > 0\}$, $A_2 \subseteq \{Jf = 0\}$, and apply Cases 1 and 2 above. $\qquad\square$

3.4.3 Change of variables formula

THEOREM 3.11 (Integration over level sets). *Let $f : \mathbb{R}^n \to \mathbb{R}^m$ be Lipschiz, $n \geq m$. Then for each $\mathcal{L}^n$-summable function $g : \mathbb{R}^n \to \mathbb{R}$,*

(i) *$g|_{f^{-1}\{y\}}$ is $\mathcal{H}^{n-m}$ summable for $\mathcal{L}^m$-a.e. y, and*

(ii)

$$\int_{\mathbb{R}^n} g\,Jf\,dx = \int_{\mathbb{R}^m}\left[\int_{f^{-1}\{y\}} g\,d\mathcal{H}^{n-m}\right]dy.$$

Remark. For each $y \in \mathbb{R}^m, f^{-1}\{y\}$ is closed and thus $\mathcal{H}^{n-m}$-measurable. $\qquad\square$

Proof. 1. *Case 1. $g \geq 0$.* Write $g = \sum_{i=1}^{\infty} \frac{1}{i}\chi_{A_i}$ for appropriate $\mathcal{L}^n$-measurable sets $\{A_i\}_{i=1}^{\infty}$; this is possible according to Theorem 1.12. Then the Monotone Convergence Theorem implies

$$\int_{\mathbb{R}^n} g\,Jf\,dx = \sum_{i=1}^{\infty} \frac{1}{i}\int_{A_i} Jf\,dx$$

$$= \sum_{i=1}^{\infty} \frac{1}{i}\int_{\mathbb{R}^m} \mathcal{H}^{n-m}(A_i \cap f^{-1}\{y\})\,dy$$

$$= \int_{\mathbb{R}^n} \sum_{i=1}^{\infty} \frac{1}{i}\mathcal{H}^{n-m}(A_i \cap f^{-1}\{y\})\,dy$$

$$= \int_{\mathbb{R}^n}\left[\int_{f^{-1}\{y\}} g\,d\mathcal{H}^{n-m}\right]dy.$$

2. *Case 2. g is any $\mathcal{L}^n$-summable function.* Write $g = g^+ - g^-$ and use Case 1. $\qquad\square$

3.4.4   Applications

A. Integrals over balls.

THEOREM 3.12 (Polar coordinates). *Let $g : \mathbb{R}^n \to \mathbb{R}$ be $\mathcal{L}^n$-summable. Then*

$$\int_{\mathbb{R}^n} g \, dx = \int_0^\infty \left(\int_{\partial B(r)} g \, d\mathcal{H}^{n-1} \right) dr.$$

In particular,

$$\frac{d}{dr} \left(\int_{B(r)} g \, dx \right) = \int_{\partial B(r)} g \, d\mathcal{H}^{n-1}$$

for $\mathcal{L}^1$-a.e. $r > 0$.

Proof. Set $f(x) = |x|$; then for $x \neq 0$ we have

$$Df(x) = \frac{x}{|x|}, \quad Jf(x) = 1. \qquad \square$$

B. Integration over level sets.

THEOREM 3.13 (Integration over level sets). *Assume $f : \mathbb{R}^n \to \mathbb{R}$ is Lipschitz continuous.*

(i) *Then*

$$\int_{\mathbb{R}^n} |Df| \, dx = \int_{-\infty}^\infty \mathcal{H}^{n-1}(\{f = t\}) \, dt.$$

(ii) *Assume also*

$$\operatorname{ess\,inf} |Df| > 0,$$

and suppose $g : \mathbb{R}^n \to \mathbb{R}$ is $\mathcal{L}^n$-summable. Then

$$\int_{\{f>t\}} g \, dx = \int_t^\infty \left(\int_{\{f=s\}} \frac{g}{|Df|} \, d\mathcal{H}^{n-1} \right) ds.$$

(iii) *In particular,*

$$\frac{d}{dt} \left(\int_{\{f>t\}} g \, dx \right) = - \int_{\{f=t\}} \frac{g}{|Df|} d\mathcal{H}^{n-1}$$

for $\mathcal{L}^1$-a.e. t.

Remark. Compare (i) with the coarea formula for BV functions, proved later in Theorem 5.9. □

Proof. 1. To prove (i), observe that $Jf = |Df|$.

2. Write $E_t := \{f > t\}$ and use Theorem 3.11 to calculate

$$\int_{\{f>t\}} g \, dx = \int_{\mathbb{R}^n} \chi_{E_t} \frac{g}{|Df|} Jf \, dx$$

$$= \int_{-\infty}^{\infty} \left(\int_{\partial E_s} \frac{g}{|Df|} \chi_{E_t} \, d\mathcal{H}^{n-1} \right) ds$$

$$= \int_t^{\infty} \left(\int_{\partial E_s} \frac{g}{|Df|} \, d\mathcal{H}^{n-1} \right) ds.$$

This gives (ii), and (iii) follows. □

C. Distance functions.

THEOREM 3.14 (Level sets of distance functions). *Assume* $K \subset \mathbb{R}^n$ *is a nonempty compact set and write*

$$d(x) := \text{dist}(x, K) \quad (x \in \mathbb{R}^n).$$

Then for each $0 < a < b$ *we have*

$$\int_a^b \mathcal{H}^{n-1}(\{d = t\}) \, dt = \mathcal{L}^n(\{a \le d \le b\}).$$

Proof. 1. Given $x \in \mathbb{R}^n$, select $c \in K$ so that $d(x) = |x - c|$. Then for any other point $y \in \mathbb{R}^n$, we have

$$d(y) - d(x) \le |y - c| - |x - c| \le |x - y|.$$

Interchanging x and y, we see that $|d(y) - d(x)| \le |x - y|$; consequently,

$$\text{Lip}(d) \le 1.$$

Rademacher's Theorem therefore implies that the distance function is differentiable $\mathcal{L}^n$-a.e..

2. Select any point $x \in \mathbb{R}^n - K$ at which $Dd(x)$ exists. Then $|Dd(x)| \leq 1$, since $\mathrm{Lip}(d) \leq 1$. As above, select $c \in K$ so that $d(x) = |x - c|$. Then

$$d(tx + (1 - t)c) = t|x - c|$$

for all $0 \leq t \leq 1$; and therefore

$$|x - c| = Dd(x) \cdot (x - c) \leq |Dd(x)||x - c|.$$

Thus $|Dd(x)| \geq 1$.

3. It follows that

$$|Dd| = 1 \qquad \mathcal{L}^n\text{-a.e. in } \mathbb{R}^n - K.$$

We may consequently invoke Theorem 3.13 to finish the proof. $\qquad\square$

3.5   References and notes

The primary reference is again Federer [F, Chapters 1 and 3]. Theorem 3.1 is from Simon [S, Section 5.1]. The proof of Rademacher's Theorem, which we took from [S, Section 5.2], is due to Morrey (cf. [My, p. 65]). Theorem 3.3 in Section 3.1 is [F, Section 3.2.8]. See Clarke [C] for more on calculus for Lipschitz continuous functions.

The discussion of linear maps and Jacobians in Section 3.2 is strongly based on Hardt [H]. S. Antman helped us with the proof of the Polar Decomposition Theorem, and A. Damlamian provided the calculations for the Binet–Cauchy formula. See also Gantmacher [Ga, pages 9–12, 276–278].

The proof of the area formula in Section 3.3, originating with [F, Sections 3.2.2–3.2.5], follows Hardt's exposition in [H]. Our proof in Section 3.4 of the coarea formula also closely follows [H], and is in turn based on [F, Sections 3.2.8–3.2.13]. Theorem 3.14 is from [F, Section 3.2.34].

Chapter 4

Sobolev Functions

In this chapter we study *Sobolev functions* on $\mathbb{R}^n$, functions with weak first partial derivatives belonging to some L^p space. The various Sobolev spaces have good completeness and compactness properties and consequently are often the proper settings for applications of functional analysis to, for instance, linear and nonlinear PDE theory.

Now, as we will see, by definition, integration-by-parts is valid for Sobolev functions. It is, however, far less obvious to what extent the other rules of calculus are valid. We intend to investigate this general question, with particular emphasis on pointwise properties of Sobolev functions.

Section 4.1 provides basic definitions. In Section 4.2 we derive various ways of approximating Sobolev functions by smooth functions. Section 4.3 interprets boundary values of Sobolev functions using traces, and Section 4.4 discusses extending such functions off Lipschitz continuous domains. We prove the fundamental Sobolev-type inequalities in Section 4.5, an immediate application of which is the compactness theorem in Section 4.6. The key to understanding the fine properties of Sobolev functions is capacity, introduced in Section 4.7 and utilized in Sections 4.8 and 4.9.

4.1 Definitions and elementary properties

Throughout this chapter, U denotes an open subset of $\mathbb{R}^n$.

DEFINITION 4.1. *Assume $f \in L^1_{\text{loc}}(U)$ and $i \in \{1, \dots, n\}$. We say $g_i \in L^1_{\text{loc}}(U)$ is the **weak partial derivative** of f with respect to x_i in U if*

$$\int_U f \phi_{x_i} \, dx = - \int_U g_i \phi \, dx \qquad (\star)$$

for all $\phi \in C^1_c(U)$.

NOTATION It is easy to check that the weak partial derivative with respect to x_i, if it exists, is uniquely defined $\mathcal{L}^n$-a.e. We write

$$f_{x_i} := g_i \quad (i = 1, \ldots, n)$$

and

$$Df := (f_{x_1}, \ldots, f_{x_n}),$$

provided the weak derivatives $f_{x_1}, \ldots, f_{x_n}$ exist.

DEFINITION 4.2. *Let $1 \leq p \leq \infty$.*

(i) *The function f belongs to the **Sobolev space***

$$W^{1,p}(U)$$

if $f \in L^p(U)$ and if for $i = 1, \ldots, n$ the weak partial derivatives f_{x_i} exist and belong to $L^p(U)$.

(ii) *The function f belongs to*

$$W_{\text{loc}}^{1,p}(U)$$

if $f \in W^{1,p}(V)$ for each open set $V \subset\subset U$.

(iii) *We say f is a **Sobolev function** if $f \in W_{\text{loc}}^{1,p}(U)$ for some $1 \leq p \leq \infty$.*

(iv) *We do not identify two Sobolev functions that agree $\mathcal{L}^n$-a.e.*

Remark. So if f is a Sobolev function, then *by definition* the integration-by-parts formula

$$\int_U f\phi_{x_i} \, dx = -\int_U f_{x_i}\phi \, dx$$

is valid for all $\phi \in C_c^1(U)$ and $i = 1, \ldots n$. $\qquad\qquad\square$

NOTATION If $f \in W^{1,p}(U)$, define

$$\|f\|_{W^{1,p}(U)} := \left(\int_U |f|^p + |Df|^p \, dx \right)^{\frac{1}{p}}$$

for $1 \leq p < \infty$, and

$$\|f\|_{W^{1,\infty}(U)} := \underset{U}{\text{ess sup}}(|f| + |Df|).$$

DEFINITION 4.3.

(i) *We say*

$$f_k \to f \quad in \ W^{1,p}(U)$$

provided

$$\|f_k - f\|_{W^{1,p}(U)} \to 0.$$

(ii) *Similarly,*

$$f_k \to f \quad in \ W^{1,p}_{loc}(U)$$

provided

$$\|f_k - f\|_{W^{1,p}(V)} \to 0$$

for each open set $V \subset\subset U$.

4.2 Approximation

4.2.1 Approximation by smooth functions

NOTATION

(i) If $\epsilon > 0$, we write

$$U_\epsilon := \{x \in U \mid \text{dist}(x, \partial U) > \epsilon\}.$$

(ii) Define the C^∞-function $\eta : \mathbb{R}^n \to \mathbb{R}$ by

$$\eta(x) := \begin{cases} c\exp\left(\dfrac{1}{|x|^2 - 1}\right) & \text{if } |x| < 1 \\ 0 & \text{if } |x| \geq 1, \end{cases}$$

the constant $c > 0$ adjusted so that

$$\int_{\mathbb{R}^n} \eta(x)\, dx = 1.$$

(iii) Write

$$\eta_\epsilon(x) := \frac{1}{\epsilon^n}\eta\left(\frac{x}{\epsilon}\right) \quad (\epsilon > 0, x \in \mathbb{R}^n);$$

η_ϵ is called the **standard mollifier**.

(iv) If $f \in L^1_{\text{loc}}(U)$, define
$$f^\epsilon := \eta_\epsilon * f;$$

that is,
$$f^\epsilon(x) := \int_U \eta_\epsilon(x - y) f(y) \, dy \quad (x \in U_\epsilon).$$

Mollification provides us with a systematic technique for approximating Sobolev functions by C^∞ functions.

THEOREM 4.1 (Properties of mollifiers).

(i) *For each* $\epsilon > 0$, $f^\epsilon \in C^\infty(U_\epsilon)$.

(ii) *If* $f \in C(U)$, *then*
$$f^\epsilon \to f$$
uniformly on compact subsets of U.

(iii) *If* $f \in L^p_{\text{loc}}(U)$ *for some* $1 \leq p < \infty$, *then*
$$f^\epsilon \to f \quad \text{in } L^p_{\text{loc}}(U).$$

(iv) *Furthermore,* $f^\epsilon(x) \to f(x)$ *if* x *is a Lebesgue point of* f; *in particular,*
$$f^\epsilon \to f \quad \mathcal{L}^n\text{-a.e.}$$

(v) *If* $f \in W^{1,p}_{\text{loc}}(U)$ *for some* $1 \leq p \leq \infty$, *then*
$$f^\epsilon_{x_i} = \eta_\epsilon * f_{x_i} \quad (i = 1, \ldots, n)$$
on U_ϵ.

(vi) *In particular, if* $f \in W^{1,p}_{\text{loc}}(U)$ *for some* $1 \leq p < \infty$, *then*
$$f^\epsilon \to f \quad \text{in } W^{1,p}_{\text{loc}}(U).$$

Proof. 1. Fix any point $x \in U_\epsilon$ and choose $i \in \{1, \ldots, n\}$. We let e_i denote the i-th coordinate vector $(0, \ldots, 1, \ldots, 0)$. Then for $|h|$ small enough, $x + he_i \in U_\epsilon$, and thus

$$\frac{f^\epsilon(x + he_i) - f^\epsilon(x)}{h}$$
$$= \frac{1}{\epsilon^n} \int_U \frac{1}{h} \left[\eta\left(\frac{x + he_i - y}{\epsilon}\right) - \eta\left(\frac{x - y}{\epsilon}\right) \right] f(y) \, dy$$
$$= \frac{1}{\epsilon^n} \int_V \frac{1}{h} \left[\eta\left(\frac{x + he_i - y}{\epsilon}\right) - \eta\left(\frac{x - y}{\epsilon}\right) \right] f(y) \, dy$$

for some $V \subset\subset U$. The difference quotient converges as $h \to 0$ to

$$\frac{1}{\epsilon} \eta_{x_i} \left(\frac{x-y}{\epsilon} \right) = \epsilon^n \eta_{\epsilon, x_i}(x-y)$$

for each $y \in V$. Furthermore, the absolute value of the integrand is bounded by

$$\frac{1}{\epsilon} \|D\eta\|_{L^\infty} |f| \in L^1(V).$$

Hence the Dominated Convergence Theorem implies

$$f_{x_i}^\epsilon(x) = \lim_{h \to 0} \frac{f^\epsilon(x+he_i) - f^\epsilon(x)}{h}$$

exists and equals

$$\int_U \eta_{\epsilon, x_i}(x-y) f(y) \, dy.$$

A similar argument demonstrates that the partial derivatives of f^ϵ of all orders exist and are continuous at each point of U_ϵ; this proves (i).

2. Given $V \subset\subset U$, we choose $V \subset\subset W \subset\subset U$. Then for $x \in V$,

$$f^\epsilon(x) = \frac{1}{\epsilon^n} \int_{B(x,\epsilon)} \eta\left(\frac{x-y}{\epsilon}\right) f(y) \, dy = \int_{B(1)} \eta(z) f(x - \epsilon z) \, dz.$$

Thus, since $\int_{B(1)} \eta(z) \, dz = 1$, we have

$$|f^\epsilon(x) - f(x)| \le \int_{B(1)} \eta(z) |f(x - \epsilon z) - f(x)| \, dz.$$

If f is uniformly continuous on W, we conclude from this estimate that $f^\epsilon \to f$ uniformly on V. Assertion (ii) follows.

3. Assume $1 \le p \le \infty$ and $f \in L_{loc}^p(U)$. Then for $V \subset\subset W \subset\subset U$, $x \in V$, and $\epsilon > 0$ small enough, we calculate in the case $1 < p < \infty$ that

$$|f^\epsilon(x)| \le \int_{B(1)} \eta(z)^{1-\frac{1}{p}} \eta(z)^{\frac{1}{p}} |f(x - \epsilon z)| \, dz$$

$$\le \left(\int_{B(1)} \eta(z) \, dz \right)^{1-\frac{1}{p}} \left(\int_{B(1)} \eta(z) |f(x - \epsilon z)|^p \, dz \right)^{\frac{1}{p}}$$

$$= \left(\int_{B(1)} \eta(z) |f(x - \epsilon z)|^p \, dz \right)^{\frac{1}{p}}.$$

Hence for $1 \leq p < \infty$ we find

$$\int_V |f^\epsilon(x)|^p \, dx \leq \int_{B(1)} \eta(z) \left(\int_V |f(x - \epsilon z)|^p \, dx \right) dz$$

$$\leq \int_W |f(y)|^p \, dy \tag{$\star$}$$

for $\epsilon > 0$ small enough.

Now fix $\delta > 0$. Since $f \in L^p(W)$, there exists $g \in C(\bar{W})$ such that

$$\|f - g\|_{L^p(W)} \leq \delta.$$

This implies, according to estimate $(\star)$, that

$$\|f^\epsilon - g^\epsilon\|_{L^p(V)} \leq \delta.$$

Consequently,

$$\|f^\epsilon - f\|_{L^p(V)} \leq 2\delta + \|g^\epsilon - g\|_{L^p(V)} \leq 3\delta$$

provided $\epsilon > 0$ is small enough, owing to assertion (ii). Assertion (iii) is proved.

4. To prove (iv), let us suppose $f \in L^1_{\text{loc}}(U)$ and assume $x \in U$ is a Lebesgue point of f. Then, by the calculation above, we see

$$|f^\epsilon(x) - f(x)| \leq \frac{1}{\epsilon^n} \int_{B(x,\epsilon)} \eta\left(\frac{x-y}{\epsilon}\right) |f(y) - f(x)| \, dy$$

$$\leq \alpha(n) \|\eta\|_{L^\infty} \fint_{B(x,\epsilon)} |f - f(x)| \, dy$$

$$= o(1) \quad \text{as } \epsilon \to 0.$$

5. Now assume $f \in W^{1,p}_{\text{loc}}(U)$ for some $1 \leq p \leq \infty$. Consequently, as computed above,

$$f^\epsilon_{x_i}(x) = \int_U \eta_{\epsilon,x_i}(x - y) f(y) \, dy = - \int_U \eta_{\epsilon,y_i}(x - y) f(y) \, dy$$

$$= \int_U \eta_\epsilon(x - y) f_{x_i}(y) \, dy = (\eta_\epsilon * f_{x_i})(x)$$

for $x \in U_\epsilon$. This establishes assertion (v), and (vi) follows at once from (iii). $\qquad \square$

THEOREM 4.2 (Local approximation by smooth functions).
*Assume that $f \in W^{1,p}(U)$ for some $1 \le p < \infty$. Then there exists a
sequence $\{f_k\}_{k=1}^{\infty} \subset W^{1,p}(U) \cap C^{\infty}(U)$ such that*

$$f_k \to f \quad in \ W^{1,p}(U).$$

Note that we do not assert $f_k \in C^{\infty}(\bar{U})$, but see Theorem 4.3 below.

Proof. 1. Fix $\epsilon > 0$ and define $U_0 := \emptyset$ and

$$U_k := \left\{ x \in U \mid \text{dist}(x, \partial U) > \frac{1}{k} \right\} \cap B^0(0, k) \quad (k = 1, 2, \dots).$$

Set

$$V_k := U_{k+1} - \bar{U}_{k-1} \quad (k = 1, 2, \dots),$$

and let $\{\zeta_k\}_{k=1}^{\infty}$ be a sequence of smooth functions such that

$$\begin{cases} \zeta_k \in C_c^{\infty}(V_k), \ 0 \le \zeta_k \le 1, \quad (k = 1, 2, \dots), \\ \sum_{k=1}^{\infty} \zeta_k \equiv 1 \text{ on } U. \end{cases}$$

For each $k = 1, 2, \dots, f \ \zeta_k \in W^{1,p}(U)$, with $\text{spt}(f\zeta_k) \subseteq V_k$. Hence
there exists $\epsilon_k > 0$ such that

$$\begin{cases} \text{spt}(\eta_{\epsilon_k} * (f\zeta_k)) \subseteq V_k \\ \left(\int_U |\eta_{\epsilon_k} * (f\zeta_k) - f\zeta_k|^p \, dx \right)^{\frac{1}{p}} < \frac{\epsilon}{2^k} \\ \left(\int_U |\eta_{\epsilon_k} * (D(f\zeta_k)) - D(f\zeta_k)|^p \, dx \right)^{\frac{1}{p}} < \frac{\epsilon}{2^k}. \end{cases} \quad (\star)$$

Define

$$f_\epsilon := \sum_{k=1}^{\infty} \eta_{\epsilon_k} * (f\zeta_k).$$

In some neighborhood of each point $x \in U$, there are only finitely many
nonzero terms in this sum; hence

$$f_\epsilon \in C^{\infty}(U).$$

2. Since

$$f = \sum_{k=1}^{\infty} f\zeta_k,$$

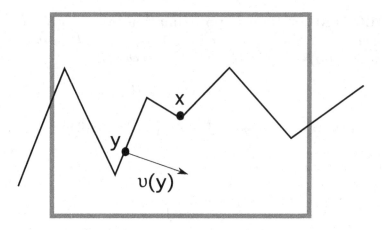

$(\star)$ implies

$$\|f_\epsilon - f\|_{L^p(U)} \leq \sum_{k=1}^{\infty} \left(\int_U |\eta_{\epsilon_k} * (f\zeta_k) - f\zeta_k|^p \, dx \right)^{\frac{1}{p}} < \epsilon$$

and

$$\|Df_\epsilon - Df\|_{L^p(U)}$$

$$\leq \sum_{k=1}^{\infty} \left(\int_U |\eta_{\epsilon_k} * (D(f\zeta_k)) - D(f\zeta_k)|^p \, dx \right)^{\frac{1}{p}} < \epsilon.$$

Consequently $f_\epsilon \in W^{1,p}(U)$ and

$$f_\epsilon \to f \quad \text{in } W^{1,p}(U)$$

as $\epsilon \to 0$. $\square$

   Our intention next is to approximate a Sobolev function by functions smooth all the way up to the boundary. This necessitates some
hypothesis on the geometric behavior of ∂U.

DEFINITION 4.4. *We say the boundary ∂U is **Lipschitz** if for each
point $x \in \partial U$, there exist $r > 0$ and a Lipschitz continuous mapping γ:
$\mathbb{R}^{n-1} \to \mathbb{R}$ such that, upon our rotating and relabeling the coordinate
axes if necessary, we have*

$$U \cap Q(x,r) = \{y \mid \gamma(y_1, \ldots, y_{n-1}) < y_n\} \cap Q(x,r),$$

where

$$Q(x,r) := \{y \mid |y_i - x_i| < r, i = 1, \ldots, n\}.$$

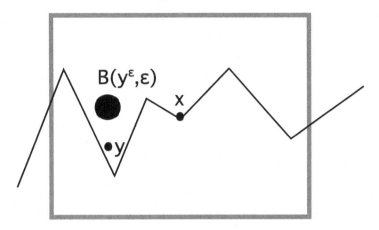

In other words, near each point $x \in \partial U$, the boundary is the graph of a Lipschitz continuous function.

Remark. By Rademacher's Theorem, the outer unit normal $\nu(y)$ to U exists for $\mathcal{H}^{n-1}$-a.e. $y \in \partial U$. $\qquad\square$

THEOREM 4.3 (Global approximation by smooth functions).
Assume U is bounded and ∂U is Lipschitz.

(i) *If $f \in W^{1,p}(U)$ for some $1 \le p < \infty$, there exists a sequence $\{f_k\}_{k=1}^\infty \subseteq W^{1,p}(U) \cap C^\infty(\bar{U})$ such that*

$$f_k \to f \quad in \ W^{1,p}(U).$$

(ii) *If in addition $f \in C(\bar{U})$, then*

$$f_k \to f \quad uniformly.$$

Proof. 1. For $x \in \partial U$, take $r > 0$ and $\gamma : \mathbb{R}^{n-1} \to \mathbb{R}$ as in the definition above. Also write $Q := Q(x, r), Q' = Q(x, \frac{r}{2})$.

2. Suppose first f vanishes near $\partial Q' \cap U$. For $y \in U \cap Q', \epsilon > 0$ and $\alpha > 0$, we define

$$y^\epsilon := y + \epsilon \alpha e_n.$$

Observe $B(y^\epsilon, \epsilon) \subset U \cap Q$ for all ϵ sufficiently small, provided α is large enough, say $\alpha := \mathrm{Lip}(\gamma) + 2$.

3. We define

$$f_\epsilon(y) := \frac{1}{\epsilon^n} \int_U \eta \left(\frac{z}{\epsilon} \right) f(y^\epsilon - z) \, dz$$

$$= \frac{1}{\epsilon^n} \int_{B(y^\epsilon, \epsilon)} \eta \left(\frac{y - w}{\epsilon} + \alpha e_n \right) f(w) \, dw$$

for $y \in U \cap Q'$. As in the proof of Theorem 4.1, we check

$$f_\epsilon \in C^\infty(U \bar\cap Q')$$

and

$$f_\epsilon \to f \quad \text{in } W^{1,p}(U \cap Q').$$

Furthermore, since $f = 0$ near $\partial Q' \cap U$, we have $f_\epsilon = 0$ near $\partial Q' \cap U$ for sufficiently small $\epsilon > 0$; we can thus extend f_ϵ to be 0 on $U - Q'$.

4. Since ∂U is compact, we can cover ∂U with finitely many cubes $Q'_i = Q(x_i, \frac{r_i}{2})(i = 1, 2, \ldots, N)$, as above. Let $\{\zeta_i\}_{i=0}^N$ be a sequence of smooth functions such that

$$\begin{cases} 0 \le \zeta_i \le 1, & \text{spt}(\zeta_i) \subseteq Q'_i \quad (i = 1, \ldots, N) \\ 0 \le \zeta_0 \le 1, & \text{spt}(\zeta_0) \subseteq U \\ \sum_{i=0}^N \zeta_i \equiv 1 & \text{on } U \end{cases}$$

and set

$$f^i := f\zeta_i \quad (i = 0, 1, 2, \ldots, N).$$

Fix $\delta > 0$. Construct as in Step 3 functions $g^i := (f^i)_{\epsilon_i} \in C^\infty(\bar{U})$ satisfying

$$\text{spt}(g^i) \subset \bar{U} \cap Q_i, \quad \|g^i - f^i\|_{W^{1,p}(U \cap Q)} < \frac{\delta}{2N}$$

for $i = 1, \ldots, N$. Mollify f^0 as in proof of Theorem 4.2 to produce $g^0 \in C_c^\infty(U)$ such that

$$\|g^0 - f^0\|_{W^{1,p}(U)} < \frac{\delta}{2}.$$

Finally, set

$$g := \sum_{i=0}^N g^i \in C^\infty(\bar{U})$$

and compute

$$\|g - f\|_{W^{1,p}(U)} \le \|g^0 - f^0\|_{W^{1,p}(U)} + \sum_{i=1}^N \|g^i - f^i\|_{W^{1,p}(U \cap Q,)} < \delta. \quad \square$$

The construction shows that if $f \in W^{1,p}(U) \cap C(\bar{U})$, then $f_k \to f$ uniformly on $\bar{U}$ as well.

4.2.2 Product and chain rules

In view of Section 4.2 we can approximate Sobolev functions by smooth functions, and consequently we can now verify that many of the usual calculus rules hold for weak derivatives.

Assume $1 \leq p < \infty$.

THEOREM 4.4 (Calculus rules for Sobolev functions).

(i) *If* $f, g \in W^{1,p}(U) \cap L^{\infty}(U)$, *then*

$$fg \in W^{1,p}(U) \cap L^{\infty}(U)$$

and

$$(fg)_{x_i} = f_{x_i} g + f g_{x_i} \quad \mathcal{L}^n\text{-}a.e.$$

for $i = 1, 2, \ldots, n$.

(ii) *If* $f \in W^{1,p}(U)$ *and* $F \in C^1(\mathbb{R}), F' \in L^{\infty}(\mathbb{R}), F(0) = 0$, *then*

$$F(f) \in W^{1,p}(U)$$

and

$$F(f)_{x_i} = F'(f) f_{x_i} \quad \mathcal{L}^n\text{-}a.e.$$

for $i = 1, 2, \ldots, n$.

(iii) *If* $f \in W^{1,p}(U)$, *then* $f^+, f^-, |f| \in W^{1,p}(U)$ *and*

$$Df^+ = \begin{cases} Df & \mathcal{L}^n\text{-}a.e. \text{ on } \{f > 0\} \\ 0 & \mathcal{L}^n\text{-}a.e. \text{ on } \{f \leq 0\}, \end{cases}$$

$$Df^- = \begin{cases} 0 & \mathcal{L}^n\text{-}a.e. \text{ on } \{f \geq 0\} \\ -Df & \mathcal{L}^n\text{-}a.e. \text{ on } \{f < 0\}, \end{cases}$$

$$D|f| = \begin{cases} Df & \mathcal{L}^n\text{-}a.e. \text{ on } \{f > 0\} \\ 0 & \mathcal{L}^n\text{-}a.e. \text{ on } \{f = 0\} \\ -Df & \mathcal{L}^n\text{-}a.e. \text{ on } \{f < 0\}. \end{cases}$$

(iv) $Df = 0 \quad \mathcal{L}^n\text{-}a.e. \text{ on } \{f = 0\}$.

Remark. If $\mathcal{L}^n(U) < \infty$, the condition $F(0) = 0$ for (ii) is unnecessary. Assertion (iv) generalizes Theorem 3.3,(i) in Section 3.1. $\qquad\square$

Proof. 1. To establish (i), choose $\phi \in C_c^1(U)$ with $\operatorname{spt} \phi \subset V \subset\subset U$. Let $f^\epsilon := \eta_\epsilon * f, g^\epsilon := \eta_\epsilon * g$ as in Section 4.2. Then

$$
\begin{aligned}
\int_U fg\phi_{x_i} dx &= \int_V fg\phi_{x_i}\, dx \\
&= \lim_{\epsilon \to 0} \int_V f^\epsilon g^\epsilon \phi_{x_i}\, dx \\
&= -\lim_{\epsilon \to 0} \int_V \left(f^\epsilon_{x_i} g^\epsilon + f^\epsilon g^\epsilon_{\partial x_i} \right) \phi\, dx \\
&= -\int_V (f_{x_i} g + f g_{x_i})\, \phi\, dx \\
&= -\int_U (f_{x_i} g + f g_{x_i})\, \phi\, dx,
\end{aligned}
$$

according to Theorem 4.1.

2. To prove (ii), choose ϕ, V, and f^ϵ as above. Then

$$
\begin{aligned}
\int_U F(f)\phi_{x_i} dx &= \int_V F(f)\phi_{x_i} dx \\
&= \lim_{\epsilon \to 0} \int_V F(f^\epsilon)\phi_{x_i} dx \\
&= -\lim_{\epsilon \to 0} \int_V F'(f^\epsilon) f^\epsilon_{x_i} \phi\, dx \\
&= -\int_V F'(f) f_{x_i} \phi\, dx \\
&= -\int_U F'(f) f_{x_i} \phi\, dx,
\end{aligned}
$$

where again we have repeatedly used Theorem 4.1.

3. Fix $\epsilon > 0$ and define

$$
F_\epsilon(r) := \begin{cases} (r^2 + \epsilon^2)^{\frac{1}{2}} - \epsilon & \text{if } r \geq 0 \\ 0 & \text{if } r < 0. \end{cases}
$$

Then $F_\epsilon \in C^1(\mathbb{R}), F_\epsilon^1 \in L^\infty(\mathbb{R})$, and so assertion (ii) implies for $\phi \in C_c^1(U)$

$$
\int_U F_\epsilon(f)\phi_{x_i}\, dx = -\int_U F_\epsilon'(f) f_{x_i} \phi\, dx.
$$

Now let $\epsilon \to 0$ to find

$$\int_U f^+ \phi_{x_i} \, dx = - \int_{U \cap \{f > 0\}} f_{x_i} \phi \, dx.$$

This proves the first part of (iii), and the other assertions follow from the formulas

$$f^- = (-f)^+, \ |f| = f^+ + f^-.$$

Assertion (iv) is a consequence of (iii), since

$$Df = Df^+ - Df^-. \qquad \qquad \square$$

4.2.3 $W^{1,\infty}$ and Lipschitz continuous functions

THEOREM 4.5 (Lipschitz continuity and $W^{1,\infty}$). *Assume* $f : U \to \mathbb{R}$. *Then*

$$f \text{ is locally Lipschitz continuous in } U$$

if and only if

$$f \in W^{1,\infty}_{\text{loc}}(U).$$

Proof. 1. First suppose f is locally Lipschitz continuous. Fix $i \in \{1, \ldots, n\}$. Then for each $V \subset\subset W \subset\subset U$, pick $0 < h < \text{dist}(V, \partial W)$, and define

$$g_i^h(x) := \frac{f(x + he_i) - f(x)}{h} \qquad (x \in V).$$

Now

$$\sup_{h > 0} |g_i^h| \leq \text{Lip}(f|_W) < \infty.$$

Then according to Theorem 1.42 there is a sequence $h_j \to 0$ and a function $g_i \in L^\infty_{\text{loc}}(U)$ such that

$$g_i^{h_j} \rightharpoonup g_i \quad \text{weakly in } L^p_{\text{loc}}(U)$$

for all $1 < p < \infty$. But if $\phi \in C^1_c(V)$, we have

$$\int_U f(x) \frac{\phi(x + he_i) - \phi(x)}{h} \, dx = - \int_U g_i^h(x) \phi(x + he_i) \, dx.$$

We set $h = h_j$ and let $j \to \infty$:

$$\int_U f \phi_{x_i} \, dx = - \int_U g_i \, \phi \, dx.$$

Hence g_i is the weak partial derivative of f with respect to x_i for $i = 1, \ldots, n$, and thus $f \in W^{1,\infty}_{\text{loc}}(U)$.

2. Conversely, suppose $f \in W^{1,\infty}_{\text{loc}}(U)$. Let $B \subset\subset U$ be any closed ball contained in U. Then by Theorem 4.1 we know

$$\sup_{0 < \epsilon < \epsilon_0} ||Df^\epsilon||_{L^\infty(B)} < \infty$$

for ϵ_0 sufficiently small, where $f^\epsilon := \eta_\epsilon * f$ is the usual mollification. Since f^ϵ is C^∞, we have

$$f^\epsilon(x) - f^\epsilon(y) = \int_0^1 Df^\epsilon(y + t(x - y))dt \cdot (x - y)$$

for $x, y \in B$; whence

$$|f^\epsilon(x) - f^\epsilon(y)| \leq C|x - y|,$$

the constant C independent of ϵ. Thus

$$|f(x) - f(y)| \leq C|x - y| \quad (x, y \in B).$$

Hence $f|_B$ is Lipschitz continuous for each ball $B \subset\subset U$, and so f is locally Lipschitz continuous in U. $\qquad\square$

4.3 Traces

THEOREM 4.6 (Traces of Sobolev functions). *Assume U is bounded, ∂U is Lipschitz, $1 \leq p < \infty$.*

(i) *There exists a bounded linear operator*

$$T : W^{1,p}(U) \to L^p(\partial U; \mathcal{H}^{n-1})$$

such that

$$Tf = f \quad \text{on } \partial U$$

for all $f \in W^{1,p}(U) \cap C(\bar{U})$.

(ii) *Furthermore, for all $\phi \in C^1(\mathbb{R}^n; \mathbb{R}^n)$ and $f \in W^{1,p}(U)$,*

$$\int_U f \operatorname{div} \phi \, dx = -\int_U Df \cdot \phi \, dx + \int_{\partial U} (\phi \cdot \nu)Tf \, d\mathcal{H}^{n-1},$$

ν denoting the unit outer normal to ∂U.

DEFINITION 4.5. *The function Tf, which is uniquely defined up to sets of $\mathcal{H}^{n-1} \llcorner \partial U$ measure zero, is called the **trace** of f on ∂U.*

We interpret Tf as providing the "boundary values" of f on ∂U.

Remark. We will see in Section 5.3 that for $\mathcal{H}^{n-1}$-a.e. point $x \in \partial U$,

$$\lim_{r \to 0} \fint_{B(x,r) \cap U} |f - Tf(x)| \, dy = 0,$$

and so

$$Tf(x) = \lim_{r \to 0} \fint_{B(x,r) \cap U} f \, dy.$$

$\square$

Proof. 1. Assume first $f \in C^1(\bar{U})$. Since ∂U is Lipschitz continuous, we can for any point $x \in \partial U$ find $r > 0$ and a Lipschitz continuous function $\gamma: \mathbb{R}^{n-1} \to \mathbb{R}$ such that, upon rotating and relabeling the coordinate axes if necessary,

$$U \cap Q(x,r) = \{y \mid \gamma(y_1, \ldots, y_{n-1}) < y_n\} \cap Q(x,r).$$

Write $Q := Q(x,r)$ and suppose temporarily $f \equiv 0$ on $U - Q$. Observe

$$-e_n \cdot \nu \geq (1 + (\mathrm{Lip}(\gamma))^2)^{-\frac{1}{2}} > 0 \quad \mathcal{H}^{n-1}\text{-a.e. on } Q \cap \partial U. \qquad (\star)$$

2. Fix $\epsilon > 0$, set

$$\beta_\epsilon(t) := (t^2 + \epsilon^2)^{\frac{1}{2}} - \epsilon \quad (t \in \mathbb{R}),$$

and compute using the Gauss–Green Theorem that

$$\int_{\partial U} \beta_\epsilon(f) \, d\mathcal{H}^{n-1} = \int_{Q \cap \partial U} \beta_\epsilon(f) \, d\mathcal{H}^{n-1}$$

$$\leq C \int_{Q \cap \partial U} \beta_\epsilon(f)(-e_n \cdot \nu) \, d\mathcal{H}^{n-1}$$

$$= -C \int_{Q \cap U} (\beta_\epsilon(f))_{y_n} \, dy$$

$$\leq C \int_{Q \cap U} |\beta'_\epsilon(f)| |Df| \, dy$$

$$\leq C \int_U |Df| \, dy,$$

since $|\beta'_\epsilon| \leq 1$. Now send $\epsilon \to 0$, to discover

$$\int_{\partial U} |f| \, d\mathcal{H}^{n-1} \leq C \int_U |Df| \, dy. \qquad (\star\star)$$

3. We have established $(\star\star)$ under the assumption that $f \equiv 0$ on $U - Q$ for some cube $Q = Q(x,r), x \in \partial U$. In the general case, we can cover ∂U by a finite number of such cubes and use a partition of unity as in the proof of Theorem 4.3 to obtain

$$\int_{\partial U} |f| \, d\mathcal{H}^{n-1} \leq C \int_U |Df| + |f| \, dy$$

for all $f \in C^1(\bar{U})$. For $1 < p < \infty$, we apply this estimate with $|f|^p$ replacing $|f|$, to obtain

$$\int_{\partial U} |f|^p \, d\mathcal{H}^{n-1} \leq C \int_U |Df||f|^{p-1} + |f|^p \, dy$$

$$\leq C \int_U |Df|^p + |f|^p \, dy \qquad (\star\star\star)$$

for all $f \in C^1(\bar{U})$.

4. Thus if we define

$$Tf := f|_{\partial U}$$

for $f \in C^1(\bar{U})$, we see from $(\star\star\star)$, Theorem 4.3 that T uniquely extends to a bounded linear operator from $W^{1,p}(U)$ to $L^p(\partial U; \mathcal{H}^{n-1})$, with

$$Tf = f|_{\partial U}$$

for all $f \in W^{1,p}(U) \cap C(\bar{U})$. This proves assertion (i); assertion (ii) follows from an approximation argument using the Gauss–Green Theorem. $\qquad \square$

4.4 Extensions

THEOREM 4.7 (Extending Sobolev functions). *Assume U is bounded, ∂U is Lipschitz, and $1 \leq p < \infty$. Let $U \subset\subset V$. There exists a bounded linear operator*

$$E : W^{1,p}(U) \to W^{1,p}(\mathbb{R}^n)$$

such that

$$Ef = f \quad \text{on } U$$

and

$$\text{spt}(Ef) \subset V$$

for all $f \in W^{1,p}(U)$.

DEFINITION 4.6. *Ef is called an **extension** of f to $\mathbb{R}^n$.*

Proof. 1. First we introduce some notation:

(a) Given $x = (x_1, \ldots, x_n) \in \mathbb{R}^n$, let us write

$$x = (x', x_n)$$

for $x' = (x_1, \ldots, x_{n-1}) \in \mathbb{R}^{n-1}, x_n \in \mathbb{R}$. Similarly, we write $y = (y', y_n)$.

(b) Given $x \in \mathbb{R}^n$, and $r, h > 0$, define the open cylinder

$$C(x, r, h) := \{y \in \mathbb{R}^n \mid |y' - x'| < r, |y_n - x_n| < h\}.$$

Since ∂U is Lipschitz continuous, for each $x \in \partial U$ there exist, upon our rotating and relabeling the coordinate axes if necessary, $r, h > 0$ and a Lipschitz continuous function $\gamma : \mathbb{R}^{n-1} \to \mathbb{R}$ such that

$$\begin{cases} \max_{|x' - y'| < r} |\gamma(y') - x_n| < \frac{h}{4}, \\ U \cap C(x, r, h) = \{y \mid |x' - y'| < r, \gamma(y') < y_n < x_n + h\}, \\ C(x, r, h) \subseteq V. \end{cases}$$

Fix $x \in \partial U$ and with r, h, γ as above, write

$$C := C(x, r, h), \ C' := C(x, \frac{r}{2}, \frac{h}{2})$$
$$U^+ := C' \cap U, \ U^- := C' - \bar{U}.$$

2. Let $f \in C^1(\bar{U})$ and suppose for the moment spt $f \subseteq C' \cap \bar{U}$. Set

$$\begin{cases} f^+(y) = f(y) & \text{if } y \in \bar{U}^+, \\ f^-(y) = f(y', 2\gamma(y') - y_n) & \text{if } y \in U^-. \end{cases}$$

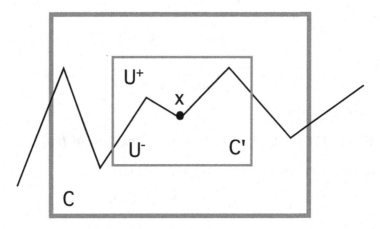

This is an "extension by reflection". Note that $f^- = f^+$ on $\partial U \cap C'$.

3. *Claim #1*: $\|f^-\|_{W^{1,p}(U^-)} \leq C\|f\|_{W^{1,p}(U)}$.

Proof of claim: Let $\phi \in C_c^1(U^-)$ and let $\{\gamma_k\}_{k=1}^\infty$ be a sequence of C^∞ functions such that

$$\begin{cases} \gamma_k \geq \gamma, \ \gamma_k \to \gamma \quad \text{uniformly} \\ D\gamma_k \to D\gamma \quad \mathcal{L}^{n-1}\text{-a.e.}, \ \sup_k \|D\gamma_k\|_{L^\infty} < \infty. \end{cases}$$

Then, for $i = 1, \ldots, n-1$,

$$\int_{U^-} f^- \phi_{y_i} dy$$
$$= \int_{U^-} f(y', 2\gamma(y') - y_n) \phi_{y_i} dy$$
$$= \lim_{k \to \infty} \int_{U^-} f(y', 2\gamma_k(y') - y_n) \phi_{y_i} dy$$
$$= -\lim_{k \to \infty} \int_{U^-} (f_{y_i}(y', 2\gamma_k(y') - y_n)$$
$$+ 2f_{y_n}(y', 2\gamma_k(y') - y_n)\gamma_{k,y_i}(y'))\phi \, dy$$
$$= -\int_{U^-} (f_{y_i}(y', 2\gamma(y') - y_n)$$
$$+ 2f_{y_n}(y', 2\gamma(y') - y_n)\gamma_{y_i}(y'))\phi \, dy.$$

Similarly,

$$\int_{U^-} f^- \phi_{y_n} dy = \int_{U^-} f_{y_n}(y', 2\gamma(y') - y_n)\phi \, dy.$$

Now recall

$$\|D\gamma\|_{L^\infty} < \infty,$$

and thus

$$\int_{U^-} |Df(y', 2\gamma(y') - y_n)|^p \, dy \le C \int_U |Df|^p \, dy < \infty$$

by the change of variables formula (Theorem 3.9).

4. Define

$$Ef := \bar{f} = \begin{cases} f^+ & \text{on } \bar{U}^+ \\ f^- & \text{on } \bar{U}^- \\ 0 & \text{on } \mathbb{R}^n - (\bar{U}^+ \cup \bar{U}^-), \end{cases}$$

and note $\bar{f}$ is continuous on $\mathbb{R}^n$.

5. *Claim #2*: $E(f) \in W^{1,p}(\mathbb{R}^n)$, spt $E(f) \subseteq C' \subseteq V$, and

$$\|E(f)\|_{W^{1,p}(\mathbb{R}^n)} \le C\|f\|_{W^{1,p}(U)}.$$

Proof of claim: Let $\phi \in C_c^1(C')$. For $1 \le i \le n$

$$\int_{C'} \bar{f}\phi_{y_i} \, dy = \int_{U^+} f^+ \phi_{y_i} \, dy + \int_{U^-} f^- \phi_{y_i} \, dy$$

$$= -\int_{U^+} f_{y_i}^+ \phi \, dy - \int_{U^-} f_{y_i}^- \phi \, dy$$

$$+ \int_{\partial U} (T(f^+) - T(f^-))\phi\nu_i \, d\mathcal{H}^{n-1}$$

by Theorem 4.6. But $T(f^+) = T(f^-) = f|_{\partial U}$, and so the last term vanishes.

This calculation and Claim #1 complete the proof in case f is C^1, with support in $C' \cap \bar{U}$.

6. Now assume $f \in C^1(\bar{U})$, but drop the restriction on its support. Since ∂U is compact, we can cover ∂U with finitely many cylinders $C_k = C(x_k, r_k, h_k)(k = 1, \ldots, N)$ for which assertions analogous to the foregoing hold. Let $\{\zeta_k\}_{k=0}^N$ be a partition of unity as in the proof of Theorem 4.3, define $E(\zeta_k f)(k = 1, 2, \ldots, N)$ as above, and set

$$Ef := \sum_{k=1}^N E(\zeta_k f) + \zeta_0 f.$$

7. Finally, if $f \in W^{1,p}(U)$, we approximate f by functions $f_k \in W^{1,p}(U) \cap C^1(\bar{U})$ and set

$$Ef := \lim_{k \to \infty} Ef_k. \qquad \square$$

4.5   Sobolev inequalities

4.5.1   Gagliardo–Nirenberg–Sobolev inequality

We prove next that if $f \in W^{1,p}(\mathbb{R}^n)$ for some $1 \le p < n$, then in fact $f \in L^{p^*}(\mathbb{R}^n)$ where $p^* > p$.

DEFINITION 4.7. *For $1 \le p < n$, define*

$$p^* := \frac{np}{n-p};$$

p^* *is called the* **Sobolev conjugate** *of p.*

Note that
$$\frac{1}{p^*} = \frac{1}{p} - \frac{1}{n}.$$

THEOREM 4.8 (Gagliardo–Nirenberg–Sobolev inequality). *Assume*

$$1 \le p < n.$$

There exists a constant C_1, depending only on p and n, such that

$$\left(\int_{\mathbb{R}^n} |f|^{p^*} \, dx \right)^{\frac{1}{p^*}} \le C_1 \left(\int_{\mathbb{R}^n} |Df|^p \, dx \right)^{\frac{1}{p}}$$

for all $f \in W^{1,p}(\mathbb{R}^n)$.

Proof. 1. According to Theorem 4.2 , we may assume $f \in C_c^1(\mathbb{R}^n)$. Then for $i = 1, \dots, n$

$$f(x_1, \dots, x_i, \dots, x_n) = \int_{-\infty}^{x_i} f_{x_i}(x_1, \dots, t_i, \dots, x_n) \, dt_i$$

and so

$$|f(x)| \le \int_{-\infty}^{\infty} |Df|(x_1, \dots, t_i, \dots, x_n) \, dt_i \quad (i = 1, \dots, n).$$

Thus

$$|f(x)|^{\frac{n}{n-1}} \le \prod_{i=1}^{n} \left(\int_{-\infty}^{\infty} |Df|(x_1,\ldots,t_i,\ldots,x_n)\, dt_i \right)^{\frac{1}{n-1}}.$$

Integrate with respect to x_1:

$$\int_{-\infty}^{\infty} |f|^{1^*}\, dx_1 \le \left(\int_{-\infty}^{\infty} |Df|\, dt_1 \right)^{\frac{1}{n-1}}$$

$$\int_{-\infty}^{\infty} \prod_{i=2}^{n} \left(\int_{-\infty}^{\infty} |Df|\, dt_i \right)^{\frac{1}{n-1}} dx_1$$

$$\le \left(\int_{-\infty}^{\infty} |Df|\, dt_1 \right)^{\frac{1}{n-1}}$$

$$\left(\prod_{i=2}^{n} \int_{-\infty}^{\infty} \int_{-\infty}^{\infty} |Df|\, dx_1 dt_i \right)^{\frac{1}{n-1}}.$$

Next integrate with respect to x_2 to find

$$\int_{-\infty}^{\infty} \int_{-\infty}^{\infty} |f|^{1^*}\, dx_1 dx_2$$

$$\le \left(\int_{-\infty}^{\infty} \int_{-\infty}^{\infty} |Df|\, dx_1 dt_2 \right)^{\frac{1}{n-1}} \left(\int_{-\infty}^{\infty} \int_{-\infty}^{\infty} |Df|\, dt_1 dx_2 \right)^{\frac{1}{n-1}}$$

$$\times \prod_{i=3}^{n} \left(\int_{-\infty}^{\infty} \int_{-\infty}^{\infty} \int_{-\infty}^{\infty} |Df|\, dx_1 dx_2 dt_i \right)^{\frac{1}{n-1}}.$$

We continue, and eventually discover

$$\int_{\mathbb{R}^n} |f|^{1^*}\, dx \le \prod_{i=1}^{n} \left(\int_{-\infty}^{\infty} \cdots \int_{-\infty}^{\infty} |Df|\, dx_1 \ldots dt_i \ldots dx_n \right)^{\frac{1}{n-1}}$$

$$= \left(\int_{\mathbb{R}^n} |Df|\, dx \right)^{\frac{n}{n-1}}.$$

This immediately gives

$$\left(\int_{\mathbb{R}^n} |f|^{1^*}\, dx \right)^{\frac{1}{1^*}} \le \int_{\mathbb{R}^n} |Df|\, dx, \qquad (\star)$$

and so proves the theorem for $p = 1$.

2. If $1 < p < n$, set $g = |f|^\gamma$ with $\gamma > 0$ as selected below. Applying $(\star)$ to g, we find

$$\left(\int_{\mathbb{R}^n} |f|^{\frac{\gamma n}{n-1}} \, dx \right)^{\frac{n-1}{n}} \leq \gamma \int_{\mathbb{R}^n} |f|^{\gamma-1} |Df| \, dx$$

$$\leq \gamma \left(\int_{\mathbb{R}^n} |f|^{\frac{(\gamma-1)p}{p-1}} \, dx \right)^{\frac{p-1}{p}} \left(\int_{\mathbb{R}} |Df|^p \, dx \right)^{\frac{1}{p}}.$$

Choose γ so that

$$\frac{\gamma n}{n-1} = (\gamma - 1)\frac{p}{p-1}.$$

Then

$$\frac{\gamma n}{n-1} = (\gamma - 1)\frac{p}{p-1} = \frac{np}{n-p} = p^*.$$

Thus

$$\left(\int_{\mathbb{R}^n} |f|^{p^*} dx \right)^{\frac{n-1}{n}} \leq C \left(\int_{\mathbb{R}^n} |f|^{p^*} dx \right)^{\frac{p-1}{p}} \left(\int_{\mathbb{R}^n} |Df|^p \, dx \right)^{\frac{1}{p}},$$

and so

$$\left(\int_{\mathbb{R}^n} |f|^{p^*} dx \right)^{\frac{1}{p^*}} \leq C \left(\int_{\mathbb{R}^n} |Df|^p \, dx \right)^{\frac{1}{p}}$$

where C depends only on n and p. $\square$

4.5.2   Poincaré's inequality on balls

Our goal next is deriving a local version of the preceding inequality. For this we will need the following technical calculation:

LEMMA 4.1. *For each $1 \leq p < \infty$ there exists a constant C, depending only on n and p, such that*

$$\int_{B(x,r)} |f(y) - f(z)|^p \, dy \leq C r^{n+p-1} \int_{B(x,r)} |Df(y)|^p |y - z|^{1-n} \, dy$$

for all $B(x,r) \subset \mathbb{R}^n$, $f \in C^1(B(x,r))$ and $z \in B(x,r)$.

Proof. If $y, z \in B(x,r)$, then

$$f(y) - f(z) = \int_0^1 \frac{d}{dt} f(z + t(y - z)) \, dt$$
$$= \int_0^1 Df(z + t(y - z)) \, dt \cdot (y - z),$$

and so

$$|f(y) - f(z)|^p \leq |y - z|^p \int_0^1 |Df|^p(z + t(y - z)) \, dt.$$

Thus for $s > 0$,

$$\int_{B(x,r) \cap \partial B(z,s)} |f(y) - f(z)|^p \, d\mathcal{H}^{n-1}(y)$$

$$\leq s^p \int_0^1 \int_{B(x,r) \cap \partial B(z,s)} |Df|^p(z + t(y - z)) \, d\mathcal{H}^{n-1}(y) dt$$

$$\leq s^p \int_0^1 \frac{1}{t^{n-1}} \int_{B(x,r) \cap \partial B(z,ts)} |Df(w)|^p \, d\mathcal{H}^{n-1}(w) dt$$

$$= s^{n+p-1} \int_0^1 \int_{B(x,r) \cap \partial B(z,ts)} |Df(w)|^p |w - z|^{1-n}$$
$$d\mathcal{H}^{n-1}(w) dt$$

$$= s^{n+p-2} \int_{B(x,r) \cap B(z,s)} |Df(w)|^p |w - z|^{1-n} \, dw.$$

We integrate in s from 0 to $2r$ and use Theorem 3.12 to deduce

$$\int_{B(x,r)} |f(y) - f(z)|^p \, dy \leq C r^{n+p-1} \int_{B(x,r)} |Df(w)|^p |w - z|^{1-n} \, dw. \quad \square$$

THEOREM 4.9 (Poincaré's inequality on balls). *For each* $1 \leq p < n$ *there exists a constant* C_2, *depending only on* p *and* n, *such that*

$$\left(\fint_{B(x,r)} |f - (f)_{x,r}|^{p^*} \, dy \right)^{\frac{1}{p^*}} \leq C_2 r \left(\fint_{B(x,r)} |Df|^p \, dy \right)^{\frac{1}{p}}$$

for all $B(x,r) \subseteq \mathbb{R}^n$, $f \in W^{1,p}(B^0(x,r))$.

Recall

$$(f)_{x,r} = \fint_{B(x,r)} f \, dy.$$

Proof. 1. Approximating if necessary, we may assume that $f \in C^1(B(x,r))$. We recall Lemma 4.1 to compute

$$\fint_{B(x,r)} |f - (f)_{x,r}|^p \, dy = \fint_{B(x,r)} \left| \fint_{B(x,r)} f(y) - f(z) \, dz \right|^p dy$$

$$\leq \fint_{B(x,r)} \fint_{B(x,r)} |f(y) - f(z)|^p \, dz dy$$

$$\leq C \fint_{B(x,r)} r^{p-1} \int_{B(x,r)} |Df(z)|^p |y - z|^{1-n} dz \, dy$$

$$\leq C r^p \fint_{B(x,r)} |Df|^p \, dz. \qquad (\star)$$

2. *Claim*: There exists a constant $C = C(n,p)$ such that

$$\left(\fint_{B(x,r)} |g|^{p^*} \, dy \right)^{\frac{1}{p^*}} \leq C \left(r^p \fint_{B(x,r)} |Dg|^p \, dy + \fint_{B(x,r)} |g|^p \, dy \right)^{\frac{1}{p}}$$

for all $g \in W^{1,p}(B^0(x,r))$.

Proof of claim: First observe that, upon replacing $g(y)$ by $\frac{1}{r}g(ry)$ if necessary, we may assume $r = 1$. Similarly we may suppose $x = 0$. We next employ Theorem 4.7 to extend g to $\bar{g} \in W^{1,p}(\mathbb{R}^n)$ satisfying

$$\|\bar{g}\|_{W^{1,p}(\mathbb{R}^n)} \leq C \|g\|_{W^{1,p}(B^0(0,1))}.$$

Then Theorem 4.8 implies

$$\left(\int_{B(1)} |g|^{p^*} \, dy \right)^{\frac{1}{p^*}} \leq \left(\int_{\mathbb{R}^n} |\bar{g}|^{p^*} \, dy \right)^{\frac{1}{p^*}}$$

$$\leq C_1 \left(\int_{\mathbb{R}^n} |D\bar{g}|^p \, dy \right)^{\frac{1}{p}}$$

$$\leq C \left(\int_{B(1)} |Dg|^p + |g|^p \, dy \right)^{\frac{1}{p}}.$$

3. We use $(\star)$ and the Claim with $g := f - (f)_{x,r}$ to complete the proof. $\qquad \square$

4.5.3 Morrey's inequality

DEFINITION 4.8. *Let* $0 < \alpha < 1$. *A function* $f : \mathbb{R}^n \to \mathbb{R}$ *is* **Hölder continuous with exponent** α *provided*

$$\sup_{\substack{x,y \in \mathbb{R}^n \\ x \neq y}} \frac{|f(x) - f(y)|}{|x - y|^\alpha} < \infty.$$

THEOREM 4.10 (Morrey's inequality).

(i) *For each* $n < p < \infty$ *there exists a constant* C_3, *depending only on* p *and* n, *such that*

$$|f(y) - f(z)| \leq C_3 r \left(\fint_{B(x,r)} |Df|^p \, dw \right)^{\frac{1}{p}}$$

for all $B(x,r) \subset \mathbb{R}^n, f \in W^{1,p}(B^0(x,r))$, *and* $\mathcal{L}^n$-*a.e.* $y, z \in B(x,r)$.

(ii) *In particular, if* $f \in W^{1,p}(\mathbb{R}^n)$, *then the limit*

$$\lim_{r \to 0} (f)_{x,r} =: f^*(x)$$

exists for all $x \in \mathbb{R}^n$, *and* f^* *is Hölder continuous with exponent* $\alpha = 1 - \frac{n}{p}$.

Remark. See Theorem 4.5 for the case $p = \infty$. $\quad\square$

Proof. 1. First assume f is C^1 and use Lemma 4.1 with $p = 1$ to calculate

$$|f(y) - f(z)| \leq \fint_{B(x,r)} |f(y) - f(w)| + |f(w) - f(z)| \, dw$$

$$\leq C \int_{B(x,r)} |Df(w)|(|y - w|^{1-n} + |z - w|^{1-n}) \, dw$$

$$\leq C \left(\int_{B(x,r)} (|y - w|^{1-n} + |z - w|^{1-n})^{\frac{p}{p-1}} \, dw \right)^{\frac{p-1}{p}}$$

$$\left(\int_{B(x,r)} |Df|^p dw \right)^{\frac{1}{p}}$$

$$\leq C r^{\left(n-(n-1)\frac{p}{p-1}\right)\frac{p-1}{p}} \left(\int_{B(x,r)} |Df|^p \, dw \right)^{\frac{1}{p}}$$

$$= C r^{1-\frac{n}{p}} \left(\int_{B(x,r)} |Df|^p \, dw \right)^{\frac{1}{p}}.$$

2. By approximation, we see that if $f \in W^{1,p}(B^0(x,r))$, the same estimate holds for $\mathcal{L}^n$-a.e. $y, z \in B(x,r)$. This proves (i).

3. Now suppose $f \in W^{1,p}(\mathbb{R}^n)$. Then for $\mathcal{L}^n$-a.e. x, y we can apply the estimate of (i) with $r = |x - y|$ to obtain

$$|f(y) - f(x)| \leq C |x - y|^{1-\frac{n}{p}} \left(\int_{B(x,r)} |Df|^p \, dw \right)^{\frac{1}{p}}$$

$$\leq C \|Df\|_{L^p(\mathbb{R}^n)} |x - y|^{1-\frac{n}{p}}.$$

Thus f is equal $\mathcal{L}^n$-a.e. to a Hölder-continuous function $\bar{f}$. Clearly $f^* = \bar{f}$ everywhere in $\mathbb{R}^n$. □

4.6 Compactness

THEOREM 4.11 (Compactness and $W^{1,p}$). *Assume U is bounded, ∂U is Lipschitz, $1 < p < n$. Suppose $\{f_k\}_{k=1}^\infty$ is a sequence in $W^{1,p}(U)$ satisfying.*

$$\sup_k \|f_k\|_{W^{1,p}(U)} < \infty.$$

Then there exists a subsequence $\{f_{k_j}\}_{j=1}^\infty$ and a function $f \in W^{1,p}(U)$ such that

$$f_{k_j} \to f \quad \text{in } L^q(U).$$

for each

$$1 \leq q < p^*.$$

Proof. 1. Fix a bounded open set V such that $U \subset\subset V$ and extend each f_k to $\bar{f}_k \in W^{1,p}(\mathbb{R}^n)$, spt $\bar{f}_k \subset V$,

$$\sup_k \|\bar{f}_k\|_{W^{1,p}(\mathbb{R}^n)} \leq C \sup_k \|f_k\|_{W^{1,p}(U)} < \infty. \tag{$\star$}$$

2. Let $\bar{f}_k^\epsilon := \eta_\epsilon * \bar{f}_k$ be the usual mollification, as described in Section 4.2.

Claim #1: $\|\bar{f}_k^\epsilon - \bar{f}_k\|_{L^p(\mathbb{R}^n)} \leq C\epsilon$, uniformly in k.

Proof of claim: First suppose the functions $\bar{f}_k$ are smooth, and calculate

$$|\bar{f}_k^\epsilon(x) - \bar{f}_k(x)| \leq \int_{B(1)} \eta(z)|\bar{f}_k(x - \epsilon z) - f_k(x)|\, dz$$

$$= \int_{B(1)} \eta(z) \left| \int_0^1 \frac{d}{dt}\bar{f}_k(x - t\epsilon z)\, dt \right| dz$$

$$\leq \epsilon \int_{B(1)} \eta(z) \int_0^1 |D\bar{f}_k(x - \epsilon t z)|\, dt dz.$$

Thus

$$\|\bar{f}_k^\epsilon - \bar{f}_k\|_{L^p(\mathbb{R}^n)}^p$$

$$\leq C\epsilon^p \int_{B(1)} \eta(z) \int_0^1 \left(\int_{\mathbb{R}^n} |D\bar{f}_k(x - \epsilon t z)|^p\, dx \right) dt dz$$

$$\leq C\epsilon^p \|\bar{f}_k\|_{W^{1,p}(\mathbb{R}_n)}^p$$

$$\leq C\epsilon^p.$$

according to $(\star)$. The general case follows by approximation.

3. *Claim #2:* For each $\epsilon > 0$, the sequence $\{\bar{f}_k^\epsilon\}_{k=1}^\infty$ is bounded and equicontinuous on $\mathbb{R}^n$.

Proof of claim: We calculate

$$|\bar{f}_k^\epsilon(x)| \leq \int_{B(x,\epsilon)} \eta_\epsilon(x - y)|\bar{f}_k(y)|\, dy$$

$$\leq C\epsilon^{-\frac{n}{p}} \|\bar{f}_k\|_{L^p(\mathbb{R}^n)}$$

$$\leq C\epsilon^{-\frac{n}{p}}$$

and

$$|D\bar{f}_k^\epsilon(x)| \leq \int_{B(x,\epsilon)} |D\eta_\epsilon(x - y)||\bar{f}_k(y)|\, dy \leq C\epsilon^{-\frac{n}{p}-1}.$$

4. *Claim #3:* For each $\delta > 0$ there exists a subsequence $\{f_{k_j}\}_{j=1}^\infty \subseteq \{f_k\}_{k=1}^\infty$ such that

$$\limsup_{i,j \to \infty} \|f_{k_i} - f_{k_j}\|_{L^p(U)} \leq \delta.$$

Proof of claim: Recalling Claim #1, we choose $\epsilon > 0$ so small that

$$\sup_k \|\bar{f}_k^\epsilon - \bar{f}_k\|_{L^p(\mathbb{R}^n)} \le \frac{\delta}{3}.$$

Next we use Claim #2 and the Arzela–Ascoli Theorem to find a subsequence $\{\bar{f}_{k_j}^\epsilon\}_{j=1}^\infty$ which converges uniformly on $\mathbb{R}^n$. Then

$$
\begin{aligned}
\|f_{k_j} - f_{k_i}\|_{L^p(U)} &\le \|\bar{f}_{k_j} - \bar{f}_{k_i}\|_{L^p(\mathbb{R}^n)} \\
&\le \|\bar{f}_{k_j} - \bar{f}_{k_j}^\epsilon\|_{L^p(\mathbb{R}^n)} + \|\bar{f}_{k_j}^\epsilon - \bar{f}_{k_i}^\epsilon\|_{L^p(\mathbb{R}^n)} \\
&\qquad\qquad\qquad\qquad + \|\bar{f}_{k_i}^\epsilon - \bar{f}_{k_i}\|_{L^p(\mathbb{R}^n)} \\
&\le \frac{2\delta}{3} + \|\bar{f}_{k_j}^\epsilon - \bar{f}_{k_i}^\epsilon\|_{L^p(\mathbb{R}^n)} \le \delta
\end{aligned}
$$

for i, j large enough.

5. We use a diagonal argument and Claim #3 with $\delta = 1, \frac{1}{2}, \frac{1}{4}$, etc. to obtain a subsequence, also denoted $\{f_{k_j}\}_{j=1}^\infty$, converging to f in $L^p(U)$. We observe also for $1 \le q < p^*$,

$$\|f_{k_j} - f\|_{L^q(U)} \le \|f_{k_j} - f\|_{L^p(U)}^\theta \|f_{k_j} - f\|_{L^{p^*}(U)}^{1-\theta},$$

where $\frac{1}{q} = \frac{\theta}{p} + \frac{1-\theta}{p^*}$ and hence $\theta > 0$. Since $\{f_k\}_{k=1}^\infty$ is bounded in $L^{p^*}(U)$, we see

$$\lim_{j\to\infty} \|f_{k_j} - f\|_{L^q(U)} = 0$$

for each $1 \le q < p^*$ Since $p > 1$, it follows from Theorem 1.42 that $f \in W^{1,p}(U)$.   □

Remark. The compactness assertion is false for the endpoint case that $q = p^*$. In case $p = 1$, the above argument shows that there is a subsequence $\{f_{k_j}\}_{j=1}^\infty$ and $f \in L^{1^*}(U)$ such that

$$\lim_{j\to\infty} \|f_{k_j} - f\|_{L^q(U)} = 0$$

for each $1 \le q < 1^*$. It follows from Theorem 5.2 that $f \in BV(U)$.   □

4.7   Capacity

We next introduce *capacity* as a way to study certain "small" subsets of $\mathbb{R}^n$. We will later see that in fact capacity is precisely suited for characterizing the fine properties of Sobolev functions. For this section, fix

$$1 \leq p < n.$$

4.7.1 Definitions and elementary properties

DEFINITION 4.9.

$$K^p := \{f : \mathbb{R}^n \to \mathbb{R} \mid f \geq 0, f \in L^{p^*}(\mathbb{R}^n), Df \in L^p(\mathbb{R}^n; \mathbb{R}^n)\}.$$

DEFINITION 4.10. *If $A \subset \mathbb{R}^n$, set*

$$\mathrm{Cap}_p(A) := \inf \left\{ \int_{\mathbb{R}^n} |Df|^p \, dx \mid f \in K^p, A \subseteq \{f \geq 1\}^0 \right\}.$$

*We call $\mathrm{Cap}_p(A)$ the **p-capacity** of A.*

Remarks.

(i) Note carefully the requirement that A must lie within the region $\{f \geq 1\}^0$, the *interior* of the set $\{f \geq 1\}$.

(ii) Using regularization, we see

$$\mathrm{Cap}_p(K) = \inf \left\{ \int_{\mathbb{R}^n} |Df|^p \, dx \mid f \in C_c^\infty(\mathbb{R}^n), f \geq \chi_K \right\}$$

for each compact set $K \subset \mathbb{R}^n$.

(iii) Clearly, $A \subseteq B$ implies

$$\mathrm{Cap}_p(A) \leq \mathrm{Cap}_p(B).$$

$\square$

THEOREM 4.12 (Approximation in K^p).

(i) *If $f \in K^p$ for some $1 \leq p < n$, there exists a sequence $\{f_k\}_{k=1}^\infty \subseteq W^{1,p}(\mathbb{R}^n)$ such that*

$$\|f - f_k\|_{L^{p^*}(\mathbb{R}^n)} \to 0$$

and

$$\|Df - Df_k\|_{L^p(\mathbb{R}^n)} \to 0$$

as $k \to \infty$.

(ii) *If $f \in K^p$, then*

$$\|f\|_{L^{p^*}(\mathbb{R}^n)} \le C_1 \|Df\|_{L^p(\mathbb{R}^n)},$$

where C_1 is the constant from the Gagliardo–Nirenberg–Sobolev inequality.

Proof. Select $\zeta \in C_c^1(\mathbb{R}^n)$ so that

$$0 \le \zeta \le 1, \ \zeta \equiv 1 \text{ on } B(1), \ \text{spt}\,\zeta \subset B(2), \ |D\zeta| \le 2.$$

For each $k = 1, 2, \ldots$, set $\zeta_k(x) := \zeta(\frac{x}{k})$.

Given $f \in K^p$, write $f_k := f\zeta_k$. Then $f_k \in W^{1,p}(\mathbb{R}^n)$,

$$\int_{\mathbb{R}^n} |f - f_k|^{p^*} \, dy \le \int_{\mathbb{R}^n - B(k)} |f|^{p^*} \, dy,$$

and

$$\int_{\mathbb{R}^n} |Df - Df_k|^p \, dy$$

$$\le 2^{p-1} \left\{ \int_{\mathbb{R}^n} |(1 - \zeta_k)Df|^p + |fD\zeta_k|^p \, dy \right\}$$

$$\le 2^{p-1} \left\{ \int_{\mathbb{R}^n - B(k)} |Df|^p \, dy + \frac{2^p}{k^p} \int_{B(2k) - B(k)} |f|^p \, dy \right\}$$

$$\le C \int_{\mathbb{R}^n - B(k)} |Df|^p \, dy + C \left(\int_{\mathbb{R}^n - B(k)} |f|^{p^*} \, dy \right)^{1 - \frac{p}{n}}.$$

This proves assertion (i). Assertion (ii) follows from (i) and the Gagliardo–Nirenberg–Sobolev inequality (Theorem 4.8). □

THEOREM 4.13 (Properties of K^p).

(i) *Assume $f, g \in K^p$. Then*

$$h := \max\{f, g\} \in K^p$$

and

$$Dh = \begin{cases} Df & \mathcal{L}^n\text{-a.e. on } \{f \ge g\} \\ Dg & \mathcal{L}^n\text{-a.e. on } \{f \le g\}. \end{cases}$$

An analogous assertion holds for min $\{f, g\}$.

(ii) *If $f \in K^p$ and $t \geq 0$, then*

$$h := \min\{f, t\} \in K^p.$$

(iii) *Given a sequence $\{f_k\}_{k=1}^{\infty} \subseteq K^p$, define*

$$g := \sup_{1 \leq k < \infty} f_k, \quad h := \sup_{1 \leq k < \infty} |Df_k|.$$

If $h \in L^p(\mathbb{R}^n)$, then $g \in K^p$ and

$$|Dg| \leq h \quad \mathcal{L}^n\text{-}a.e.$$

Proof. 1. To prove (i) we note

$$h = \max\{f, g\} = f + (g - f)^+.$$

Hence Theorem 4.4 implies

$$Dh = \begin{cases} Df & \mathcal{L}^n\text{-a.e. on } \{f \geq g\} \\ Dg & \mathcal{L}^n\text{-a.e. on } \{f \leq g\}. \end{cases}$$

Thus $Dh \in L^p$. Since $0 \leq h \leq f + g$, we have $h \in L^{p^*}$ as well.

2. The proof of (ii) is similar; we need only observe

$$0 \leq h = \min\{f, t\} \leq f,$$

and so $h \in L^{p^*}$.

3. To prove (iii) let us set

$$g_l := \sup_{1 \leq k \leq l} f_k.$$

Using assertion (i), we see $g_l \in K^p$ and

$$|Dg_l| \leq \sup_{1 \leq k \leq l} |Df_k| \leq h.$$

Since $g_l \to g$ monotonically, we can use Theorem 4.12 to calculate that

$$\begin{aligned}
\|g\|_{L^{p^*}(\mathbb{R}^n)} &= \lim_{l \to \infty} \|g_l\|_{L^{p^*}(\mathbb{R}^n)} \\
&\leq C_1 \liminf_{l \to \infty} \|Dg_l\|_{L^p(\mathbb{R}^n)} \\
&\leq C_1 \|h\|_{L^p(\mathbb{R}^n)}.
\end{aligned}$$

Thus $g \in L^{p^*}$. Now, for each $\phi \in C_c^1(\mathbb{R}^n; \mathbb{R}^n)$,

$$\int_{\mathbb{R}^n} g \, \mathrm{div} \, \phi \, dy = \lim_{l \to \infty} \int_{\mathbb{R}^n} g_l \, \mathrm{div} \, \phi \, dy$$

$$= -\lim_{l \to \infty} \int_{\mathbb{R}^n} \phi \cdot Dg_l \, dy$$

$$\leq \int_{\mathbb{R}^n} |\phi| h \, dy.$$

It follows that the linear functional L defined by

$$L(\phi) := \int_{\mathbb{R}^n} g \, \mathrm{div} \, \phi \, dy \qquad (\phi \in C_c^1(\mathbb{R}^n; \mathbb{R}^n))$$

has a unique extension $\bar{L}$ to $C_c(\mathbb{R}^n; \mathbb{R}^n)$ such that

$$\bar{L}(\phi) \leq \int_{\mathbb{R}^n} |\phi| h \, dy,$$

for $\phi \in C_c(\mathbb{R}^n; \mathbb{R}^n)$. We apply Riesz Representation Theorem 1.38 and note the measure μ constructed in its proof satisfies

$$\mu(A) \leq \int_A h \, dy$$

for each Lebesgue measurable set $A \subseteq \mathbb{R}^n$. It follows that

$$\bar{L}(\phi) = \int_{\mathbb{R}^n} \phi \cdot k \, dy$$

where $k \in L^p(\mathbb{R}^n; \mathbb{R}^n)$ and $|k| \leq h \; \mathcal{L}^n$-a.e. Thus $g \in K^p$ and $|Dg| = |k| \leq h \; \mathcal{L}^n$-a.e. $\qquad \square$

THEOREM 4.14 (Capacity as measure). Cap_p *is a measure on* $\mathbb{R}^n$.

Warning: Cap_p is *not* a Borel measure. In fact, if $A \subseteq \mathbb{R}^n$ and $0 < \mathrm{Cap}_p(A) < \infty$, then A is *not* Cap_p-measurable. Remember also that what we call a measure in these notes is usually called an "outer measure" in other texts.

Proof. Assume $A \subseteq \bigcup_{k=1}^\infty A_k$, $\sum_{k=1}^\infty \mathrm{Cap}_p(A_k) < \infty$. Fix $\epsilon > 0$. For each $k = 1, \ldots$, choose $f_k \in K^p$ so that

$$A_k \subseteq \{f_k \geq 1\}$$

and

$$\int_{\mathbb{R}^n} |Df_k|^p \, dx \le \operatorname{Cap}_p(A_k) + \frac{\epsilon}{2^k}.$$

Define $g := \sup_{1 \le k < \infty} f_k$. Then $A \subseteq \{g \ge 1\}^0, g \in K^p$ by Theorem 4.13, and

$$\int_{\mathbb{R}^n} |Dg|^p \, dx \le \int_{\mathbb{R}^n} \sup_{1 \le k < \infty} |Df_k|^p \, dx$$

$$\le \sum_{k=1}^{\infty} \int_{\mathbb{R}^n} |Df_k|^p \, dx$$

$$\le \sum_{k=1}^{\infty} \operatorname{Cap}_p(A_k) + \epsilon.$$

Thus

$$\operatorname{Cap}_p(A) \le \sum_{k=1}^{\infty} \operatorname{Cap}_p(A_k) + \epsilon. \qquad \square$$

THEOREM 4.15 (Properties of capacity). *Assume $A, B \subseteq \mathbb{R}^n$.*

(i) $\operatorname{Cap}_p(A) = \inf\{\operatorname{Cap}_p(U) \mid U \ open , A \subseteq U\}.$

(ii) $\operatorname{Cap}_p(\lambda A) = \lambda^{n-p} \operatorname{Cap}_p(A) \quad (\lambda > 0).$

(iii) $\operatorname{Cap}_p(L(A)) = \operatorname{Cap}_p(A)$ *for each affine isometry $L : \mathbb{R}^n \to \mathbb{R}^n$.*

(iv) $\operatorname{Cap}_p(B(x,r)) = r^{n-p} \operatorname{Cap}_p(B(1)).$

(v) $\operatorname{Cap}_p(A) \le C \mathcal{H}^{n-p}(A),$ *for some constant C depending only on p and n.*

(vi) $\mathcal{L}^n(A) \le C \operatorname{Cap}_p(A)^{\frac{n}{n-p}}$ *for some constant C depending only on p and n.*

(vii) $\operatorname{Cap}_p(A \cup B) + \operatorname{Cap}_p(A \cap B) \le \operatorname{Cap}_p(A) + \operatorname{Cap}_p(B).$

(viii) *If $A_1 \subseteq \ldots A_k \subseteq A_{k+1} \ldots$, then*

$$\lim_{k \to \infty} \operatorname{Cap}_p(A_k) = \operatorname{Cap}_p \left(\bigcup_{k=1}^{\infty} A_k \right).$$

(ix) *If $A_1 \supset \ldots A_k \supset A_{k+1} \ldots$ are compact, then*

$$\lim_{k \to \infty} \operatorname{Cap}_p(A_k) = \operatorname{Cap}_p \left(\bigcap_{k=1}^{\infty} A_k \right).$$

Remark. Assertion (ix) may be false if the sets $\{A_k\}_{k=1}^{\infty}$ are not compact. See Theorem 4.16 for an improvement of (v). □

Proof. 1. Clearly $\mathrm{Cap}_p(A) \le \inf\{\mathrm{Cap}_p(U) \mid U \text{ open}, U \supset A\}$. On the other hand, for each $\epsilon > 0$, there exists $f \in K^p$ such that $A \subseteq \{f \ge 1\}^0 =: U$ and

$$\int_{\mathbb{R}^n} |Df|^p \, dx \le \mathrm{Cap}_p(A) + \epsilon.$$

But then

$$\mathrm{Cap}_p(U) \le \int_{\mathbb{R}^n} |Df|^p \, dx,$$

and so statement (i) holds.

2. Fix $\epsilon > 0$ and choose $f \in K^p$ as above. Let $g(x) := f(\frac{x}{\lambda})$. Then $g \in K^p, \lambda A \subseteq \{g \ge 1\}^0$ and

$$\int_{\mathbb{R}^n} |Dg|^p \, dx = \lambda^{n-p} \int_{\mathbb{R}^n} |Df|^p \, dx.$$

Thus $\mathrm{Cap}_p(\lambda A) \le \lambda^{n-p}(\mathrm{Cap}_p(A) + \epsilon)$. The other inequality is similar, and so (ii) is verified.

3. Assertion (iii) is clear, and statement (iv) is a consequence of (ii), (iii).

4. To prove (v), fix $\delta > 0$ and suppose

$$A \subseteq \bigcup_{k=1}^{\infty} B(x_k, r_k)$$

where $2r_k < \delta$ $(k = 1, \dots)$. Then

$$\mathrm{Cap}_p(A) \le \sum_{k=1}^{\infty} \mathrm{Cap}_p(B(x_k, r_k)) = \mathrm{Cap}_p(B(1)) \sum_{k=1}^{\infty} r_k^{n-p}.$$

Hence

$$\mathrm{Cap}_p(A) \le C\mathcal{H}^{n-p}(A).$$

Choose $\epsilon > 0, f \in K^p$ as in Part 1 of the proof. Then by Theorem 4.12

$$\mathcal{L}^n(A)^{\frac{1}{p^*}} \le \left(\int_{\mathbb{R}^n} f^{p^*} \, dx \right)^{\frac{1}{p^*}}$$

$$\leq C_1 \left(\int_{\mathbb{R}^n} |Df|^p \, dx \right)^{\frac{1}{p}}$$

$$\leq C_1 (\mathrm{Cap}_p(A) + \epsilon)^{\frac{1}{p}}.$$

Consequently,

$$\mathcal{L}^n(A) \leq C \mathrm{Cap}_p(A)^{\frac{p^*}{p}};$$

this is (vi).

5. Fix $\epsilon > 0$, select $f \in K^p$ as above, and choose also $g \in K^p$ so that

$$B \subseteq \{g \geq 1\}^0, \int_{\mathbb{R}^n} |Dg|^p \, dx \leq \mathrm{Cap}_p(B) + \epsilon.$$

Then $\max\{f, g\}, \min\{f, g\} \in K^p$ and

$$|D(\max\{f, g\})|^p + |D(\min\{f, g\})|^p = |Df|^p + |Dg|^p \quad \mathcal{L}^n\text{-a.e.},$$

according to Theorem 4.13. Furthermore,

$$A \cup B \subseteq \{\max\{f, g\} \geq 1\}^0,$$

$$A \cap B \subseteq \{\min\{f, g\} \geq 1\}^0.$$

Thus

$$\mathrm{Cap}_p(A \cup B) + \mathrm{Cap}_p(A \cap B) \leq \int_{\mathbb{R}^n} |D(\max\{f, g\})|^p$$

$$+ |D(\min\{f, g\})|^p \, dx$$

$$= \int_{\mathbb{R}^n} |Df|^p + |Dg|^p \, dx$$

$$\leq \mathrm{Cap}_p(A) + \mathrm{Cap}_p(B) + 2\epsilon$$

and assertion (vii) is proved.

6. We will prove statement (viii) for the case $1 < p < n$ only; see Federer and Ziemer [FZ] for $p = 1$. Assume $\lim_{k \to \infty} \mathrm{Cap}_p(A_k) < \infty$ and $\epsilon > 0$. Then for each $k = 1, 2, \ldots$, choose $f_k \in K^p$ such that

$$A_k \subseteq \{x \mid f_k(x) \geq 1\}^0$$

and

$$\int_{\mathbb{R}^n} |Df_k|^p \, dx < \mathrm{Cap}_p(A_k) + \frac{\epsilon}{2^k}.$$

Define

$$h_m := \max\{f_k \mid 1 \le k \le m\}, \quad h_0 := 0$$

and notice from Theorem 4.13 that $h_m = \max(h_{m-1}, f_m) \in K^p$ and

$$A_{m-1} \subseteq \{x \mid \min(h_{m-1}, f_m) \ge 1\}^0.$$

We compute

$$\int_{\mathbb{R}^n} |Dh_m|^p \, dx + \mathrm{Cap}_p(A_{m-1}) \le \int_{\mathbb{R}^n} |D(\max(h_{m-1}, f_m))|^p \, dx$$
$$+ \int_{\mathbb{R}^n} |D(\min(h_{m-1}, f_m))|^p \, dx$$
$$= \int_{\mathbb{R}^n} |Dh_{m-1}|^p + |Df_m|^p \, dx$$
$$\le \int_{\mathbb{R}^n} |Dh_{m-1}|^p \, dx + \mathrm{Cap}_p(A_m)$$
$$+ \frac{\epsilon}{2^m}.$$

Consequently,

$$\int_{\mathbb{R}^n} |Dh_m|^p \, dx - \int_{\mathbb{R}^n} |Dh_{m-1}|^p \, dx$$
$$\le \mathrm{Cap}_p(A_m) - \mathrm{Cap}_p(A_{m-1}) + \frac{\epsilon}{2^m};$$

from which it follows by adding that

$$\int_{\mathbb{R}^n} |Dh_m|^p \, dx \le \mathrm{Cap}_p(A_m) + \epsilon \quad (m = 1, 2, \dots).$$

Set $f := \lim_{m \to \infty} h_m$. Then $\bigcup_{k=1}^{\infty} A_k \subseteq \{x \mid f(x) \ge 1\}^0$. Furthermore,

$$\|f\|_{L^{p^*}(\mathbb{R}^n)} = \lim_{m \to \infty} \|h_m\|_{L^{p^*}(\mathbb{R}^n)}$$
$$\le C_1 \liminf_{m \to \infty} \|Dh_m\|_{L^p(\mathbb{R}^n)}$$
$$\le C \left(\lim_{m \to \infty} \mathrm{Cap}_p(A_m) + \epsilon \right)^{\frac{1}{p}}.$$

Since $p > 1$, a subsequence of $\{Dh_m\}_{m=1}^{\infty}$ converges weakly to Df in $L^p(\mathbb{R}^n)$ (cf. Theorem 1.42); thus $f \in K^p$. Consequently,

$$\mathrm{Cap}_p(\cup_{k=1}^{\infty} A_k) \le \|Df\|_{L^p(\mathbb{R}^n)}^p \le \lim_{m \to \infty} \mathrm{Cap}_p(A_m) + \epsilon.$$

7. We prove (ix) by first noting

$$\text{Cap}_p\left(\cap_{k=1}^{\infty} A_k\right) \le \lim_{k \to \infty} \text{Cap}_p(A_k).$$

On the other hand, choose any open set U with $\cap_{k=1}^{\infty} A_k \subseteq U$. As $\cap_{k=1}^{\infty} A_k$ is compact, there exists a positive integer m such that $A_k \subset U$ for $k \ge m$. Thus

$$\lim_{k \to \infty} \text{Cap}_p(A_k) \le \text{Cap}_p(U).$$

Recall (i) to complete the proof of (ix). □

4.7.2 Capacity and Hausdorff dimension

As noted earlier, we are interested in capacity as a way of characterizing certain "very small" subsets of $\mathbb{R}^n$. Obviously Hausdorff measures provide another approach, and so it is important to understand the relationships between capacity and Hausdorff measure.

We begin with a refinement of assertion (v) from Theorem 4.15:

THEOREM 4.16 (Capacity and Hausdorff measure).
Assume $1 < p < n$. If $\mathcal{H}^{n-p}(A) < \infty$, then

$$\text{Cap}_p(A) = 0.$$

Proof. 1. According to Theorem 4.15, (viii), we may assume A is compact.

Claim: There exists a constant C, depending only on n and A, such that if V is any open set containing A, there exists an open set W and $f \in K^p$ such that

$$\begin{cases} A \subseteq W \subset \{f = 1\}, \ \text{spt}(f) \subset V, \\ \int_{\mathbb{R}^n} |Df|^p \, dx \le C. \end{cases}$$

Proof of claim: Let V be an open set containing A and let $\delta := \frac{1}{2} \text{dist}(A, \mathbb{R}^n - V)$. Since $\mathcal{H}^{n-p}(A) < \infty$ and A is compact, there exists a finite collection $\{B^0(x_i, r_i)\}_{i=1}^m$ of open balls such that $2r_i < \delta$, $B^0(x_i, r_i) \cap A \ne \emptyset$, $A \subseteq \bigcup_{i=1}^m B^0(x_i, r_i)$, and

$$\sum_{i=1}^m \alpha(n-p) r_i^{n-p} \le C\mathcal{H}^{n-p}(A) + 1.$$

for some constant C.

Now set $W := \bigcup_{i=1}^{m} B^0(x_i, r_i)$ and define $f_i \in K^p$ by

$$f_i(x) = \begin{cases} 1 & \text{if } |x - x_i| \le r_i \\ 2 - \frac{|x - x_i|}{r_i} & \text{if } r_i \le |x - x_i| \le 2r_i \\ 0 & \text{if } 2r_i \le |x - x_i|. \end{cases}$$

Then

$$\int_{\mathbb{R}^n} |Df_i|^p \, dx \le C r_i^{n-p}.$$

Let $f := \max_{1 \le i \le m} f_i$. Then $f \in K^p$, $W \subseteq \{f = 1\}$, $\text{spt}(f) \subseteq V$, and

$$\int_{\mathbb{R}^n} |Df|^p \, dx \le \sum_{i=1}^{m} \int_{\mathbb{R}^n} |Df_i|^p \, dx \le C \sum_{i=1}^{m} r_i^{n-p} \le C(\mathcal{H}^{n-p}(A) + 1).$$

2. Using the claim inductively, we can find open sets $\{V_k\}_{k=1}^{\infty}$ and functions $f_k \in K^p$ such that

$$\begin{cases} A \subseteq V_{k+1} \subset V_k, \bar{V}_{k+1} \subset \{f_k = 1\}^0, \\ \text{spt}(f_k) \subseteq V_k, \ \int_{\mathbb{R}^n} |Df_k|^p \, dx \le C. \end{cases}$$

Set

$$S_j := \sum_{k=1}^{j} \frac{1}{k}$$

and

$$g_j := \frac{1}{S_j} \sum_{k=1}^{j} \frac{f_k}{k}.$$

Then $g_j \in K^p$, $g_j \ge 1$ on V_{j+1}. Since $\text{spt} |Df_k| \subseteq V_k - \bar{V}_{k+1}$, we see that

$$\text{Cap}_p(A) \le \int_{\mathbb{R}^n} |Dg_j|^p \, dx = \frac{1}{S_j^p} \sum_{k=1}^{j} \frac{1}{k^p} \int_{\mathbb{R}^n} |Df_k|^p \, dx$$

$$\le \frac{C}{S_j^p} \sum_{k=1}^{j} \frac{1}{k^p} \to 0 \quad \text{as } j \to \infty,$$

since $p > 1$. $\qquad\qquad\qquad\qquad\qquad\qquad\qquad\qquad\qquad\qquad\qquad\qquad\square$

THEOREM 4.17 (More on capacity and Hausdorff measure).
Assume $A \subset \mathbb{R}^n$ and $1 \leq p < \infty$. If $\mathrm{Cap}_p(A) = 0$, then

$$\mathcal{H}^s(A) = 0 \quad \text{for all } s > n - p.$$

Remark. We will prove later in Section 5.6 that $\mathrm{Cap}_1(A) = 0$ if and only if $\mathcal{H}^{n-1}(A) = 0$. $\qquad\square$

Proof. 1. Let $\mathrm{Cap}_p(A) = 0$ and $n - p < s < \infty$. Then for all $i \geq 1$, there exists $f_i \in K^p$ such that $A \subseteq \{f_i \geq 1\}^0$ and

$$\int_{\mathbb{R}^n} |Df_i|^p \, dx \leq \frac{1}{2^i}.$$

Let $g := \sum_{i=1}^{\infty} f_i$. Then

$$\left(\int_{\mathbb{R}^n} |Dg|^p \, dx \right)^{\frac{1}{p}} \leq \sum_{i=1}^{\infty} \left(\int_{\mathbb{R}^n} |Df_i|^p \, dx \right)^{\frac{1}{p}} < \infty,$$

and by the Gagliardo–Nirenberg–Sobolev inequality (Theorem 4.8),

$$\left(\int_{\mathbb{R}^n} |g|^{p^*} \, dx \right)^{\frac{1}{p^*}} \leq \sum_{i=1}^{\infty} \left(\int_{\mathbb{R}^n} |f_i|^{p^*} \, dx \right)^{\frac{1}{p^*}}$$

$$\leq \sum_{i=1}^{\infty} C_1 \left(\int_{\mathbb{R}^n} |Df_i|^p \, dx \right)^{\frac{1}{p}} < \infty.$$

Thus $g \in K^p$.

2. Note $A \subseteq \{g \geq m\}^0$ for all $m \geq 1$. Fix any $a \in A$. Then for r small enough that $B(a, r) \subseteq \{g \geq m\}^0$, we have $(g)_{a,r} \geq m$; therefore $(g)_{a,r} \to \infty$ as $r \to 0$.

3. *Claim:* For each $a \in A$,

$$\limsup_{r \to 0} \frac{1}{r^s} \int_{B(a,r)} |Dg|^p \, dx = +\infty.$$

Proof of claim: Let $a \in A$ and suppose

$$\limsup_{r \to 0} \frac{1}{r^s} \int_{B(a,r)} |Dg|^p \, dx < \infty.$$

Then there exists a constant $M < \infty$ such that

$$\frac{1}{r^s} \int_{B(a,r)} |Dg|^p \, dx \leq M$$

for all $0 < r \leq 1$. Then for $0 < r \leq 1$,

$$\fint_{B(a,r)} |g - (g)_{a,r}|^p \, dx \leq C_2 r^p \fint_{B(a,r)} |Dg|^p \, dx \leq C r^\theta,$$

where $\theta := s - (n - p) > 0$. Thus

$$\left| (g)_{a,\frac{r}{2}} - (g)_{a,r} \right| = \frac{1}{\mathcal{L}^n(B(a,\frac{r}{2}))} \left| \int_{B(a,\frac{r}{2})} g - (g)_{a,r} \, dx \right|$$

$$\leq 2^n \fint_{B(a,r)} |g - (g)_{a,r}| \, dx$$

$$\leq 2^n \left(\fint_{B(a,r)} |g - (g)_{a,r}|^p \, dx \right)^{\frac{1}{p}}$$

$$= C r^{\frac{\theta}{p}}.$$

Hence if $k > j$,

$$\left| (g)_{a,\frac{1}{2^k}} - (g)_{a,\frac{1}{2^j}} \right| \leq \sum_{l=j+1}^{k} \left| (g)_{a,\frac{1}{2^l}} - (g)_{a,\frac{1}{2^{l-1}}} \right| \leq C \sum_{l=j+1}^{k} \left(\frac{1}{2^{l-1}} \right)^{\frac{\theta}{p}}.$$

This last sum is the tail of a geometric series, and so $\{(g)_{a,\frac{1}{2^k}}\}_{k=1}^\infty$ is a Cauchy sequence. Thus $(g)_{a,\frac{1}{2^k}} \not\to \infty$, a contradiction since $(g)_{a,r} \to \infty$ as $r \to 0$.

Consequently,

$$A \subseteq \left\{ a \in \mathbb{R}^n \mid \limsup_{r \to 0} \frac{1}{r^s} \int_{B(a,r)} |Dg|^p \, dx = +\infty \right\}$$

$$\subseteq \left\{ a \in \mathbb{R}^n \mid \limsup_{r \to 0} \frac{1}{r^s} \int_{B(a,r)} |Dg|^p \, dx > 0 \right\} =: \Lambda_s.$$

But since $|Dg|^p$ is $\mathcal{L}^n$-summable, $\mathcal{H}^s(\Lambda_s) = 0$, according to Theorem 2.10. $\qquad \square$

4.8 Quasicontinuity, precise representatives of Sobolev functions

This section studies the fine properties of Sobolev functions.

THEOREM 4.18 (Capacity estimate). *Assume $f \in K^p$ and $\epsilon > 0$. Let*

$$A := \{x \in \mathbb{R}^n \mid (f)_{x,r} > \epsilon \text{ for some } r > 0\}.$$

Then

$$\mathrm{Cap}_p(A) \leq \frac{C}{\epsilon^p} \int_{\mathbb{R}^n} |Df|^p \, dx, \qquad (\star)$$

where C depends only on n and p.

Remark. This is a capacity variant of the simple estimate

$$\mathcal{L}^n(\{x \in \mathbb{R}^n \mid f(x) > \epsilon\}) \leq \frac{1}{\epsilon^p} \int_{\mathbb{R}^n} |f|^p \, dx.$$

$\square$

Proof. 1. For the moment we set $\epsilon = 1$ and observe that if $x \in A$ and $(f)_{x,r} > 1$, then

$$\alpha(n)r^n \leq \int_{B(x,r)} f \, dy \leq (\alpha(n)r^n)^{1-\frac{1}{p^*}} \left(\int_{B(x,r)} f^{p^*} \, dy \right)^{\frac{1}{p^*}}.$$

Therefore

$$r \leq C$$

for some constant C.

2. According to the Besicovitch Covering Theorem 1.27, there exist an integer N_n and countable collections $\mathcal{F}_1, \ldots, \mathcal{F}_{N_n}$ of disjoint closed balls such that

$$A \subseteq \bigcup_{i=1}^{N_n} \bigcup_{B \in \mathcal{F}_i} B$$

and

$$(f)_B > 1 \quad \text{for each } B \in \bigcup_{i=1}^{N_n} \mathcal{F}_i.$$

Denote by B_i^j the elements of $\mathcal{F}_i$ $(i = 1, \ldots, N_n; j = 1, \ldots)$. Choose $h_{ij} \in K^p$ such that

$$h_{ij} = ((f)_{B_i^j} - f)^+ \quad \text{on } B_i^j$$

and

$$\int_{\mathbb{R}^n} |Dh_{ij}|^p \, dx \le C \int_{B_i^j} |Df|^p \, dx \quad (i = 1, \ldots N_n; j = 1, 2, \ldots),$$

where C depends only on n and p. This is possible according to Theorem 4.7 and Poincaré's inequality. Note that

$$f + h_{ij} \ge (f)_{B_i^j} \ge 1 \quad \text{in } B_i^j.$$

Hence, setting

$$h := \sup\{h_{ij} \mid i = 1, \ldots, N_n, j = 1, \ldots\} \in K^p,$$

we observe that

$$f + h \ge 1 \quad \text{on } A. \tag{$\star\star$}$$

3. Now

$$\int_{\mathbb{R}^n} |D(f + h)|^p \, dx \le C \left\{ \int_{\mathbb{R}^n} |Df|^p \, dx + \sum_{i=1}^{N_n} \sum_{j=1}^{\infty} \int_{\mathbb{R}^n} |Dh_{ij}|^p \, dx \right\}$$

$$\le C \int_{\mathbb{R}^n} |Df|^p \, dx.$$

Consequently, since A is open and so $(\star\star)$ implies

$$A \subseteq \{f + h \ge 1\}^0,$$

we have

$$\text{Cap}_p(A) \le \int_{\mathbb{R}^n} |D(f + h)|^p \, dx \le C \int_{\mathbb{R}^n} |Df|^p \, dx.$$

4. In case $0 < \epsilon \ne 1$, we set $g := \epsilon^{-1} f \in K^p$; so that

$$A := \{x \mid (f)_{x,r} > \epsilon \text{ for some } r > 0\}$$
$$= \{x \mid (g)_{x,r} > 1 \text{ for some } r > 0\}.$$

Thus

$$\text{Cap}_p(A) \le C \int_{\mathbb{R}^n} |Dg|^p \, dx = \frac{C}{\epsilon^p} \int_{\mathbb{R}^n} |Df|^p \, dx. \qquad \square$$

We now study the fine structure properties of Sobolev functions, using capacity to measure the size of the "bad" sets.

DEFINITION 4.11. *A function f is* **p-quasicontinuous** *if for each $\epsilon > 0$, there exists an open set V such that*

$$\text{Cap}_p(V) \leq \epsilon$$

and

$$f|_{\mathbb{R}^n - V} \text{ is continuous.}$$

THEOREM 4.19 (Fine properties of Sobolev functions). *Suppose $f \in W^{1,p}(\mathbb{R}^n), 1 \leq p < n.$*

(i) *There is a Borel set $E \subset \mathbb{R}^n$ such that*

$$\text{Cap}_p(E) = 0$$

and

$$\lim_{r \to 0}(f)_{x,r} =: f^*(x)$$

exists for each $x \in \mathbb{R}^n - E.$

(ii) *In addition,*

$$\lim_{r \to 0} \fint_{B(x,r)} |f - f^*(x)|^{p^*} \, dy = 0$$

for each $x \in \mathbb{R}^n - E.$

(iii) *The precise representative f^* is p-quasicontinuous.*

Remark. Notice that if f is a Sobolev function and $f = g$ $\mathcal{L}^n$-a.e., then g is also a Sobolev function. Consequently if we wish to study the fine properties of f, we must turn our attention to the precise representative f^*, defined in Section 1.7. $\qquad\square$

Proof. 1. Set

$$A := \left\{ x \in \mathbb{R}^n \mid \limsup_{r \to 0} \frac{1}{r^{n-p}} \int_{B(x,r)} |Df|^p \, dy > 0 \right\}.$$

By Theorem 2.10 and Theorem 4.16,

$$\mathcal{H}^{n-p}(A) = 0, \ \text{Cap}_p(A) = 0.$$

Now, according to Poincaré's inequality,

$$\lim_{r \to 0} \fint_{B(x,r)} |f - (f)_{x,r}|^{p^*} \, dy = 0 \tag{$\star$}$$

for each $x \notin A$. Choose functions $f_i \in W^{1,p}(\mathbb{R}^n) \cap C^\infty(\mathbb{R}^n)$ such that

$$\int_{\mathbb{R}^n} |Df - Df_i|^p \, dy \leq \frac{1}{2^{(p+1)i}} \quad (i = 1, 2, \dots),$$

and set

$$B_i := \left\{ x \in \mathbb{R}^n \mid \fint_{B(x,r)} |f - f_i| \, dy > \frac{1}{2^i} \text{ for some } r > 0 \right\}.$$

According to Theorem 4.18,

$$\frac{\mathrm{Cap}_p(B_i)}{2^{pi}} \leq C \int_{\mathbb{R}^n} |Df - Df_i|^p \, dy \leq \frac{C}{2^{(p+1)i}}.$$

Consequently, $\mathrm{Cap}_p(B_i) \leq \frac{C}{2^i}$. Furthermore,

$$|(f)_{x,r} - f_i(x)| \leq \fint_{B(x,r)} |f - (f)_{x,r}| \, dy + \fint_{B(x,r)} |f - f_i| \, dy$$

$$+ \fint_{B(x,r)} |f_i - f_i(x)| \, dy.$$

Thus $(\star)$ and the definition of B_i imply

$$\limsup_{r \to 0} |(f)_{x,r} - f_i(x)| \leq \frac{1}{2^i} \quad (x \notin A \cup B_i). \tag{$\star\star$}$$

Set $E_k := A \cup (\cup_{j=k}^\infty B_j)$. Then

$$\mathrm{Cap}_p(E_k) \leq \mathrm{Cap}_p(A) + \sum_{j=k}^\infty \mathrm{Cap}_p(B_j) \leq C \sum_{j=k}^\infty \frac{1}{2^j}.$$

Furthermore, if $x \in \mathbb{R}^n - E_k$ and $i, j \geq k$, then

$$|f_i(x) - f_j(x)| \leq \limsup_{r \to 0} |(f)_{x,r} - f_i(x)|$$

$$+ \limsup_{r \to 0} |(f)_{x,r} - f_j(x)|$$

$$\leq \frac{1}{2^i} + \frac{1}{2^j}$$

by ($\star\star$). Hence $\{f_j\}_{j=1}^{\infty}$ converges uniformly on $\mathbb{R}^n - E_k$ to a continuous function g. Furthermore,

$$\limsup_{r \to 0} |g(x) - (f)_{x,r}| \leq |g(x) - f_i(x)| + \limsup_{r \to 0} |f_i(x) - (f)_{x,r}|;$$

so that ($\star\star$) implies

$$g(x) = \lim_{r \to 0} (f)_{x,r} = f^*(x) \quad (x \in \mathbb{R}^n - E_k).$$

Now set $E := \cap_{k=1}^{\infty} E_k$. Then $\mathrm{Cap}_p(E) \leq \lim_{k \to \infty} \mathrm{Cap}_p(E_k) = 0$ and

$$f^*(x) = \lim_{r \to 0} (f)_{x,r} \quad \text{exists for each } x \in \mathbb{R}^n - E.$$

This proves (i).

2. To prove (ii), note $A \subseteq E$ and so ($\star$) implies for $x \in \mathbb{R}^n - E$ that

$$\lim_{r \to 0} \left(\fint_{B(x,r)} |f - f^*(x)|^{p^*} \, dy \right)^{\frac{1}{p^*}}$$

$$\leq \lim_{r \to 0} |(f)_{x,r} - f^*(x)| + \lim_{r \to 0} \left(\fint_{B(x,r)} |f - (f)_{x,r}|^{p^*} \, dy \right)^{\frac{1}{p^*}}$$

$$= 0.$$

3. Finally, we prove (iii) by fixing $\epsilon > 0$ and then choosing k such that $\mathrm{Cap}_p(E_k) < \frac{\epsilon}{2}$. According to Theorem 4.15, there exists an open set $U \supset E_k$ with $\mathrm{Cap}_p(U) < \epsilon$. Since the $\{f_i\}_{i=1}^{\infty}$ converge uniformly to f^* on $\mathbb{R}^n - U$, we see that $f^*|_{\mathbb{R}^n - U}$ is continuous. $\square$

4.9 Differentiability on lines

We will study in this section the properties of a Sobolev function f, or more exactly its precise representative f^*, restricted to lines.

4.9.1   Sobolev functions of one variable

NOTATION If $h : \mathbb{R} \to \mathbb{R}$ is absolutely continuous on each compact subinterval, we write h' to denote its derivative (which exists $\mathcal{L}^1$-a.e.).

THEOREM 4.20 (Sobolev functions of one variable). *Let* $1 \leq p < \infty$.

(i) *If* $f \in W^{1,p}_{loc}(\mathbb{R})$, *its precise representative* f^* *is absolutely continuous on each compact subinterval of* $\mathbb{R}$ *and* $(f^*)' \in L^p_{loc}(\mathbb{R})$.

(ii) *Conversely, suppose* $f \in L^p_{loc}(\mathbb{R})$ *and* $f = g$ $\mathcal{L}^1$-*a.e., where* g *is absolutely continuous on each compact subinterval of* $\mathbb{R}$ *and* $g' \in L^p_{loc}(\mathbb{R})$. *Then* $f \in W^{1,p}_{loc}(\mathbb{R})$.

Proof. 1. First assume $f \in W^{1,p}_{loc}(\mathbb{R})$ and let f' denote its weak derivative. For $0 < \epsilon \leq 1$ define $f^\epsilon := \eta_\epsilon * f$, as before. Then

$$f^\epsilon(y) = f^\epsilon(x) + \int_x^y (f^\epsilon)'(t)\,dt. \qquad (\star)$$

Let x_0 be a Lebesgue point of f and $\epsilon, \delta \in (0,1)$. Since

$$|f^\epsilon(x) - f^\delta(x)| \leq \int_{x_0}^x |(f^\epsilon)'(t) - (f^\delta)'(t)|dt + |f^\epsilon(x_0) - f^\delta(x_0)|$$

for $x \in \mathbb{R}$, it follows from Theorem 4.1 that $\{f^\epsilon\}_{\epsilon>0}$ converges uniformly on compact subsets of $\mathbb{R}$ to a continuous function g with $g = f$ $\mathcal{L}^1$-a.e. From $(\star)$ we see

$$g(x) = g(x_0) + \int_{x_0}^x f'(t)dt;$$

and hence g is locally absolutely continuous with $g' = f'$ $\mathcal{L}^1$-a.e.

Finally, since $(f)_{x,r} = (g)_{x,r} \to g(x)$ for each $x \in \mathbb{R}$, we see $g = f^*$. This proves (i).

2. On the other hand, assume $f = g$ $\mathcal{L}^1$-a.e., g is absolutely continuous and $g' \in L^p_{loc}(\mathbb{R})$. Then for each $\phi \in C^1_c(\mathbb{R})$,

$$\int_{-\infty}^\infty f\phi'\,dx = \int_{-\infty}^\infty g\phi'\,dx = -\int_{-\infty}^\infty g'\phi dx,$$

and thus g' is the weak derivative of f. Since $g' \in L^p_{loc}(\mathbb{R})$, we conclude $f \in W^{1,p}_{loc}(\mathbb{R})$. □

4.9.2   Differentiability on a.e. line

THEOREM 4.21 (Sobolev functions restricted to lines).

(i) *If $f \in W_{loc}^{1,p}(\mathbb{R}^n)$, then for each $k = 1, \ldots, n$ the functions*

$$f_k^*(x', t) := f^*(\ldots, x_{k-1}, t, x_{k+1}, \ldots)$$

are absolutely continuous in t on compact subsets of $\mathbb{R}$, for $\mathcal{L}^{n-1}$-a.e. point $x' = (x_1, \ldots, x_{k-1}, x_{k+1}, \ldots, x_n) \in \mathbb{R}^{n-1}$. In addition,

$$(f_k^*)' \in L_{loc}^p(\mathbb{R}^n).$$

(ii) *Conversely, suppose $f \in L_{loc}^p(\mathbb{R}^n)$ and $f = g$ $\mathcal{L}^n$-a.e., where for each $k = 1, \ldots, n$, the functions*

$$g_k(x', t) := g(x_1, \ldots, x_{k-1}, t, x_{k+1}, \ldots, x_n)$$

are absolutely continuous in t on compact subsets of $\mathbb{R}$ for $\mathcal{L}^{n-1}$-a.e. point $x = (x_1, \ldots, x_{k-1}, x_{k+1}, \ldots x_n) \in \mathbb{R}^{n-1}$, and $g_k' \in L_{loc}^p(\mathbb{R}^n)$. Then $f \in W_{loc}^{1,p}(\mathbb{R}^n)$.

Proof. 1. It suffices to prove assertion (i) for the case $k = n$. Define $f^\epsilon := \eta_\epsilon * f$ as before, and recall

$$f^\epsilon \to f \quad \text{in } W_{loc}^{1,p}(\mathbb{R}^n).$$

According to Fubini's Theorem, for each $L > 0$ and $\mathcal{L}^{n-1}$-a.e. $x' = (x_1, \ldots, x_{n-1})$, the expression

$$\int_{-L}^{L} |f^\epsilon(x', t) - f(x', t)|^p + |f_{x_n}^\epsilon(x', t) - f_{x_n}(x', t)|^p \, dt$$

goes to zero as $\epsilon \to 0$. Thus the functions

$$f_n^\epsilon(t) := f^\epsilon(x', t) \quad (t \in \mathbb{R})$$

converge in $W_{loc}^{1,p}(\mathbb{R})$, and so locally uniformly, to a locally absolutely continuous function f_n, with $f_n'(t) = f_{x_n}(x', t)$ for $\mathcal{L}^1$-a.e. $t \in \mathbb{R}$.

On the other hand, Theorem 4.19, Theorem 5.12 (to be proved later), and Theorem 4.17 imply

$$f^\epsilon \to f^* \quad \mathcal{H}^{n-1}\text{-a.e.}$$

In view of Theorem 2.8, for $\mathcal{L}^{n-1}$-a.e. point x', we have

$$f_n^\epsilon(t) \to f^*(x',t)$$

for all $t \in \mathbb{R}$ Hence for $\mathcal{L}^{n-1}$-a.e. x' and all $t \in \mathbb{R}$,

$$f_n(t) = f^*(x',t).$$

This proves statement (i).

2. Assume now the hypothesis of assertion (ii). Then for each $\phi \in C_c^1(\mathbb{R}^n)$,

$$
\begin{aligned}
\int_{\mathbb{R}^n} f\phi_{x_k}\, dx &= \int_{\mathbb{R}^n} g\phi_{x_k}\, dx \\
&= \int_{\mathbb{R}^{n-1}} \left(\int_{-\infty}^{\infty} g_k(x',t)\phi'(x',t)\, dt \right) dx' \\
&= -\int_{\mathbb{R}^{n-1}} \left(\int_{-\infty}^{\infty} g_k'(x',t)\phi(x',t)\, dt \right) dx' \\
&= -\int_{\mathbb{R}^n} g_k'\phi\, dx.
\end{aligned}
$$

Thus $f_{x_k} = g_k'$ $\mathcal{L}^n$-a.e. for $k = 1,\ldots,n,$, and consequently $f \in W_{\text{loc}}^{1,p}(\mathbb{R}^n)$. $\square$

4.10 References and notes

Our main sources for Sobolev functions are Gilbarg–Trudinger [G-T, Chapter 7] and Federer–Ziemer [F-Z]. Many of these calculations appear also in [E2].

See [G-T, Sections 7.2 and 7.3] for mollification and local approximation by smooth functions. Theorem 4.2 is from [G-T, Section 7.6] and Theorem 4.3 is based upon [G-T, Theorem 7.25]. The product and chain rules are in [G-T, Section 7.4]. See also [G-T, Section 7.12] for extensions. Various Sobolev-type inequalities are in [G-T, Section 7.7]. Lemma 4.1 in Section 4.5 is a variant of [G-T, Lemma 7.16]. Compactness assertions are in [G-T, Section 7.10].

We follow [F-Z] (cf. Maz'ja [M] and Ziemer [Z]) in our treatment of capacity. Theorems 4.14–4.17 in Section 4.7 are from [F-Z], as are all the results in Section 4.8.

Much more information about capacity is available in the comprehensive books [Z] and [M]. Maly–Ziemer [M-Z] provides applications to regularity issues for solutions of elliptic PDE. Maly–Swanson–Ziemer [M-S-Z] discuss the coarea formula for Sobolev functions, and Figalli [Fg] presents a fairly simple proof of the Morse-Sard Theorem in Sobolev spaces.

Chapter 5

Functions of Bounded Variation, Sets of Finite Perimeter

We introduce and study next functions on $\mathbb{R}^n$ of *bounded variation*, which is to say functions whose weak first partial derivatives are Radon measures. This is essentially the weakest measure theoretic sense in which a function can be differentiable. We also investigate sets E having *finite perimeter*, meaning that the indicator function χ_E is BV.

It is not so obvious that any of the usual rules of calculus apply to functions whose first derivatives are merely measures. The principal goal of this chapter is therefore to study this problem, investigating in particular the extent to which a BV function is "measure theoretically C^1" and a set of finite perimeter has "a C^1 boundary measure theoretically."

Our study initially, in Sections 5.1 through 5.4, parallels the corresponding investigation of Sobolev functions in Chapter 4. Section 5.5 extends the coarea formula to the BV setting and Section 5.6 generalizes the Gagliardo–Nirenberg–Sobolev inequality. Sections 5.7, 5.8, and 5.11 analyze the measure theoretic boundary of a set of finite perimeter, and most importantly establish a version of the Gauss–Green Theorem. This investigation is carried over in Sections 5.9 and 5.10 to study the fine, pointwise properties of BV functions.

5.1 Definitions, Structure Theorem

Throughout this chapter, U denotes an open subset of $\mathbb{R}^n$.

DEFINITION 5.1.

(i) *A function* $f \in L^1(U)$ *has* **bounded variation** *in U if*

$$\sup \left\{ \int_U f \operatorname{div} \phi \, dx \mid \phi \in C_c^1(U; \mathbb{R}^n), |\phi| \leq 1 \right\} < \infty.$$

We write

$$BV(U)$$

to denote the space of functions of bounded variation in U. We do not identify two BV functions that agree $\mathcal{L}^n$-a.e.

(ii) *An $\mathcal{L}^n$-measurable subset $E \subset \mathbb{R}^n$ has* **finite perimeter** *in U if*

$$\chi_E \in BV(U).$$

It is convenient to introduce also local versions of these concepts:

DEFINITION 5.2.

(i) *A function $f \in L_{\text{loc}}^1(U)$ has* **locally bounded variation** *in U if for each open set $V \subset\subset U$,*

$$\sup \left\{ \int_V f \operatorname{div} \phi \, dx \mid \phi \in C_c^1(V; \mathbb{R}^n), |\phi| \leq 1 \right\} < \infty.$$

We write

$$BV_{\text{loc}}(U)$$

to denote the space of such functions.

(ii) *An $\mathcal{L}^n$-measurable subset $E \subset \mathbb{R}^n$ has* **locally finite perimeter** *in U if*

$$\chi_E \in BV_{\text{loc}}(U).$$

Some examples will be presented later, after we establish this general structure assertion.

THEOREM 5.1 (Structure Theorem for BV_{loc} functions). *Assume that $f \in BV_{\text{loc}}(U)$.*

Then there exist a Radon measure μ on U and a μ-measurable function

$$\sigma : U \to \mathbb{R}^n$$

such that

(i) $|\sigma(x)| = 1 \quad \mu$-a.e., and

(ii) for all $\phi \in C_c^1(U; \mathbb{R}^n)$, we have

$$\int_U f \operatorname{div} \phi \, dx = - \int_U \phi \cdot \sigma \, d\mu.$$

As we will discuss in detail later, the Structure Theorem asserts that the weak first partial derivatives of a BV function are Radon measures.

Proof. 1. Define the linear functional $L : C_c^1(U; \mathbb{R}^n) \to \mathbb{R}$ by

$$L(\phi) := - \int_U f \operatorname{div} \phi \, dx$$

for $\phi \in C_c^1(U; \mathbb{R}^n)$. Since $f \in BV_{\mathrm{loc}}(U)$, we have

$$\sup \left\{ L(\phi) \mid \phi \in C_c^1(V; \mathbb{R}^n), |\phi| \le 1 \right\} =: C(V) < \infty$$

for each open set $V \subset\subset U$, and consequently

$$|L(\phi)| \le C(V) \|\phi\|_{L^\infty} \qquad (\star)$$

for $\phi \in C_c^1(V; \mathbb{R}^n)$.

2. Select any compact set $K \subset U$, and then choose an open set V such that $K \subset V \subset\subset U$. For each $\phi \in C_c(U; \mathbb{R}^n)$ with $\operatorname{spt} \phi \subseteq K$, choose $\phi_k \in C_c^1(V; \mathbb{R}^n)$ $(k = 1, \dots)$ so that $\phi_k \to \phi$ uniformly on V. Define

$$\bar{L}(\phi) := \lim_{k \to \infty} L(\phi_k);$$

according to $(\star)$ this limit exists and is independent of the choice of the sequence $\{\phi_k\}_{k=1}^\infty$ converging to ϕ. Thus L uniquely extends to a linear functional

$$\bar{L} : C_c(U; \mathbb{R}^n) \to \mathbb{R}$$

and

$$\sup \left\{ L(\phi) \mid \phi \in C_c(U; \mathbb{R}^n), |\phi| \le 1, \operatorname{spt} \phi \subseteq K \right\} < \infty$$

for each compact set $K \subset U$. The Riesz Representation Theorem now completes the proof. $\qquad \square$

NOTATION

(i) If $f \in BV_{\mathrm{loc}}(U)$, we will henceforth write

$$\|Df\|$$

for the measure μ, and

$$[Df] := \|Df\| \llcorner \sigma.$$

Hence assertion (ii) in Theorem 5.1 reads

$$\int_U f \operatorname{div} \phi \, dx = -\int_U \phi \cdot \sigma \, d\|Df\| = -\int_U \phi \cdot d[Df]$$

for all $\phi \in C_c^1(U; \mathbb{R}^n)$.

(ii) Similarly, if $f = \chi_E$ and E is a set of locally finite perimeter in U, we will hereafter write

$$\|\partial E\|$$

for the measure μ, and

$$\nu_E := -\sigma.$$

Consequently,

$$\int_E \operatorname{div} \phi \, dx = \int_U \phi \cdot \nu_E \, d\|\partial E\|$$

for all $\phi \in C_c^1(U; \mathbb{R}^n)$.

MORE NOTATION If $f \in BV_{\mathrm{loc}}(U)$, we write

$$\mu^i = \|Df\| \llcorner \sigma^i \quad (i = 1, \dots, n)$$

for $\sigma = (\sigma^1, \dots, \sigma^n)$. By Lebesgue's Decomposition Theorem 1.31, we may further set

$$\mu^i = \mu_{\mathrm{ac}}^i + \mu_{\mathrm{s}}^i,$$

where

$$\mu_{\mathrm{ac}}^i \ll \mathcal{L}^n, \qquad \mu_{\mathrm{s}}^i \perp \mathcal{L}^n.$$

Then

$$\mu_{\mathrm{ac}}^i = \mathcal{L}^n \llcorner f_i$$

for functions $f_i \in L^1_{loc}(U)$ $(i = 1, \ldots, n)$. We write

$$
\begin{aligned}
f_{x_i} &:= f_i \quad (i = 1, \ldots, n) \\
Df &:= (f_{x_1}, \ldots, f_{x_n}), \\
[Df]_{ac} &:= (\mu^1_{ac}, \ldots, \mu^n_{ac}) = \mathcal{L}^n \llcorner Df, \\
[Df]_s &:= (\mu^1_s, \ldots, \mu^n_s).
\end{aligned}
$$

Thus

$$
[Df] = [Df]_{ac} + [Df]_s = \mathcal{L}^n \llcorner Df + [Df]_s;
$$

so that $Df \in L^1_{loc}(U; \mathbb{R}^n)$ is the density of the absolutely continuous part of $[Df]$.

Remark. Compare this with the notation for convex functions set forth in Section 6.3. □

Remarks.

(i) $\|Df\|$ is the **variation measure** of f; $\|\partial E\|$ is the **perimeter measure** of E; and $\|\partial E\|(U)$ is the **perimeter of E in U.**

(ii) If $f \in BV_{loc}(U) \cap L^1(U)$, then $f \in BV(U)$ if and only if $\|Df\|(U) < \infty$. In this case we define

$$
\|f\|_{BV(U)} := \|f\|_{L^1(U)} + \|Df\|(U).
$$

(iii) From the proof of the Riesz Representation Theorem 1.38, we see

$$
\|Df\|(V) = \sup \left\{ \int_V f \operatorname{div} \phi \, dx \mid \phi \in C^1_c(V; \mathbb{R}^n), |\phi| \leq 1 \right\},
$$

$$
\|\partial E\|(V) = \sup \left\{ \int_E f \operatorname{div} \phi \, dx \mid \phi \in C^1_c(V; \mathbb{R}^n), |\phi| \leq 1 \right\}
$$

for each open $V \subset\subset U$.

□

EXAMPLE. Assume $f \in W^{1,1}_{loc}(U)$. Then for each open set $V \subset\subset U$ and each $\phi \in C^1_c(V; \mathbb{R}^n)$, with $|\phi| \leq 1$, we have

$$
\int_U f \operatorname{div} \phi \, dx = - \int_U Df \cdot \phi \, dx \leq \int_V |Df| \, dx < \infty.
$$

Thus $f \in BV_{\text{loc}}(U)$. Furthermore,

$$\|Df\| = \mathcal{L}^n \llcorner |Df|;$$

and $\mathcal{L}^n$-a.e. we have

$$\sigma = \begin{cases} \frac{Df}{|Df|} & \text{if } Df \neq 0 \\ 0 & \text{if } Df = 0. \end{cases}$$

Hence

$$W_{\text{loc}}^{1,1}(U) \subset BV_{\text{loc}}(U),$$

and similarly

$$W^{1,1}(U) \subset BV(U).$$

In particular,

$$W_{\text{loc}}^{1,p}(U) \subset BV_{\text{loc}}(U) \quad \text{for } 1 \leq p \leq \infty.$$

Hence, *each Sobolev function has locally bounded variation.* □

EXAMPLE. Assume E is a smooth, open subset of $\mathbb{R}^n$ and $\mathcal{H}^{n-1}(\partial E \cap K) < \infty$ for each compact set $K \subset U$. Then for V and ϕ as above,

$$\int_E \operatorname{div} \phi \, dx = \int_{\partial E} \phi \cdot \nu \, d\mathcal{H}^{n-1},$$

ν denoting the outward unit normal along ∂E.

Hence

$$\int_E \operatorname{div} \phi \, dx = \int_{\partial E \cap V} \phi \cdot \nu \, d\mathcal{H}^{n-1} \leq \mathcal{H}^{n-1}(\partial E \cap V) < \infty.$$

Thus E has locally finite perimeter in U. Furthermore,

$$\|\partial E\|(U) = \mathcal{H}^{n-1}(\partial E \cap U)$$

and

$$\nu_E = \nu \quad \mathcal{H}^{n-1}\text{-a.e. on } \partial E \cap U.$$

Thus $\|\partial E\|(U)$ measures the "size" of ∂E in U. Since $\chi_E \notin W_{\text{loc}}^{1,1}(U)$ (according, for instance, to Theorem 4.21), we see

$$W_{\text{loc}}^{1,1}(U) \subsetneq BV_{\text{loc}}(U), \ W^{1,1}(U) \subsetneq BV(U).$$

So *not every function of locally bounded variation is a Sobolev function.*
□

Remark. Indeed, if $f \in BV_{\mathrm{loc}}(U)$, we can write as above

$$[Df] = [Df]_{\mathrm{ac}} + [Df]_{\mathrm{s}} = \mathcal{L}^n \llcorner Df + [Df]_{\mathrm{s}}.$$

Consequently, $f \in BV_{\mathrm{loc}}(U)$ belongs to $W^{1,p}_{\mathrm{loc}}(U)$ if and only if

$$f \in L^p_{\mathrm{loc}}(U), \quad [Df]_{\mathrm{s}} = 0, \quad Df \in L^p_{\mathrm{loc}}(U).$$

The study of BV functions is rather more subtle than the study of Sobolev functions, since we must always keep track of the singular part $[Df]_{\mathrm{s}}$ of the vector measure $[Df]$. □

5.2 Approximation and compactness

5.2.1 Lower semicontinuity

THEOREM 5.2 (Lower semicontinuity of variation measure).
Suppose $f_k \in BV(U)$ $(k = 1, \dots)$ and

$$f_k \to f \quad in \ L^1_{\mathrm{loc}}(U).$$

Then

$$\|Df\|(U) \le \liminf_{k \to \infty} \|Df_k\|(U).$$

Proof. Let $\phi \in C^1_c(U; \mathbb{R}^n)$, $|\phi| \le 1$. Then

$$\int_U f \operatorname{div} \phi \, dx = \lim_{k \to \infty} \int_U f_k \operatorname{div} \phi \, dx$$

$$= -\lim_{k \to \infty} \int_U \phi \cdot \sigma_k \, d\|Df_k\|$$

$$\le \liminf_{k \to \infty} \|Df_k\|(U).$$

Thus

$$\|Df\|(U) = \sup \left\{ \int_U f \operatorname{div} \phi \, dx \ \middle| \ \phi \in C^1_c(U; \mathbb{R}^n), |\phi| \le 1 \right\}$$

$$\le \liminf_{k \to \infty} \|Df_k\|(U).$$

□

5.2.2 Approximation by smooth functions

THEOREM 5.3 (Local approximation by smooth functions).
Assume $f \in BV(U)$.

Then there exist functions $\{f_k\}_{k=1}^{\infty} \subset BV(U) \cap C^{\infty}(U)$ *such that*

(i) $f_k \to f$ *in* $L^1(U)$ *and*

(ii) $\|Df_k\|(U) \to \|Df\|(U)$ *as* $k \to \infty$.

Remark. Compare with Theorem 4.2 in Section 4.2. Note very carefully that we do *not* assert $\|D(f_k - f)\|(U) \to 0$. $\qquad\qquad\square$

Proof. 1. Fix $\epsilon > 0$. Given a positive integer m, define for $k = 1, \ldots$ the open sets

$$U_k := \left\{ x \in U \mid \text{dist}(x, \partial U) > \frac{1}{m+k} \right\} \cap B^0(0, k+m).$$

Next, choose m so large that

$$\|Df\|(U - U_1) < \epsilon. \qquad\qquad (\star)$$

Set $U_0 := \emptyset$ and define

$$V_k := U_{k+1} - \bar{U}_{k-1} \qquad (k = 1, \ldots).$$

Let $\{\zeta_k\}_{k=1}^{\infty}$ be a sequence of smooth functions such that

$$\zeta_k \in C_c^{\infty}(V_k), \ 0 \le \zeta_k \le 1 \ (k = 1, \ldots), \ \sum_{k=1}^{\infty} \zeta_k \equiv 1 \quad \text{on } U.$$

Fix the mollifier η_ϵ, as described in Section 4.2. Then for each k, select $\epsilon_k > 0$ so small that

$$\begin{cases} \text{spt}(\eta_{\epsilon_k} * (f\zeta_k)) \subseteq V_k \\ \int_U |\eta_{\epsilon_k} * (f\zeta_k) - f\zeta_k| \, dx < \frac{\epsilon}{2^k}, \\ \int_U |\eta_{\epsilon_k} * (fD\zeta_k) - fD\zeta_k| \, dx < \frac{\epsilon}{2^k}. \end{cases} \qquad (\star\star)$$

Define

$$f_\epsilon := \sum_{k=1}^{\infty} \eta_{\epsilon_k} * (f\zeta_k).$$

In some neighborhood of each point $x \in U$ there are only finitely many nonzero terms in this sum; hence $f_\epsilon \in C^{\infty}(U)$.

2. Since also

$$f = \sum_{k=1}^{\infty} f\zeta_k,$$

(⋆⋆) implies

$$\|f_\epsilon - f\|_{L^1(U)} \le \sum_{k=1}^{\infty} \int_U |\eta_{\epsilon_k} * (f\zeta_k) - f\zeta_k| \, dx < \epsilon.$$

Consequently, $f_\epsilon \to f$ in $L^1(U)$ as $\epsilon \to 0$; and therefore Theorem 5.2 implies

$$\|Df\|(U) \le \liminf_{\epsilon \to 0} \|Df_\epsilon\|(U). \qquad (\star\star\star)$$

3. Now let $\phi \in C_c^1(U; \mathbb{R}^n), |\phi| \le 1$. Then

$$
\begin{aligned}
\int_U f_\epsilon \operatorname{div} \phi \, dx &= \sum_{k=1}^{\infty} \int_U \eta_{\epsilon_k} * (f\zeta_k) \operatorname{div} \phi \, dx \\
&= \sum_{k=1}^{\infty} \int_U f\zeta_k \operatorname{div}(\eta_{\epsilon_k} * \phi) \, dx \\
&= \sum_{k=1}^{\infty} \int_U f \operatorname{div}(\zeta_k(\eta_{\epsilon_k} * \phi)) \, dx \\
&\quad - \sum_{k=1}^{\infty} \int_U f \, D\zeta_k \cdot (\eta_{\epsilon_k} * \phi) \, dx \\
&= \sum_{k=1}^{\infty} \int_U f \operatorname{div}(\zeta_k(\eta_{\epsilon_k} * \phi)) \, dx \\
&\quad - \sum_{k=1}^{\infty} \int_U \phi \cdot (\eta_{\epsilon_k} * (f \, D\zeta_k) - f \, D\zeta_k) \, dx \\
&=: I_1^\epsilon + I_2^\epsilon.
\end{aligned}
$$

Here we used the fact $\sum_{k=1}^{\infty} D\zeta_k \equiv 0$ in U.

4. Note that

$$|\zeta_k(\eta_{\epsilon_k} * \phi)| \le 1 \quad (k = 1, \dots),$$

and that each point in U belongs to at most three of the sets $\{V_k\}_{k=1}^{\infty}$. Thus

$$|I_1^\epsilon| = \left| \int_U f \operatorname{div}(\zeta_1(\eta_{\epsilon_1} * \phi)) \, dx + \sum_{k=2}^{\infty} \int_U f \operatorname{div}(\zeta_k \eta_{\epsilon_k} * \phi) \, dx \right|$$

$$\leq |Df|(U) + \sum_{k=2}^{\infty} |Df|(V_k)$$

$$\leq |Df|(U) + 3|Df|(U - U_1)$$

$$\leq |Df|(U) + 3\epsilon \quad \text{by } (\star) .$$

On the other hand, $(\star\star)$ implies

$$|I_2^\epsilon| < \epsilon.$$

Therefore

$$\int_U f_\epsilon \operatorname{div} \phi \, dx \leq \|Df\|(U) + 4\epsilon,$$

and so

$$\|Df_\epsilon\|(U) \leq \|Df\|(U) + 4\epsilon.$$

This estimate and $(\star\star\star)$ complete the proof. $\qquad\square$

THEOREM 5.4 (Weak approximation of derivatives). *For f_k in the statement of Theorem 5.3, define the (vector-valued) Radon measure*

$$\mu_k(B) := \int_{B \cap U} Df_k \, dx$$

for each Borel set $B \subseteq \mathbb{R}^n$. Set also

$$\mu(B) := \int_{B \cap U} d[Df].$$

Then

$$\mu_k \rightharpoonup \mu$$

weakly in the sense of (vector-valued) Radon measures on $\mathbb{R}^n$.

Proof. Fix $\phi \in C_c^1(\mathbb{R}^n; \mathbb{R}^n)$ and $\epsilon > 0$. Define $U_1 \subset\subset U$ as in the previous proof and choose a smooth cutoff function ζ satisfying

$$\zeta \equiv 1 \text{ on } U_1, \ \operatorname{spt} \zeta \subset U, \ 0 \leq \zeta \leq 1.$$

Then

$$\int_{\mathbb{R}^n} \phi \, d\mu_k = \int_U \phi \cdot Df_k \, dx$$

$$= \int_U \zeta \phi \cdot Df_k \, dx + \int_U (1 - \zeta) \phi \cdot Df_k \, dx \qquad (\star)$$

$$= -\int_U \operatorname{div}(\zeta \phi) f_k \, dx + \int_U (1 - \zeta) \phi \cdot Df_k \, dx.$$

Since $f_k \to f \in L^1(U)$, the first term in $(\star)$ converges to

$$
\begin{aligned}
-\int_U \operatorname{div}(\zeta\phi) f \, dx &= \int_U \zeta\phi \cdot d[Df] \\
&= \int_U \phi \cdot d[Df] + \int_U (\zeta - 1)\phi \cdot d[Df].
\end{aligned}
\tag{$\star\star$}
$$

The last term in $(\star\star)$ is estimated by

$$
\|\phi\|_{L^\infty} \|Df\|(U - U_1) \le C\epsilon.
$$

Using Theorem 5.3, we see that for k large enough, we control the last term in $(\star)$ by

$$
\|\phi\|_{L^\infty} \|Df_k\|(U - U_1) \le C\epsilon.
$$

Hence

$$
\left| \int_{\mathbb{R}^n} \phi \, d\mu_k - \int_{\mathbb{R}^n} \phi \, d\mu \right| \le C\epsilon
$$

for all sufficiently large k. $\square$

5.2.3 Compactness

THEOREM 5.5 (Compactness for BV functions). *Let $U \subset \mathbb{R}^n$ be open and bounded, with Lipschitz boundary ∂U. Assume $\{f_k\}_{k=1}^\infty$ is a sequence in $BV(U)$ satisfying*

$$
\sup_k \|f_k\|_{BV(U)} < \infty.
$$

Then there exists a subsequence $\{f_{k_j}\}_{j=1}^\infty$ and a function $f \in BV(U)$ such that

$$
f_{k_j} \to f \quad in \ L^1(U)
$$

as $j \to \infty$.

Proof. For $k = 1, 2, \ldots$, choose $g_k \in C^\infty(U)$ so that

$$
\int_U |f_k - g_k| \, dx < \frac{1}{k}, \ \sup_k \int_U |Dg_k| \, dx < \infty;
\tag{$\star$}
$$

such functions exist according to Theorem 5.3. By the remark following Theorem 4.11 in Section 4.6 there exist $f \in L^1(U)$ and a subsequence $\{g_{k_j}\}_{j=1}^\infty$ such that $g_{k_j} \to f$ in $L^1(U)$. But then $(\star)$ implies also that $f_{k_j} \to f$ in $L^1(U)$. According to Theorem 5.2, $f \in BV(U)$. $\square$

5.3   Traces

Assume for this section that U is open and bounded, with Lipschitz boundary ∂U. Observe that since each part of ∂U is locally the graph of a Lipschitz continuous function γ, the outer unit normal ν exists $\mathcal{H}^{n-1}$ almost everywhere on ∂U, according to Rademacher's Theorem.

We now extend to BV functions the notion of trace, defined in Section 4.3 for Sobolev functions.

THEOREM 5.6 (Traces of BV functions). *Assume U is open and bounded, with ∂U Lipschitz continuous. There exists a bounded linear mapping*

$$T : BV(U) \to L^1(\partial U; \mathcal{H}^{n-1})$$

such that

$$\int_U f \operatorname{div} \phi \, dx = - \int_U \phi \cdot d\,[Df] + \int_{\partial U} (\phi \cdot \nu)\, Tf \, d\mathcal{H}^{n-1} \qquad (\star)$$

for all $f \in BV(U)$ and $\phi \in C^1(\mathbb{R}^n; \mathbb{R}^n)$.

The point is that we do not now require ϕ to vanish near ∂U.

DEFINITION 5.3. *The function Tf, which is uniquely defined up to sets of $\mathcal{H}^{n-1} \llcorner \partial U$ measure zero, is called the **trace** of f on ∂U.*

We interpret Tf as the "boundary values" of f on ∂U.

Remark. If $f \in W^{1,1}(U) \subset BV(U)$, the definition of trace above and that from Section 4.3 agree.   □

Proof. 1. First we introduce some notation:

(a) Given $x = (x_1, \ldots, x_n) \in \mathbb{R}^n$, let us write $x = (x', x_n)$ for $x' := (x_1, \ldots, x_{n-1}) \in \mathbb{R}^{n-1}$, $x_n \in \mathbb{R}$. Similarly, we write $y = (y', y_n)$.

(b) Given $x \in \mathbb{R}^n$ and $r, h > 0$, define the open cylinder

$$C(x, r, h) := \{y \in \mathbb{R}^n \mid |y' - x'| < r, |y_n - x_n| < h\}.$$

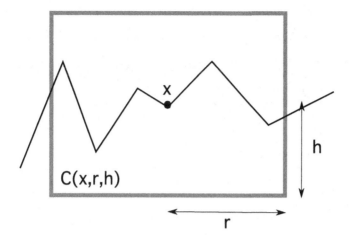

Now since ∂U is Lipschitz continuous, for each point $x \in \partial U$ there exist r, $h > 0$ and a Lipschitz continuous function $\gamma : \mathbb{R}^{n-1} \to \mathbb{R}$ such that

$$\max_{|x'-y'| \leq r} |\gamma(y') - x_n| \leq \frac{h}{4};$$

and, upon rotation and relabeling the coordinate axes if necessary, we have

$$U \cap C(x, r, h) = \{y \mid |x' - y'| < r, \gamma(y') < y_n < x_n + h\}.$$

2. Assume for the time being $f \in BV(U) \cap C^\infty(U)$. Pick $x \in \partial U$ and choose r, h, γ, etc., as above. Write

$$C := C(x, r, h).$$

If $0 < \epsilon < \frac{h}{2}$ and $y \in \partial U \cap C$, we define

$$f_\epsilon(y) := f(y', \gamma(y') + \epsilon).$$

Let us also set

$$C_{\epsilon,\delta} := \{y \in C \mid \gamma(y') + \delta < y_n < \gamma(y') + \epsilon\}$$

for $0 \leq \delta < \epsilon < \frac{h}{2}$, and define $C_\epsilon := C_{\epsilon,0}$. Write $C^\epsilon := (C \cap U) - C_\epsilon$.
Then

$$|f_\delta(y) - f_\epsilon(y)| \leq \int_\delta^\epsilon |f_{x_n}(y', \gamma(y') + t)| \, dt$$

$$\leq \int_\delta^\epsilon |Df(y', \gamma(y') + t)| \, dt.$$

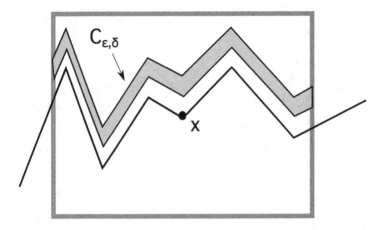

Consequently, since γ is Lipschitz continuous, the area formula from Section 3.3 implies

$$\int_{\partial U \cap C} |f_\delta - f_\epsilon| \, d\mathcal{H}^{n-1} \leq C \int_{C_{\epsilon,\delta}} |Df| \, dy = C|Df|(C_{\epsilon,\delta}).$$

Therefore $\{f_\epsilon\}_{\epsilon > 0}$ is Cauchy in $L^1(\partial U \cap C; \mathcal{H}^{n-1})$, and thus the limit

$$Tf := \lim_{\epsilon \to 0} f_\epsilon$$

exists in this space. Furthermore, our passing to limits as $\delta \to 0$ in the foregoing inequality yields the bound

$$\int_{\partial U \cap C} |Tf - f_\epsilon| \, d\mathcal{H}^{n-1} \leq C\|Df\|(C_\epsilon). \qquad (\star\star)$$

Next fix $\phi \in C_c^1(C; \mathbb{R}^n)$. Then

$$\int_{C^\epsilon} f \operatorname{div} \phi \, dy = -\int_{C^\epsilon} \phi \cdot Df \, dy + \int_{\partial U \cap C} f_\epsilon \phi_\epsilon \cdot \nu \, d\mathcal{H}^{n-1}.$$

Let $\epsilon \to 0$ to find

$$\int_{U \cap C} f \operatorname{div} \phi \, dy = -\int_{U \cap C} \phi \cdot \sigma \, d\|Df\| + \int_{\partial U \cap C} Tf \phi \cdot \nu \, d\mathcal{H}^{n-1}. \qquad (\star\star\star)$$

3. Since ∂U is compact, we can cover ∂U with finitely many cylinders $C_i = C(x_i, r_i, h_i)$ $(i = 1, \ldots, N)$ for which assertions analogous to

$(\star\star)$ and $(\star\star\star)$ hold. An argument using a partition of unity subordinate to the $\{C_i\}_{i=1}^{\infty}$ then establishes formula $(\star)$.

Observe also that $(\star\star\star)$ shows the definition of "Tf" to be the same (up to sets of $\mathcal{H}^{n-1}\llcorner\partial U$ measure zero) on any part of ∂U that happens to lie in two or more of the cylinders C_i.

4. Now assume only $f \in BV(U)$. In this general case, choose $f_k \in BV(U) \cap C^{\infty}(U)(k = 1, 2, \dots)$ such that

$$f_k \to f \text{ in } L^1(U), \|Df_k\|(U) \to \|Df\|(U)$$

and

$$\mu_k \rightharpoonup \mu \quad \text{weakly,}$$

where the measures $\{\mu_k\}_{k=1}^{\infty}$, μ are defined as in Theorem 5.4.

5. *Claim*: $\{Tf_k\}_{k=1}^{\infty}$ is a Cauchy sequence in $L^1(\partial U; \mathcal{H}^{n-1})$.

Proof of claim: Choose a cylinder C as in the previous part of the proof. Fix $\epsilon > 0$, $y \in \partial U \cap C$, and then define

$$f_k^\epsilon(y) := \frac{1}{\epsilon}\int_0^\epsilon f_k(y', \gamma(y') + t)\, dt = \frac{1}{\epsilon}\int_0^\epsilon (f_k)^t(y)\, dt.$$

Then $(\star\star)$ implies

$$\int_{\partial U \cap C} |Tf_k - f_k^\epsilon|\, d\mathcal{H}^{n-1} \le \frac{1}{\epsilon}\int_0^\epsilon \int_{\partial U \cap C} |Tf_k - (f_k)^t|\, d\mathcal{H}^{n-1} dt$$

$$\le C\|Df_k\|(C_\epsilon).$$

Thus

$$\int_{\partial U \cap C} |Tf_k - Tf_l|\, d\mathcal{H}^{n-1} \le \int_{\partial U \cap C} |Tf_k - f_k^\epsilon|\, d\mathcal{H}^{n-1}$$

$$+ \int_{\partial U \cap C} |Tf_l - f_l^\epsilon|\, d\mathcal{H}^{n-1}$$

$$+ \int_{\partial U \cap C} |f_k^\epsilon - f_l^\epsilon|\, d\mathcal{H}^{n-1}$$

$$\le C(\|Df_k\| + \|Df_l\|)(C_\epsilon)$$

$$+ \frac{C}{\epsilon}\int_{C_\epsilon} |f_k - f_l|\, dy,$$

and so

$$\limsup_{k,l\to\infty} \int_{\partial U \cap C} |T f_k - T f_l|\, d\mathcal{H}^{n-1} \le C\|Df\|\,(\bar{C}_\epsilon \cap U).$$

Since the quantity on the right-hand side goes to zero as $\epsilon \to 0$, the claim is proved.

6. In view of the claim, we may define

$$Tf := \lim_{k\to\infty} T f_k;$$

this definition does not depend on the particular choice of approximating sequence. Finally, formula $(\star)$ holds for each f_k and thus also holds in the limit for f. $\qquad\square$

THEOREM 5.7 (Local properties of traces). *Assume U is open, bounded, with ∂U Lipschitz continuous. Suppose also $f \in BV(U)$. Then for $\mathcal{H}^{n-1}$-a.e. $x \in \partial U$,*

$$\lim_{r\to 0} \fint_{B(x,r)\cap U} |f - Tf(x)|\, dy = 0,$$

and so

$$Tf(x) = \lim_{r\to 0} \fint_{B(x,r)\cap U} f\, dy.$$

Remark. Thus in particular if $f \in BV(U) \cap C(\bar{U})$, then

$$Tf = f|_{\partial U} \quad \mathcal{H}^{n-1}\text{-a.e.}$$

$\qquad\square$

Proof. 1. *Claim*: For $\mathcal{H}^{n-1}$-a.e. $x \in \partial U$,

$$\lim_{r\to 0} \frac{\|Df\|(B(x,r)\cap U)}{r^{n-1}} = 0.$$

Proof of claim: Fix $\sigma > 0$, $\delta > \epsilon > 0$, and let

$$A_\sigma := \left\{ x \in \partial U \mid \limsup_{r\to 0} \frac{\|Df\|(B(x,r))\cap U)}{r^{n-1}} > \sigma \right\}.$$

Then for each $x \in A_\sigma$, there exists $0 < r < \epsilon$ such that

$$\frac{\|Df\|(B(x,r)\cap U)}{r^{n-1}} \ge \sigma. \qquad (\star)$$

Using Vitali's Covering Theorem, we obtain a countable collection of disjoint balls $\{B(x_i, r_i)\}_{i=1}^{\infty}$ satisfying $(\star)$, such that

$$A_\sigma \subseteq \bigcup_{i=1}^{\infty} B(x_i, 5r_i).$$

Then

$$H_{10\delta}^{n-1}(A_\sigma) \leq \sum_{i=1}^{\infty} \alpha(n-1)(5r_i)^{n-1}$$

$$\leq \frac{C}{\gamma} \sum_{i=1}^{\infty} \|Df\|(B(x_i, r_i) \cap U)$$

$$\leq C\|Df\|(U^\epsilon),$$

where

$$U^\epsilon := \{x \in U \mid \operatorname{dist}(x, \partial U) < \epsilon\}.$$

Send $\epsilon \to 0$ to find $H_{10\delta}^{n-1}(A_\sigma) = 0$ for all $\delta > 0$.

2. Now fix a point $x \in \partial U$ such that

$$\lim_{r \to 0} \frac{\|Df\|(B(x, r) \cap U)}{r^{n-1}} = 0, \qquad (\star\star)$$

$$\lim_{r \to 0} \fint_{B(x,r) \cap \partial U} |Tf - Tf(x)| \, d\mathcal{H}^{n-1} = 0.$$

According to the claim and the Lebesgue–Besicovitch Differentiation Theorem, $\mathcal{H}^{n-1}$-a.e. $x \in \partial U$ will do. Let $h = h(r) := 2 \max(1, 4\operatorname{Lip}(\gamma))r$, and consider the cylinders

$$C(r) = C(x, r, h).$$

Observe that for sufficiently small r, the cylinders $C(r)$ work in place of the cylinder C in the previous proof. Thus estimates similar to those developed in that proof show

$$\int_{\partial U \cap C(r)} |Tf - f_\epsilon| \, d\mathcal{H}^{n-1} \leq C\|Df\|(C(r) \cap U),$$

where

$$f_\epsilon(y) := f(y', \gamma(y') + \epsilon) \quad \left(y \in C(r) \cap \partial U, 0 < \epsilon < \frac{h(r)}{2}\right).$$

Consequently, we may employ the coarea formula to estimate

$$\int_{B(x,r)\cap U} |Tf(y',\gamma(y')) - f(y)| \, dy \leq Cr\|Df\|(C(r)\cap U).$$

Hence we compute

$$\fint_{B(x,r)\cap U} |f(y) - Tf(x)| \, dy \leq \frac{C}{r^{n-1}} \int_{C(r)\cap \partial U} |Tf - Tf(x)| \, d\mathcal{H}^{n-1}$$

$$+ \frac{C}{r^n} \int_{B(x,r)\cap U} |Tf(y',\gamma(y')) - f(y)| \, dy$$

$$\leq o(1) + \frac{C}{r^{n-1}}\|Df\|(C(r)\cap U)$$

$$= o(1) \quad \text{as } r \to 0$$

by $(\star\star)$. □

5.4　Extensions

THEOREM 5.8 (Extensions of BV functions). *Assume $U \subset \mathbb{R}^n$ is open and bounded, with ∂U Lipschitz continuous. Let $f_1 \in BV(U)$, $f_2 \in BV(\mathbb{R}^n - \bar{U})$.*
　　Define

$$\bar{f}(x) := \begin{cases} f_1(x) & x \in U \\ f_2(x) & x \in \mathbb{R}^n - \bar{U}. \end{cases}$$

Then

$$\bar{f} \in BV(\mathbb{R}^n)$$

and

$$\|D\bar{f}\|(\mathbb{R}^n) = \|Df_1\|(U) + \|Df_2\|(\mathbb{R}^n - \bar{U}) + \int_{\partial U} |Tf_1 - Tf_2| \, d\mathcal{H}^{n-1}.$$

Remark. In particular, under the stated assumptions on U, the extension

$$Ef := \begin{cases} f & \text{on } U \\ 0 & \text{on } \mathbb{R}^n - U \end{cases}$$

belongs to $BV(\mathbb{R}^n)$ provided $f \in BV(U)$ and the set U has finite perimeter, with $\|\partial U\|(\mathbb{R}^n) = \mathcal{H}^{n-1}(\partial U)$. $\square$

Proof. 1. Let $\phi \in C_c^1(\mathbb{R}^n, \mathbb{R}^n), |\phi| \leq 1$. Then

$$\int_{\mathbb{R}^n} \bar{f} \operatorname{div} \phi \, dx = \int_U f_1 \operatorname{div} \phi \, dx + \int_{\mathbb{R}^n - \bar{U}} f_2 \operatorname{div} \phi \, dx$$

$$= -\int_U \phi \cdot d[Df_1] - \int_{\mathbb{R}^n - \bar{U}} \phi \cdot d[Df_2]$$

$$+ \int_{\partial U} (Tf_1 - Tf_2)\phi \cdot \nu \, d\mathcal{H}^{n-1}$$

$$\leq \|Df_1\|(U) + \|Df_2\|(\mathbb{R}^n - \bar{U})$$

$$+ \int_{\partial U} |Tf_1 - Tf_2| \, d\mathcal{H}^{n-1}.$$

Thus $\bar{f} \in BV(\mathbb{R}^n)$ and

$$\|D\bar{f}\|(\mathbb{R}^n) \leq \|Df_1\|(U) + \|Df_2\|(\mathbb{R}^n - \bar{U}) + \int_{\partial U} |Tf_1 - Tf_2| \, d\mathcal{H}^{n-1}.$$

2. To show equality, observe that

$$-\int_{\mathbb{R}^n} \phi \cdot d[D\bar{f}] = -\int_U \phi \cdot d[Df_1] - \int_{\mathbb{R}^n - \bar{U}} \phi \cdot d[Df_2]$$

$$+ \int_{\partial U} (Tf_1 - Tf_2)\phi \cdot \nu \, d\mathcal{H}^{n-1} \qquad (\star)$$

for all $\phi \in C_c^1(\mathbb{R}^n; \mathbb{R}^n)$. Thus

$$[D\bar{f}] = \begin{cases} [Df_1] & \text{on } U \\ [Df_2] & \text{on } \mathbb{R}^n - \bar{U}. \end{cases}$$

Consequently, $(\star)$ implies

$$-\int_{\partial U} \phi \cdot d[D\bar{f}] = \int_{\partial U} (Tf_1 - Tf_2)\phi \cdot \nu \, d\mathcal{H}^{n-1},$$

and so

$$\|D\bar{f}\|(\partial U) = \int_{\partial U} |Tf_1 - Tf_2| \, d\mathcal{H}^{n-1}.$$ $\square$

5.5   Coarea formula for BV functions

Next we relate the variation measure of f and the perimeters of its level sets.

NOTATION For $f \colon U \to \mathbb{R}$ and $t \in \mathbb{R}$, define

$$E_t := \{x \in U \mid f(x) > t\}.$$

LEMMA 5.1. *If $f \in BV(U)$, the mapping*

$$t \mapsto \|\partial E_t\|(U) \quad (t \in \mathbb{R})$$

is $\mathcal{L}^1$-measurable.

Proof. The mapping

$$(x,t) \mapsto \chi_{E_t}(x)$$

is $\mathcal{L}^n \times \mathcal{L}^1$-measurable; and thus for each $\phi \in C_c^1(U; \mathbb{R}^n)$, the function

$$t \mapsto \int_{E_t} \operatorname{div} \phi \, dx = \int_U \chi_{E_t} \operatorname{div} \phi \, dx$$

is $\mathcal{L}^1$-measurable. Let D denote any countable dense subset of $C_c^1(U : \mathbb{R}^n)$. Then

$$t \mapsto \|\partial E_t\|(U) = \sup_{\phi \in D, |\phi| \leq 1} \int_{E_t} \operatorname{div} \phi \, dx$$

is $\mathcal{L}^1$-measurable. $\qquad\square$

THEOREM 5.9 (Coarea formula for BV functions).

(i) *If $f \in BV(U)$, then E_t has finite perimeter for $\mathcal{L}^1$-a.e. point $t \in \mathbb{R}$, and*

$$\|Df\|(U) = \int_{-\infty}^{\infty} \|\partial E_t\|(U) \, dt.$$

(ii) *Conversely, if $f \in L^1(U)$ and*

$$\int_{-\infty}^{\infty} \|\partial E_t\|(U) \, dt < \infty,$$

then $f \in BV(U)$.

Remark. Compare this with Theorem 3.13 in Section 3.4. $\qquad\square$

Proof. 1. Let $f \in L^1(U)$ and $\phi \in C_c^1(U; \mathbb{R}^n), |\phi| \leq 1$.

Claim #1: We have

$$\int_U f \operatorname{div} \phi \, dx = \int_{-\infty}^{\infty} \left(\int_{E_t} \operatorname{div} \phi \, dx \right) dt.$$

Proof of claim: First suppose $f \geq 0$; so that

$$f(x) = \int_0^{\infty} \chi_{E_t}(x) \, dt$$

for a.e. $x \in U$. Thus

$$\begin{aligned}
\int_U f \operatorname{div} \phi \, dx &= \int_U \left(\int_0^{\infty} \chi_{E_t}(x) \, dt \right) \operatorname{div} \phi(x) \, dx \\
&= \int_0^{\infty} \left(\int_U \chi_{E_t}(x) \operatorname{div} \phi(x) \, dx \right) dt \\
&= \int_0^{\infty} \left(\int_{E_t} \operatorname{div} \phi \, dx \right) dt.
\end{aligned}$$

Similarly, if $f \leq 0$,

$$f(x) = \int_{-\infty}^0 (\chi_{E_t}(x) - 1) \, dt;$$

whence

$$\begin{aligned}
\int_U f \operatorname{div} \phi \, dx &= \int_U \left(\int_{-\infty}^0 (\chi_{E_t}(x) - 1) \, dt \right) \operatorname{div} \phi(x) \, dx \\
&= \int_{-\infty}^0 \left(\int_U (\chi_{E_t}(x) - 1) \operatorname{div} \phi(x) \, dx \right) dt \\
&= \int_{-\infty}^0 \left(\int_{E_t} \operatorname{div} \phi \, dx \right) dt.
\end{aligned}$$

For the general case, write $f = f^+ - f^-$.

2. From Claim #1 we see that for all ϕ as above,

$$\int_U f \operatorname{div} \phi \, dx \leq \int_{-\infty}^{\infty} \|\partial E_t\|(U) \, dt.$$

Hence

$$\|Df\|(U) \leq \int_{-\infty}^{\infty} \|\partial E_t\|(U) \, dt. \qquad (\star)$$

This proves (ii).

2. *Claim #2*: Assertion (i) holds for all $f \in BV(U) \cap C^{\infty}(U)$.

Proof of claim: Let

$$m(t) := \int_{U - E_t} |Df| \, dx = \int_{\{f \leq t\}} |Df| \, dx.$$

Then the function m is nondecreasing, and thus m' exists $\mathcal{L}^1$-a.e., with

$$\int_{-\infty}^{\infty} m'(t) \, dt \leq \int_U |Df| \, dx. \qquad (\star\star)$$

Now fix any $-\infty < t < \infty$, $r > 0$, and define $\eta \colon \mathbb{R} \to \mathbb{R}$ this way:

$$\eta(s) := \begin{cases} 0 & \text{if } s \leq t \\ \frac{s-t}{r} & \text{if } t \leq s \leq t + r \\ 1 & \text{if } s \geq t + r. \end{cases}$$

Then

$$\eta'(s) = \begin{cases} \frac{1}{r} & \text{if } t < s < t + r \\ 0 & \text{if } s < t \text{ or } s > t + r. \end{cases}$$

Hence, for all $\phi \in C_c^1(U; \mathbb{R}^n)$,

$$-\int_U \eta(f(x)) \operatorname{div} \phi \, dx = \int_U \eta'(f(x)) Df \cdot \phi \, dx$$

$$= \frac{1}{r} \int_{E_t - E_{t+r}} Df \cdot \phi \, dx. \qquad (\star\star\star)$$

Now

$$\frac{m(t+r) - m(t)}{r} = \frac{1}{r} \left[\int_{U - E_{t+r}} |Df| \, dx - \int_{U - E_t} |Df| \, dx \right]$$

$$= \frac{1}{r} \int_{E_t - E_{t+r}} |Df| \, dx$$

$$\geq \frac{1}{r} \int_{E_t - E_{t+r}} Df \cdot \phi \, dx$$

$$= -\int_U \eta(f(x)) \operatorname{div} \phi \, dx$$

by $(\star\star\star)$. For those t such that $m'(t)$ exists, we then let $r \to 0$:

$$m'(t) \geq -\int_{E_t} \operatorname{div} \phi \, dx$$

for $\mathcal{L}^1$-a.e. t. Take the supremum over all ϕ as above:

$$\|\partial E_t\|(U) \leq m'(t),$$

and recall (★★) to find

$$\int_{-\infty}^{\infty} \|\partial E_t\|(U) \, dt \le \int_U |Df| \, dx = \|Df\|(U).$$

This estimate and (★) complete the proof.

3. *Claim #3*: Assertion (i) holds for each function $f \in BV(U)$.

Proof of claim: Fix $f \in BV(U)$ and choose $\{f_k\}_{k=1}^{\infty}$ as in Theorem 5.3. Then

$$f_k \to f \quad \text{in } L^1(U).$$

as $k \to \infty$. Define

$$E_t^k := \{x \in U \mid f_k(x) > t\}.$$

Now

$$\int_{-\infty}^{\infty} |\chi_{E_t^k}(x) - \chi_{E_t}(x)| \, dt = \int_{\min\{f(x),f_k(x)\}}^{\max\{f(x),f_k(x)\}} dt = |f_k(x) - f(x)|;$$

consequently,

$$\int_U |f_k(x) - f(x)| \, dx = \int_{-\infty}^{\infty} \left(\int_U |\chi_{E_t^k}(x) - \chi_{E_t}(x)| \, dx \right) dt.$$

Since $f_k \to f$ in $L^1(U)$, there exists a subsequence which, upon reindexing by k if need be, satisfies

$$\chi_{E_t^k} \to \chi_{E_t} \quad \text{in } L^1(U)$$

for $\mathcal{L}^1$-a.e. t. Then the lower semicontinuity Theorem 5.2 implies

$$\|\partial E_t\|(U) \le \liminf_{k \to \infty} \|\partial E_t^k\|(U).$$

Thus Fatou's Lemma implies

$$\int_{-\infty}^{\infty} \|\partial E_t\|(U) \, dt \le \liminf_{k \to \infty} \int_{-\infty}^{\infty} \|\partial E_t^k\|(U) \, dt$$

$$= \lim_{k \to \infty} \|Df_k\|(U) = \|Df\|(U).$$

This calculation and (★) complete the proof. $\qquad \square$

5.6   Isoperimetric inequalities

We now develop certain inequalities relating the $\mathcal{L}^n$-measure of a set and its perimeter.

5.6.1   Sobolev's and Poincaré's inequalities for BV

THEOREM 5.10 (Inequalities for BV functions).

(i) *There exists a constant C_1 such that*

$$\|f\|_{L^{1*}(\mathbb{R}^n)} \le C_1 \|Df\|(\mathbb{R}^n)$$

for all $f \in BV(\mathbb{R}^n)$, where

$$1^* = \frac{n}{n-1}.$$

(ii) *There exists a constant C_2 such that*

$$\|f - (f)_{x,r}\|_{L^{1*}(B(x,r))} \le C_2 \|Df\|(B^0(x,r))$$

for all balls $B(x,r) \subset \mathbb{R}^n$ and $f \in BV_{\mathrm{loc}}(\mathbb{R}^n)$, where

$$(f)_{x,r} := \fint_{B(x,r)} f \, dy.$$

(iii) *For each $0 < \alpha \le 1$, there exists a constant $C_3(\alpha)$ such that*

$$\|f\|_{L^{1*}(B(x,r))} \le C_3(\alpha) \|Df\|(B^0(x,r))$$

for all $B(x,r) \subseteq \mathbb{R}^n$ and all $f \in BV_{\mathrm{loc}}(\mathbb{R}^n)$ satisfying

$$\frac{\mathcal{L}^n(B(x,r) \cap \{f = 0\})}{\mathcal{L}^n(B(x,r))} \ge \alpha.$$

Proof. 1. Choose $f_k \in C_c^\infty(\mathbb{R}^n)$ $(k = 1, \dots)$ so that

$$f_k \to f \text{ in } L^1(\mathbb{R}^n), \quad f_k \to f \ \mathcal{L}^n\text{-a.e.}, \quad \|Df_k\|(\mathbb{R}^n) \to \|Df\|(\mathbb{R}^n).$$

Then Fatou's Lemma and the Gagliardo–Nirenberg–Sobolev inequality imply

$$\|f\|_{L^{1*}(\mathbb{R}^n)} \le \liminf_{k \to \infty} \|f_k\|_{L^{1*}(\mathbb{R}^n)}$$

$$\le \lim_{k \to \infty} C_1 \|Df_k\|_{L^1(\mathbb{R}^n)}$$

$$= C_1 \|Df\|(\mathbb{R}^n).$$

This proves (i).

2. Statement (ii) follows similarly from Poincaré's inequality, Section 4.5.

3. Suppose

$$\frac{\mathcal{L}^n(B(x,r) \cap \{f = 0\})}{\mathcal{L}^n(B(x,r))} \geq \alpha > 0. \qquad (\star)$$

Then

$$\|f\|_{L^{1*}(B(x,r))} \leq \|f - (f)_{x,r}\|_{L^{1*}(B(x,r))} + \|(f)_{x,r}\|_{L^{1*}(B(x,r))}$$
$$\leq C_2 \|Df\|(B^0(x,r)) + |(f)_{x,r}|(\mathcal{L}^n(B(x,r)))^{1-\frac{1}{n}}. \qquad (\star\star)$$

But

$$|(f)_{x,r}|(\mathcal{L}^n(B(x,r)))^{1-\frac{1}{n}}$$

$$\leq \frac{1}{\mathcal{L}^n(B(x,r))^{\frac{1}{n}}} \int_{B(x,r) \cap \{f \neq 0\}} |f| \, dy$$

$$\leq \left(\int_{B(x,r)} |f|^{1*} \, dy \right)^{1-\frac{1}{n}} \left(\frac{\mathcal{L}^n(B(x,r)) \cap \{f \neq 0\}}{\mathcal{L}^n(B(x,r))} \right)^{\frac{1}{n}}$$

$$\leq \|f\|_{L^{1*}(B(x,r))} (1 - \alpha)^{\frac{1}{n}},$$

by $(\star)$. We employ this estimate in $(\star\star)$ to compute

$$\|f\|_{L^{1*}(B(x,r))} \leq \frac{C_2}{(1 - (1 - \alpha)^{\frac{1}{n}})} \|Df\|(B^0(x,r)). \qquad \square$$

5.6.2 Isoperimetric inequalities

THEOREM 5.11 (Isoperimetric inequalities). *Let E be a bounded set of finite perimeter in $\mathbb{R}^n$.*

(i) *Then*

$$\mathcal{L}^n(E)^{1-\frac{1}{n}} \leq C_1 \|\partial E\|(\mathbb{R}^n),$$

and

(ii) *for each ball $B(x,r) \subset \mathbb{R}^n$,*

$$\min \{\mathcal{L}^n(B(x,r) \cap E), \mathcal{L}^n(B(x,r) - E)\}^{1-\frac{1}{n}}$$
$$\leq 2C_2 \|\partial E\|(B^0(x,r)).$$

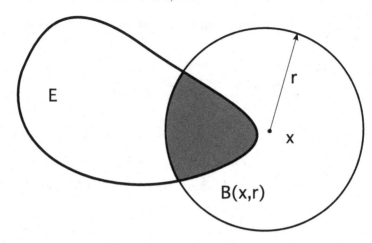

The constants C_1 and C_2 are those from Theorems 4.8 and 4.9 in Section 4.5.

Remark. Statement (i) is the **isoperimetric inequality** and (ii) is the **relative isoperimetric inequality**. □

Proof: 1. Let $f = \chi_E$ in Theorem 5.10,(i) to prove (i).

2. Let $f = \chi_{B(x,r) \cap E}$ in Theorem 5.10,(ii), in which case

$$(f)_{x,r} = \frac{\mathcal{L}^n(B(x,r) \cap E)}{\mathcal{L}^n(B(x,r))}.$$

Thus

$$\int_{B(x,r)} |f - (f)_{x,r}|^{1^*} dy = \left(\frac{\mathcal{L}^n(B(x,r) - E)}{\mathcal{L}^n(B(x,r))}\right)^{1^*} \mathcal{L}^n(B(x,r) \cap E)$$
$$+ \left(\frac{\mathcal{L}^n(B(x,r) \cap E)}{\mathcal{L}^n(B(x,r))}\right)^{1^*} \mathcal{L}^n(B(x,r) - E).$$

Now if $\mathcal{L}^n(B(x,r) \cap E) \leq \mathcal{L}^n(B(x,r) - E)$, then

$$\left(\int_{B(x,r)} |f - (f)_{x,r}|^{1^*} dy\right)^{1 - \frac{1}{n}}$$
$$\geq \left[\frac{\mathcal{L}^n(B(x,r) - E)}{\mathcal{L}^n(B(x,r))}\right] \mathcal{L}^n(B(x,r) \cap E)^{1 - \frac{1}{n}}$$
$$\geq \frac{1}{2} \min\{\mathcal{L}^n(B(x,r) \cap E), \mathcal{L}^n(B(x,r) - E)\}^{1 - \frac{1}{n}}.$$

The other case is similar. □

Remark. We have shown that the Gagliardo–Nirenberg–Sobolev inequality implies the isoperimetric inequality. In fact, *the converse is true as well*: the isoperimetric inequality implies the Gagliardo–Nirenberg–Sobolev inequality.

To see this, assume $f \in C_c^1(\mathbb{R}^n)$, $f \geq 0$. We calculate

$$\int_{\mathbb{R}^n} |Df| \, dx = \|Df\|(\mathbb{R}^n)$$

$$= \int_{-\infty}^{\infty} \|\partial E_t\|(\mathbb{R}^n) \, dt$$

$$\geq \frac{1}{C_1} \int_{-\infty}^{\infty} \mathcal{L}^n(E_t)^{1-\frac{1}{n}} \, dt.$$

Now let

$$f_t := \min\{t, f\}, \quad \chi(t) := \left(\int_{\mathbb{R}^n} f_t^{1^*} \, dx \right)^{1-\frac{1}{n}} \quad (t \in \mathbb{R}).$$

Then χ is nondecreasing on $(0, \infty)$ and

$$\lim_{t \to \infty} \chi(t) = \left(\int_{\mathbb{R}^n} |f|^{1^*} \, dx \right)^{1-\frac{1}{n}}.$$

Also, for $h > 0$, we have

$$0 \leq \chi(t+h) - \chi(t) \leq \left(\int_{\mathbb{R}^n} |f_{t+h} - f_t|^{1^*} \, dx \right)^{1-\frac{1}{n}} \leq h \mathcal{L}^n(E_t)^{1-\frac{1}{n}}.$$

Thus χ is locally Lipschitz continuous, and

$$\chi'(t) \leq \mathcal{L}^n(E_t)^{1-\frac{1}{n}}$$

for $\mathcal{L}^1$-a.e. t. Integrate from 0 to ∞:

$$\left(\int_{\mathbb{R}^n} |f|^{1^*} \, dx \right)^{1-\frac{1}{n}} = \int_0^{\infty} \chi'(t) \, dt$$

$$\leq \int_0^{\infty} \mathcal{L}^n(E_t)^{1^*} \, dt$$

$$\leq C_1 \int_{\mathbb{R}^n} |Df| \, dx. \qquad \square$$

5.6.3 $\mathcal{H}^{n-1}$ and $\mathbf{Cap_1}$

As a first application of the isoperimetric inequalities, we establish this refinement of Theorem 4.17 in Section 4.7:

THEOREM 5.12 ($\mathcal{H}^{n-1}$and $\mathbf{Cap_1}$). *Assume $n \geq 2$ and $A \subset \mathbb{R}^n$ is compact. Then*

$$\mathrm{Cap_1}(A) = 0 \text{ if and only if } \mathcal{H}^{n-1}(A) = 0.$$

Proof. According to Theorem 4.15, $\mathrm{Cap_1}(A) = 0$ if $\mathcal{H}^{n-1}(A) = 0$.

Now suppose $\mathrm{Cap_1}(A) = 0$. If $f \in K^1$ and $A \subset \{f \geq 1\}^0$, then by Theorem 5.9,

$$\int_0^1 \|\partial E_t\|(\mathbb{R}^n)\, dt \leq \int_{\mathbb{R}^n} |Df|\, dx,$$

where $E_t := \{f > t\}$. Thus for some $t \in (0, 1)$,

$$\|\partial E_t\|(\mathbb{R}^n) \leq \int_{\mathbb{R}^n} |Df|\, dx.$$

Clearly $A \subseteq E_t^0$; and by the isoperimetric inequality, $\mathcal{L}^n(E_t) < \infty$. Thus for each $x \in A$, there exists $r > 0$ such that

$$\frac{\mathcal{L}^n(E_t \cap B(x, r))}{\alpha(n)r^n} = \frac{1}{4}.$$

In light of the relative isoperimetric inequality, we have for each such ball $B(x, r)$ that

$$\left(\frac{1}{4}\alpha(n)r^n\right)^{\frac{n-1}{n}} = (\mathcal{L}^n(E_t \cap B(x, r)))^{\frac{n-1}{n}} \leq C\|\partial E_t\|(B(x, r));$$

that is,

$$r^{n-1} \leq C\|\partial E_t\|(B(x, r)).$$

By Vitali's Covering Theorem, there exists a disjoint collection of balls $\{B(x_j, r_j)\}_{j=1}^\infty$ as above, with $x_j \in A$ and

$$A \subseteq \bigcup_{j=1}^\infty B(x_j, 5r_j).$$

Thus

$$\sum_{j=1}^\infty (5r_j)^{n-1} \leq C\|\partial E_t\|(\mathbb{R}^n) \leq C\int_{\mathbb{R}^n} |Df|\, dx.$$

Since $\text{Cap}_1(A) = 0$, given $\epsilon > 0$, the function f can be chosen so that

$$\int_{\mathbb{R}^n} |Df| \, dx \le \epsilon.$$

Thus for each j,

$$r_j \le (C\|\partial E_t\|(\mathbb{R}^n))^{\frac{1}{n-1}} \le C\epsilon^{\frac{1}{n-1}}.$$

This implies $\mathcal{H}^{n-1}(A) = 0$. □

5.7 The reduced boundary

In this and the next section we study the detailed structure of sets of locally finite perimeter. Our goal is to verity that such a set has "a C^1 boundary measure theoretically."

5.7.1 Estimates

We hereafter assume

E is a set of locally finite perimeter in $\mathbb{R}^n$.

Recall the definitions of ν_E, $\|\partial E\|$, etc., from Section 5.1.

DEFINITION 5.4. *Let $x \in \mathbb{R}^n$. We say $x \in \partial^* E$, the **reduced boundary** of E, if*

(i) $\|\partial E\|(B(x,r)) > 0$ *for all $r > 0$,*

(ii)

$$\lim_{r \to 0} \fint_{B(x,r)} \nu_E \, d\|\partial E\| = \nu_E(x),$$

and

(iii) $|\nu_E(x)| = 1.$

Remark. According to Theorem 1.32,

$$\|\partial E\|(\mathbb{R}^n - \partial^* E) = 0.$$ □

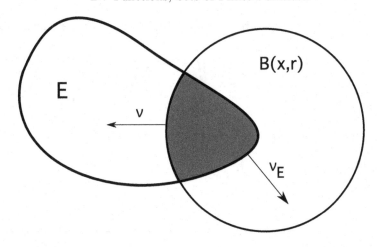

LEMMA 5.2. *Let $\phi \in C_c^1(\mathbb{R}^n; \mathbb{R}^n)$. Then for each $x \in \mathbb{R}^n$,*

$$\int_{E \cap B(x,r)} \operatorname{div} \phi \, dy = \int_{B(x,r)} \phi \cdot \nu_E \, d\|\partial E\| + \int_{E \cap \partial B(x,r)} \phi \cdot \nu \, d\mathcal{H}^{n-1}$$

for $\mathcal{L}^1$-a.e. $r > 0$, ν denoting the outward unit normal to $\partial B(x,r)$.

Proof. Assume $h : \mathbb{R}^n \to \mathbb{R}$ is smooth; then

$$\int_E \operatorname{div}(h\phi) \, dy = \int_E h \, \operatorname{div} \phi \, dy + \int_E Dh \cdot \phi \, dy.$$

Thus

$$\int_{\mathbb{R}^n} h\phi \cdot \nu_E \, d\|\partial E\| = \int_E h \operatorname{div} \phi \, dy + \int_E Dh \cdot \phi \, dy. \qquad (\star)$$

By approximation, $(\star)$ holds also for

$$h_\epsilon(y) := g_\epsilon(|y - x|),$$

where

$$g_\epsilon(s) := \begin{cases} 1 & \text{if } 0 \leq s \leq r \\ \frac{r-s+\epsilon}{\epsilon} & \text{if } r \leq s \leq r+\epsilon \, . \\ 0 & \text{if } s \geq r+\epsilon. \end{cases}$$

Then

$$g_\epsilon'(s) = \begin{cases} 0 & \text{if } 0 \leq s < r \text{ or } s > r+\epsilon \\ -\frac{1}{\epsilon} & \text{if } r < s < r+\epsilon; \end{cases}$$

and therefore

$$Dh_\epsilon(y) = \begin{cases} 0 & \text{if } |y - x| < r \text{ or } |y - x| > r + \epsilon \\ -\frac{1}{\epsilon} \frac{y-x}{|y-x|} & \text{if } r < |y - x| < r + \epsilon. \end{cases}$$

Set $h = h_\epsilon$ in $(\star)$:

$$\int_{\mathbb{R}^n} h_\epsilon \phi \cdot \nu_E \, d\|\partial E\| = \int_E h_\epsilon \, \text{div} \, \phi \, dy$$
$$- \frac{1}{\epsilon} \int_{E \cap \{y | r < |y-x| < r+\epsilon\}} \phi \cdot \frac{y - x}{|y - x|} \, dy.$$

Let $\epsilon \to 0$ and recall Theorem 3.12 in Section 3.4 to find

$$\int_{B(x,r)} \phi \cdot \nu_E \, d\|\partial E\| = \int_{E \cap B(x,r)} \text{div} \, \phi \, dy - \int_{E \cap \partial B(x,r)} \phi \cdot \nu \, d\mathcal{H}^{n-1}$$

for $\mathcal{L}^1$-a.e. $r > 0$. $\qquad\qquad\qquad\qquad\qquad\qquad\qquad\qquad\qquad\qquad\square$

LEMMA 5.3. *There exist positive constants $A_1, \ldots, A_5$, depending only on n, such that for each $x \in \partial^* E$,*

(i) $\liminf_{r \to 0} \frac{\mathcal{L}^n(B(x,r) \cap E)}{r^n} > A_1 > 0,$

(ii) $\liminf_{r \to 0} \frac{\mathcal{L}^n(B(x,r) - E)}{r^n} > A_2 > 0,$

(iii) $\liminf_{r \to 0} \frac{\|\partial E\|(B(x,r))}{r^{n-1}} > A_3 > 0,$

(iv) $\limsup_{r \to 0} \frac{\|\partial E\|(B(x,r))}{r^{n-1}} \leq A_4,$

(v) $\limsup_{r \to 0} \frac{\|\partial(E \cap B(x,r))\|(\mathbb{R}^n)}{r^{n-1}} \leq A_5.$

Proof. 1. Fix $x \in \partial^* E$. According to Lemma 5.2, for $\mathcal{L}^1$-a.e. $r > 0$

$$\|\partial(E \cap B(x,r))\|(\mathbb{R}^n) \leq \|\partial E\|(B(x,r)) + \mathcal{H}^{n-1}(E \cap \partial B(x,r)). \quad (\star)$$

Now choose $\phi \in C_c^1(\mathbb{R}^n; \mathbb{R}^n)$ such that

$$\phi \equiv \nu_E(x) \quad \text{on } B(x,r).$$

Then the formula from Lemma 5.2 reads

$$\int_{B(x,r)} \nu_E(x) \cdot \nu_E \, d\|\partial E\| = - \int_{E \cap \partial B(x,r)} \nu_E(x) \cdot \nu \, d\mathcal{H}^{n-1}. \quad (\star\star)$$

Since $x \in \partial^* E$,

$$\lim_{r \to 0} \nu_E(x) \cdot \fint_{B(x,r)} \nu_E \, d\|\partial E\| = |\nu_E(x)|^2 = 1;$$

thus for $\mathcal{L}^1$-a.e. sufficiently small $r > 0$, say $0 < r < r_0 = r_0(x)$, $(\star\star)$ implies

$$\frac{1}{2}\|\partial E\|(B(x,r)) \leq \mathcal{H}^{n-1}(E \cap \partial B(x,r)). \qquad (\star\star\star)$$

This and $(\star)$ give

$$\|\partial(E \cap B(x,r))\|(\mathbb{R}^n) \leq 3\mathcal{H}^{n-1}(E \cap \partial B(x,r)) \qquad (\star\star\star\star)$$

for a.e. $0 < r < r_0$.

2. Write $g(r) := \mathcal{L}^n(B(x,r) \cap E)$. Then

$$g(r) = \int_0^r \mathcal{H}^{n-1}(\partial B(x,s) \cap E) \, ds,$$

whence g is absolutely continuous; and

$$g'(r) = \mathcal{H}^{n-1}(\partial B(x,r) \cap E) \quad \text{for a.e. } r > 0.$$

Using now the isoperimetric inequality and $(\star\star\star\star)$, we compute

$$\begin{aligned} g(r)^{1-\frac{1}{n}} &= \mathcal{L}^n(B(x,r) \cap E)^{1-\frac{1}{n}} \\ &\leq C\|\partial(B(x,r) \cap E)\|(\mathbb{R}^n) \\ &\leq C\mathcal{H}^{n-1}(\partial B(x,r) \cap E) \\ &= C_1 g'(r) \end{aligned}$$

for a.e. $r \in (0, r_0)$. Thus

$$\frac{1}{C_1} \leq g(r)^{\frac{1}{n}-1} g'(r) = n(g^{\frac{1}{n}}(r))',$$

and so

$$g^{\frac{1}{n}}(r) \geq \frac{r}{C_1 n}.$$

Then

$$g(r) \geq \frac{r^n}{(C_1 n)^n}$$

for $0 < r \leq r_0$. This proves assertion (i).

3. Since for all $\phi \in C_c^1(\mathbb{R}^n; \mathbb{R}^n)$

$$\int_E \operatorname{div} \phi \, dx + \int_{\mathbb{R}^n - E} \operatorname{div} \phi \, dx = \int_{\mathbb{R}^n} \operatorname{div} \phi \, dx = 0,$$

it is easy to check that

$$\|\partial E\| = \|\partial(\mathbb{R}^n - E)\|, \quad \nu_E = -\nu_{\mathbb{R}^n - E}.$$

Consequently, statement (ii) follows from (i).

4. According to the relative isoperimetric inequality,

$$\frac{\|\partial E\|(B(x,r))}{r^{n-1}} \geq C \min \left\{ \frac{\mathcal{L}^n(B(x,r) \cap E)}{r^n}, \frac{\mathcal{L}^n(B(x,r) - E)}{r^n} \right\}^{\frac{n-1}{n}}$$

and thus assertion (iii) follows from (i), (ii).

5. By $(\star\star\star)$,

$$\|\partial E\|(B(x,r)) \leq 2\mathcal{H}^{n-1}(E \cap \partial B(x,r)) \leq Cr^{n-1} \quad (0 < r < r_0);$$

this is (iv).

6. Statement (v) is a consequence of $(\star)$ and (iv). $\square$

5.7.2   Blow-up

DEFINITION 5.5. *For each $x \in \partial^* E$, define the hyperplane*

$$H(x) := \{y \in \mathbb{R}^n \mid \nu_E(x) \cdot (y - x) = 0\}$$

and the half-spaces

$$H^+(x) := \{y \in \mathbb{R}^n \mid \nu_E(x) \cdot (y - x) \geq 0\},$$
$$H^-(x) := \{y \in \mathbb{R}^n \mid \nu_E(x) \cdot (y - x) \leq 0\}.$$

NOTATION Fix $x \in \partial^* E$, $r > 0$, and set

$$E_r := \{y \in \mathbb{R}^n \mid r(y - x) + x \in E\}.$$

Observe that $y \in E \cap B(x,r)$ if and only if $g_r(y) \in E_r \cap B(x,1)$, where $g_r(y) := \frac{y-x}{r} + x$.

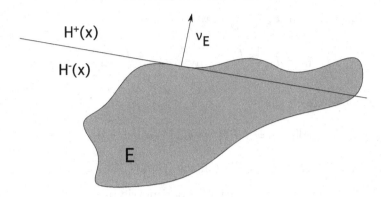

THEOREM 5.13 (Blow-up of reduced boundary). *Assume* $x \in \partial^* E$. *Then*

$$\chi_{E_r} \to \chi_{H^-(x)} \quad in \ L^1_{\text{loc}}(\mathbb{R}^n)$$

as $r \to 0$.

Thus for small enough $r > 0$, $E \cap B(x,r)$ approximately equals the half ball $H^-(x) \cap B(x,r)$.

Proof. 1. First of all, we may as well assume:

$$\begin{cases} x = 0, \ \nu_E(0) = e_n = (0,\ldots,0,1), \\ H(0) = \{y \in \mathbb{R}^n \mid y_n = 0\}, \\ H^+(0) = \{y \in \mathbb{R}^n \mid y_n \geq 0\}, \\ H^-(0) = \{y \in \mathbb{R}^n \mid y_n \leq 0\}. \end{cases}$$

2. Choose any sequence $r_k \to 0$. It will be enough to show there exists a subsequence $\{s_j\}_{j=1}^\infty \subseteq \{r_k\}_{k=1}^\infty$ for which

$$\chi_{E_{s_j}} \to \chi_{H^-(0)} \quad in \ L^1_{\text{loc}}(\mathbb{R}^n).$$

3. Fix $L > 0$ and let

$$D_r := E_r \cap B(L), \ g_r(y) = \frac{y}{r}.$$

Then for any $\phi \in C^1_c(\mathbb{R}^n; \mathbb{R}^n)$, with $|\phi| \leq 1$, we have

$$\int_{D_r} \text{div} \, \phi \, dz = \frac{1}{r^{n-1}} \int_{E \cap B(rL)} \text{div}(\phi \circ g_r) \, dy.$$

$$= \frac{1}{r^{n-1}} \int_{\mathbb{R}^n} (\phi \circ g_r) \cdot \nu_{E \cap B(rL)} \, d\|\partial(E \cap B(rL))\|$$

$$\leq \frac{\|\partial(E \cap B(rL)\|(\mathbb{R}^n)}{r^{n-1}}$$

$$\leq C < \infty$$

for all $r \in (0, 1]$, according to (v) of Lemma 5.3. Consequently,

$$\|\partial D_r\|(\mathbb{R}^n) \leq C < \infty \quad (0 < r \leq 1).$$

Furthermore,

$$\|\chi_{D_r}\|_{L^1(\mathbb{R}^n)} = \mathcal{L}^n(D_r) \leq \mathcal{L}^n(B(L)) < \infty \quad (r > 0).$$

Hence

$$\|\chi_{D_r}\|_{BV(\mathbb{R}^n)} \leq C < \infty$$

for all $0 < r \leq 1$.

In view of this estimate and the compactness Theorem 5.5, there exists a subsequence $\{s_j\}_{j=1}^\infty \subseteq \{r_k\}_{k=1}^\infty$ and $f \in BV_{\text{loc}}(\mathbb{R}^n)$ such that

$$\chi_{E_j} \to f \quad \text{in } L^1_{\text{loc}}(\mathbb{R}^n)$$

for $E_j := E_{s_j}$. We may assume also $\chi_{E_j} \to f$ $\mathcal{L}^n$-a.e.. Hence $f(x) \in \{0, 1\}$ for $\mathcal{L}^n$-a.e. x and so

$$f = \chi_F \quad \mathcal{L}^n \text{-a.e.},$$

where $F \subset \mathbb{R}^n$ has locally finite perimeter. So if $\phi \in C_c^1(\mathbb{R}^n; \mathbb{R}^n)$,

$$\int_F \text{div} \, \phi \, dy = \int_{\mathbb{R}_n} \phi \cdot \nu_F \, d\|\partial F\| \qquad (\star)$$

for some $\|\partial F\|$-measurable function ν_F, with $|\nu_F| = 1$ $\|\partial F\|$-a.e.. We must prove $F = H^-(0)$.

4. *Claim #1*: $\nu_F = e_n$ $\|\partial F\|$-a.e.

Proof of claim: Let us write $\nu_j := \nu_{E_j}$. Then if $\phi \in C_c^1(\mathbb{R}^n; \mathbb{R}^n)$, we have

$$\int_{\mathbb{R}^n} \phi \cdot \nu_j \, d\|\partial E_j\| = \int_{E_j} \text{div} \, \phi \, dy \quad (j = 1, 2, \dots).$$

Since

$$\chi_{E_j} \to \chi_F \quad \text{in } L^1_{\text{loc}},$$

we see from the above and $(\star)$ that

$$\int_{\mathbb{R}^n} \phi \cdot \nu_j \, d\|\partial E_j\| \to \int_{\mathbb{R}^n} \phi \cdot \nu_F \, d\|\partial F\|$$

as $j \to \infty$. Thus

$$\nu_j \|\partial E_j\| \rightharpoonup \nu_F \|\partial F\|$$

weakly in the sense of Radon measures. Consequently, for each $L > 0$ for which $\|\partial F\|(\partial B(L)) = 0$, and hence for all but at most countably many $L > 0$,

$$\int_{B(L)} \nu_j \, d\|\partial E_j\| \to \int_{B(L)} \nu_F \, d\|\partial F\|. \tag{$\star\star$}$$

On the other hand, for all ϕ as above,

$$\int_{\mathbb{R}^n} \phi \cdot \nu_j \, d\|\partial E_j\| = \frac{1}{s_j^{n-1}} \int_{\mathbb{R}^n} (\phi \circ g_{sj}) \cdot \nu_E \, d\|\partial E\|;$$

whence

$$\begin{cases} \|\partial E_j\|(B^0(0,L)) = \frac{1}{s_j^{n-1}} \|\partial E\|(B(0, s_j L)) \\ \int_{B(L)} \nu_j \, d\|\partial E_j\| = \frac{1}{s_j^{n-1}} \int_{B(0, s_j L)} \nu_E \, d\|\partial E\|. \end{cases} \tag{$\star\star\star$}$$

Therefore

$$\lim_{j \to \infty} \fint_{B(L)} \nu_j \, d\|\partial E_j\| = \lim_{j \to \infty} \fint_{B(0, s_j L)} \nu_E \, d\|\partial E\| = \nu_E(0) = e_n,$$

since $0 \in \partial^* E$. If $\|\partial F\|(\partial B(L)) = 0$, the lower semicontinuity Theorem 5.2 implies

$$\|\partial F\|(B(L)) \le \liminf_{j \to \infty} \|\partial E_j\|(B(L))$$

$$= \lim_{j \to \infty} \int_{B(L)} e_n \cdot \nu_j \, d\|\partial E_j\|$$

$$= \int_{B(L)} e_n \cdot \nu_F \, d\|\partial F\|,$$

by $(\star\star)$. Since $|\nu_F| = 1$ $\|\partial F\|$-a.e., the above inequality forces

$$\nu_F = e_n \ \|\partial F\|\text{-a.e.}$$

It also follows from the above inequality that

$$\|\partial F\|(B(L)) = \lim_{j\to\infty} \|\partial E_j\|(B(L))$$

whenever $\|\partial F\|(\partial B(0, L)) = 0$.

5. *Claim #2.* F is a half space.

Proof of claim: By Claim #1, for all $\phi \in C_c^1(\mathbb{R}^n; \mathbb{R}^n)$,

$$\int_F \operatorname{div} \phi \, dz = \int_{\mathbb{R}^n} \phi \cdot e_n \, d\|\partial F\|.$$

Fix $\epsilon > 0$ and let $f^\epsilon := \eta_\epsilon * \chi_F$, where η_ϵ is the usual mollifier. Then $f^\epsilon \in C^\infty(\mathbb{R}^n)$, and so

$$\int_{\mathbb{R}^n} f^\epsilon \operatorname{div} \phi \, dz = \int_F \operatorname{div}(\eta_\epsilon * \phi) \, dz$$

$$= \int_{\mathbb{R}^n} \eta_\epsilon * (\phi \cdot e_n) \, d\|\partial F\|.$$

But also

$$\int_{\mathbb{R}^n} f^\epsilon \operatorname{div} \phi \, dz = -\int_{\mathbb{R}^n} Df^\epsilon \cdot \phi \, dz.$$

Thus

$$f^\epsilon_{z_i} = 0 \quad (i = 1, \ldots, n-1), \quad f^\epsilon_{z_n} \leq 0.$$

As $f^\epsilon \to \chi_F$ $\mathcal{L}^n$-a.e. as $\epsilon \to 0$, we conclude that, up to a set of $\mathcal{L}^n$-measure zero,

$$F = \{y \in \mathbb{R}^n \mid y_n \leq \gamma\} \quad \text{for some } \gamma \in \mathbb{R}.$$

6. *Claim #3:* $F = H^-(0)$.

Proof of claim: we must show $\gamma = 0$ above. Assume instead $\gamma > 0$. Since $\chi_{E_j} \to \chi_F$ in $L^1_{\text{loc}}(\mathbb{R}^n)$,

$$\alpha(n)\gamma^n = \mathcal{L}^n(B(0, \gamma) \cap F) = \lim_{j\to\infty} \mathcal{L}^n(B(0, \gamma) \cap E_j)$$

$$= \lim_{j\to\infty} \frac{\mathcal{L}^n(B(0, \gamma s_j) \cap E)}{s_j^n},$$

a contradiction to Lemma 5.3,(ii).

Similarly, the case $\gamma < 0$ leads to a contradiction to Lemma 5.3,(i).

$\square$

We can now read off more detailed information concerning the blow-up of E around a point $x \in \partial^* E$:

THEOREM 5.14 (More on blow-up of reduced boundary).
Assume that $x \in \partial^* E$. Then

(i) $\displaystyle \lim_{r \to 0} \frac{\mathcal{L}^n(B(x,r) \cap E \cap H^+(x))}{r^n} = 0$

(ii) $\displaystyle \lim_{r \to 0} \frac{\mathcal{L}^n((B(x,r) - E) \cap H^-(x))}{r^n} = 0,$

(iii) $\displaystyle \lim_{r \to 0} \frac{\|\partial E\|(B(x,r))}{\alpha(n-1)r^{n-1}} = 1.$

DEFINITION 5.6. *A unit vector $\nu_E(x)$ for which (i) holds (with $H^\pm(x)$ as defined above) is called the* **measure theoretic unit outer normal** *to E at x.*

Proof. 1. We have

$$\frac{\mathcal{L}^n(B(x,r) \cap E \cap H^+(x))}{r^n} = \mathcal{L}^n(B(x,1) \cap E_r \cap H^+(x))$$

$$\to \mathcal{L}^n(B(x,1) \cap H^-(x) \cap H^+(x)) = 0$$

as $r \to 0$. The limit (ii) has a similar proof.

2. Assume $x = 0$. By $(\star\star\star)$ in the proof of Theorem 5.13,

$$\frac{\|\partial E\|(B(r))}{r^{n-1}} = \|\partial E_r\|(B(1)).$$

Since $\|\partial H^-(0)\|(\partial B(1)) = \mathcal{H}^{n-1}(\partial B(1) \cap H(0)) = 0$, Step 2 of the proof of Theorem 5.13 implies

$$\lim_{r \to 0} \frac{\|\partial E\|(B(r))}{r^{n-1}} = \|\partial H^-(0)\|(B(1))$$

$$= \mathcal{H}^{n-1}(B(1) \cap H(0))$$

$$= \alpha(n-1). \qquad \square$$

5.7.3   Structure Theorem for sets of finite perimeter

LEMMA 5.4. *There exists a constant C, depending only on n, such that*

$$\mathcal{H}^{n-1}(B) \leq C\|\partial E\|(B)$$

for all $B \subseteq \partial^* E$.

Proof. Let $\epsilon, \delta > 0$ and $B \subseteq \partial^* E$. Since $\|\partial E\|$ is a Radon measure, there exists an open set $U \supseteq B$ such that

$$\|\partial E\|(U) \leq \|\partial E\|(B) + \epsilon.$$

According to Lemma 5.3, if $x \in \partial^* E$, then

$$\liminf_{r \to 0} \frac{\|\partial E\|(B(x,r))}{r^{n-1}} > A_3 > 0.$$

Let

$$\mathcal{F} :=$$

$$\left\{ B(x,r) \mid x \in B, \, B(x,r) \subseteq U, r < \frac{\delta}{10}, \|\partial E\|(B(x,r)) > A_3 r^{n-1} \right\}.$$

According to Vitali's Covering Theorem, there exist disjoint balls $\{B(x_i, r_i)\}_{i=1}^{\infty} \subset \mathcal{F}$ such that

$$B \subseteq \bigcup_{i=1}^{\infty} B(x_i, 5r_i).$$

Since diam $B(x_i, 5r_i) \leq \delta$ for $i = 1, \ldots,$

$$\mathcal{H}_{\delta}^{n-1}(B) \leq \sum_{i=1}^{\infty} \alpha(n-1)(5r_i)^{n-1} \leq C \sum_{i=1}^{\infty} \|\partial E\|(B(x_i, r_i))$$

$$\leq C\|\partial E\|(U) \leq C(\|\partial E\|(B) + \epsilon).$$

Let $\epsilon \to 0$ and then $\delta \to 0$. $\qquad\qquad\square$

Now we show that a set of locally finite perimeter has "measure theoretically a C^1 boundary."

THEOREM 5.15 (Structure Theorem for sets of finite perimeter). *Assume E has locally finite perimeter in $\mathbb{R}^n$.*

(i) *Then*

$$\partial^* E = \bigcup_{k=1}^{\infty} K_k \cup N,$$

where

$$\|\partial E\|(N) = 0$$

and K_k is a compact subset of a C^1 hypersurface S_k ($k = 1, 2, \ldots$).

(ii) $\nu_E|_{S_k}$ *is normal to S_k for $k = 1, \ldots$.*

(iii) *Furthermore,*

$$\|\partial E\| = \mathcal{H}^{n-1} \llcorner \partial^* E.$$

Proof. 1. For each $x \in \partial^* E$, we have according to Theorem 5.14

$$\begin{cases} \lim_{r \to 0} \dfrac{\mathcal{L}^n(B(x,r) \cap E \cap H^+(x))}{r^n} = 0, \\ \lim_{r \to 0} \dfrac{\mathcal{L}^n((B(x,r) - E) \cap H^-(x))}{r^n} = 0. \end{cases} \qquad (\star)$$

Using Egoroff's Theorem, we see that there exist disjoint $\|\partial E\|$-measurable sets $\{F_i\}_{i=1}^{\infty} \subseteq \partial^* E$ such that

$$\begin{cases} \|\partial E\| \left(\partial^* E - \bigcup_{i=1}^{\infty} F_i \right) = 0, \ \|\partial E\|(F_i) < \infty, \text{ and} \\ \text{the convergence in } (\star) \text{ is uniform for } x \in F_i \ (i = 1, \ldots). \end{cases}$$

Then, by Lusin's Theorem, for each i there exist disjoint compact sets $\{E_i^j\}_{j=1}^{\infty} \subset F_i$ such that

$$\|\partial E\| \left(F_i - \bigcup_{j=1}^{\infty} E_i^j \right) = 0, \ \nu_E|_{E_i^j} \text{ is continuous.}$$

Reindex the sets $\{E_i^j\}_{i,j=1}^{\infty}$ and call them $\{K_k\}_{k=1}^{\infty}$. Then

$$\begin{cases} \partial^* E = \bigcup_{k=1}^{\infty} K_k \cup N, \|\partial E\|(N) = 0, \\ \text{the convergence in } (\star) \text{ is uniform on } K_k, \\ \nu_E|_{K_k} \text{ is continuous } (k = 1, 2, \ldots). \end{cases} \qquad (\star\star)$$

2. Define for $\delta > 0$

$$\rho_k(\delta) := \sup \left\{ \frac{|\nu_E(x) \cdot (y - x)|}{|y - x|} \mid 0 < |x - y| \le \delta, x, y \in K_k \right\}.$$

3. *Claim:* For each $k = 1, 2, \ldots$, we have $\rho_k(\delta) \to 0$ as $\delta \to 0$.

Proof of claim: We may as well assume $k = 1$. Fix $0 < \epsilon < 1$. By $(\star)$, $(\star\star)$ there exists $0 < \delta < 1$ such that if $z \in K_1$ and $r < 2\delta$, then

$$\begin{cases} \mathcal{L}^n(E \cap B(z, r) \cap H^+(z)) < \dfrac{\epsilon^n}{2^{n+2}} \alpha(n) r^n \\ \mathcal{L}^n(E \cap B(z, r) \cap H^-(z)) > \alpha(n) \left(\dfrac{1}{2} - \dfrac{\epsilon^n}{2^{n+2}} \right) r^n. \end{cases} \qquad (\star\star\star)$$

Assume now $x, y \in K_1$, with $0 < |x - y| \le \delta$.

Case 1. $\nu_E(x) \cdot (y - x) > \epsilon |x - y|$. Then, since $\epsilon < 1$,

$$B(y, \epsilon |x - y|) \subseteq H^+(x) \cap B(x, 2|x - y|). \qquad (\star\star\star\star)$$

To see this, observe that if $z \in B(y, \epsilon |x - y|)$, then $z = y + w$, where $|w| \le \epsilon |x - y|$. Consequently,

$$\nu_E(x) \cdot (z - x) = \nu_E(x) \cdot (y - x) + \nu_E(x) \cdot w > \epsilon |x - y| - |w| \ge 0.$$

On the other hand, $(\star\star\star)$ with $z = x$ implies

$$\begin{aligned} \mathcal{L}^n(E \cap B(x, 2|x - y|) \cap H^+(x)) &< \frac{\epsilon^n}{2^{n+2}} \alpha(n)(2|x - y|)^n \\ &= \frac{\epsilon^n \alpha(n)}{4} |x - y|^n; \end{aligned}$$

whereas $(\star\star\star)$ with $z = y$ implies

$$\begin{aligned} \mathcal{L}^n(E \cap B(y, \epsilon |x - y|)) &\ge \mathcal{L}^n(E \cap B(y, \epsilon |x - y|) \cap H^-(y)) \\ &\ge \frac{\epsilon^n \alpha(n) |x - y|^n}{2} \left(1 - \frac{\epsilon^n}{2^{n+1}} \right) \\ &> \frac{\epsilon^n \alpha(n)}{4} |x - y|^n. \end{aligned}$$

However, our applying $\mathcal{L}^n \llcorner E$ to both sides of $(\star\star\star\star)$ yields an estimate contradicting the above inequalities.

Case 2. $\nu_E(x) \cdot (y - x) \leq -\epsilon |x - y|$. This similarly leads to a contradiction.

4. Now apply Whitney's Extension Theorem (proved in Section 6.5) with

$$f = 0, \; d = \nu_E \text{ on } K_k.$$

We conclude that there exist C^1-functions $\bar{f}_k : \mathbb{R}^n \to \mathbb{R}$ such that

$$\bar{f}_k = 0, \; D\bar{f}_k = \nu_E \text{ on } K_k.$$

Let

$$S_k := \left\{ x \in \mathbb{R}^n \mid \bar{f}_k = 0, |D\bar{f}_k| > \frac{1}{2} \right\} \quad (k = 1, 2, \dots).$$

According to the Implicit Function Theorem, S_k is a C^1, $(n-1)$-dimensional submanifold of $\mathbb{R}^n$. Clearly $K_k \subseteq S_k$. This proves (i) and (ii).

5. Choose a Borel set $B \subseteq \partial^* E$. In view of Lemma 5.4,

$$\mathcal{H}^{n-1}(B \cap N) \leq C \|\partial E\|(B \cap N) = 0.$$

Thus we may as well assume $B \subseteq \cup_{k=1}^\infty K_k$, and in fact $B \subseteq K_1$. By (ii) there exists a C^1-hypersurface $S_1 \supset K_1$. Let

$$\nu := \mathcal{H}^{n-1} \llcorner S_1.$$

Since S_1 is C^1,

$$\lim_{r \to 0} \frac{\nu(B(x,r))}{\alpha(n-1)r^{n-1}} = 1 \quad (x \in B).$$

Thus Theorem 5.14,(ii) implies

$$\lim_{r \to 0} \frac{\nu(B(x,r))}{\|\partial E\|(B(x,r))} = 1 \quad (x \in B).$$

Since ν and $\|\partial E\|$ are Radon measures, Theorem 1.30 implies

$$\|\partial E\|(B) = \nu(B) = \mathcal{H}^{n-1}(B). \qquad \square$$

5.8   Gauss–Green Theorem

As above, we continue to assume E is a set of locally finite perimeter in $\mathbb{R}^n$. We next refine Theorem 1.35 in Section 1.7.

DEFINITION 5.7. *Let* $x \in \mathbb{R}^n$. *We say* $x \in \partial_* E$, *the* **measure theoretic boundary** *of* E, *if*

$$\limsup_{r \to 0} \frac{\mathcal{L}^n(B(x,r) \cap E)}{r^n} > 0$$

and

$$\limsup_{r \to 0} \frac{\mathcal{L}^n(B(x,r) - E)}{r^n} > 0.$$

LEMMA 5.5. *We have*

(i) $\partial^* E \subseteq \partial_* E$, *and*

(ii) $\mathcal{H}^{n-1}(\partial_* E - \partial^* E) = 0$.

Proof. 1. Assertion (i) follows from Lemma 5.3 in Section 5.7.

2. Since the mapping

$$r \mapsto \frac{\mathcal{L}^n(B(x,r) \cap E)}{r^n}$$

is continuous, if $x \in \partial_* E$, there exist $0 < \alpha < 1$ and $r_j \to 0$ such that

$$\frac{\mathcal{L}^n(B(x,r_j) \cap E)}{\alpha(n)r_j^n} = \alpha.$$

Thus $\min\{\mathcal{L}^n(B(x,r_j) \cap E), \mathcal{L}^n(B(x,r_j) - E)\} = \min\{\alpha, 1-\alpha\}\alpha(n)r_j^n$, and so the relative isoperimetric inequality implies

$$\limsup_{r \to 0} \frac{\|\partial E\|(B(x,r))}{r^{n-1}} > 0.$$

Since $\|\partial E\|(\mathbb{R}^n - \partial^* E) = 0$, standard covering arguments imply

$$\mathcal{H}^{n-1}(\partial_* E - \partial^* E) = 0. \qquad \square$$

Now we prove that if E has locally finite perimeter, then the usual Gauss–Green formula holds, provided we consider the measure theoretic boundary of E.

THEOREM 5.16 (Gauss–Green Theorem). *Suppose $E \subset \mathbb{R}^n$ has locally finite perimeter.*

(i) *Then $\mathcal{H}^{n-1}(\partial_* E \cap K) < \infty$ for each compact set $K \subset \mathbb{R}^n$.*

(ii) *Furthermore, for $\mathcal{H}^{n-1}$-a.e. $x \in \partial_* E$, there is a unique measure theoretic unit outer normal $\nu_E(x)$ such that*

$$\int_E \operatorname{div} \phi \, dx = \int_{\partial_* E} \phi \cdot \nu_E \, d\mathcal{H}^{n-1}$$

for all $\phi \in C_c^1(\mathbb{R}^n; \mathbb{R}^n)$.

Proof. By the foregoing theory,

$$\int_E \operatorname{div} \phi \, dx = \int_{\mathbb{R}^n} \phi \cdot \nu_E \, d\|\partial E\|.$$

But

$$\|\partial E\|(\mathbb{R}^n - \partial^* E) = 0;$$

and, by Theorem 5.15 and Lemma 5.2,

$$\|\partial E\| = \mathcal{H}^{n-1} \llcorner \partial_* E.$$

Thus (ii) follows from Lemma 5.5.   □

Remark. We will see in Section 5.11 below that if $E \subseteq \mathbb{R}^n$ is $\mathcal{L}^n$-measurable and $\mathcal{H}^{n-1}(\partial_* E \cap K) < \infty$ for all compact $K \subseteq \mathbb{R}^n$, then E has locally finite perimeter. In particular, we see that *the Gauss–Green Theorem is valid for $E = U$, an open set with Lipschitz boundary.*   □

5.9   Pointwise properties of BV functions

We next extend our analysis of sets of finite perimeter to general BV functions. The goal will be to demonstrate that a BV function is "measure theoretically piecewise continuous," with "jumps along a measure theoretically C^1 surface."

We hereafter assume $f \in BV(\mathbb{R}^n)$ and investigate the approximate limits of $f(y)$ as y approaches a typical point $x \in \mathbb{R}^n$.

DEFINITION 5.8. *If f is $\mathcal{L}$-measurable, we define*

$$\mu(x) := \operatorname{ap} \limsup_{y \to x} f(y)$$

$$= \inf \left\{ t \mid \lim_{r \to 0} \frac{\mathcal{L}^n(B(x,r) \cap \{f > t\})}{r^n} = 0 \right\}$$

and

$$\lambda(x) := \operatorname{ap} \liminf_{y \to x} f(y)$$

$$= \sup \left\{ t \mid \lim_{r \to 0} \frac{\mathcal{L}^n(B(x,r) \cap \{f < t\})}{r^n} = 0 \right\}.$$

Remark. Clearly $-\infty \le \lambda(x) \le \mu(x) \le \infty$ for all $x \in \mathbb{R}^n$. □

LEMMA 5.6. *The functions λ and μ are Borel measurable.*

Proof. For each $t \in \mathbb{R}$, the set $E_t := \{x \in \mathbb{R}^n \mid f(x) > t\}$ is $\mathcal{L}^n$-measurable, and so for each $r > 0$, $t \in \mathbb{R}$, the mapping

$$x \mapsto \frac{\mathcal{L}^n(B(x,r) \cap E_t)}{r^n}$$

is continuous. This implies

$$\mu_t(x) := \limsup_{r \to 0, \ r \text{ rational}} \frac{\mathcal{L}^n(B(x,r) \cap E_t)}{r^n}$$

is a Borel measurable function of x for each $t \in \mathbb{R}$. Now, for each $s \in R$,

$$\{x \in \mathbb{R}^n \mid \mu(x) \le s\} = \bigcap_{k=1}^{\infty} \{x \in \mathbb{R}^n \mid \mu_{s+\frac{1}{k}}(x) = 0\},$$

and so μ is a Borel measurable function.

The proof that λ is Borel measurable is similar. □

DEFINITION 5.9. *Let*

$$J := \{x \in \mathbb{R}^n \mid \lambda(x) < \mu(x)\},$$

denote the set of points at which the approximate limit does not exist.

According to Theorem 1.37,

$$\mathcal{L}^n(J) = 0.$$

We will see below that for $\mathcal{H}^{n-1}$-a.e. point $x \in J$, f has a "measure theoretic jump" across a hyperplane through x.

THEOREM 5.17 (Approximating by hypersurfaces). *There exist countably many C^1-hypersurfaces $\{S_k\}_{k=1}^\infty$ such that*

$$\mathcal{H}^{n-1}\left(J - \bigcup_{k=1}^\infty S_k \right) = 0.$$

Proof. Define, as in Section 5.5,

$$E_t := \{x \in \mathbb{R}^n \mid f(x) > t\}$$

for $t \in \mathbb{R}$. According to the coarea formula for BV functions (Theorem 5.9), E_t is a set of finite perimeter in $\mathbb{R}^n$ for $\mathcal{L}^1$-a.e. t. Furthermore, observe that if $x \in J$ and $\lambda(x) < t < \mu(x)$, then

$$\limsup_{r \to 0} \frac{\mathcal{L}^n(B(x,r) \cap \{f > t\})}{r^n} > 0$$

and

$$\limsup_{r \to 0} \frac{\mathcal{L}^n(B(x,r) \cap \{f < t\})}{r^n} > 0.$$

Thus

$$\{x \in J \mid \lambda(x) < t < \mu(x)\} \subseteq \partial_* E_t. \qquad (\star)$$

Choose $D \subset \mathbb{R}^1$ to be a countable, dense set such that E_t is of finite perimeter for each $t \in D$. For each $t \in D$, $\mathcal{H}^{n-1}$-almost all of $\partial_* E_t$ is contained in a countable union of C^1 hypersurfaces; this is a consequence of the Structure Theorem 5.15.

Now, according to $(\star)$

$$J \subseteq \bigcup_{t \in D} \partial_* E_t,$$

and the theorem follows. □

THEOREM 5.18 (Approximate lim sup and lim inf). *We have*

$$-\infty < \lambda(x) \leq \mu(x) < +\infty$$

for $\mathcal{H}^{n-1}$-a.e. $x \in \mathbb{R}^n$.

Proof. 1. *Claim #1*: We have $\mathcal{H}^{n-1}(\{x \mid \lambda(x) = +\infty\}) = 0$ and $\mathcal{H}^{n-1}(\{x \mid \mu(x) = -\infty\}) = 0$.

Proof of claim: We may assume $\mathrm{spt}(f)$ is compact. Let

$$F_t := \{x \in \mathbb{R}^n \mid \lambda(x) > t\}.$$

Since $\mu(x) = \lambda(x) = f(x)$ $\mathcal{L}^n$-a.e., E_t and F_t differ at most by a set of $\mathcal{L}^n$-measure zero; whence

$$\|\partial E_t\| = \|\partial F_t\|.$$

Consequently, the coarea formula for BV functions implies

$$\int_{-\infty}^{\infty} \|\partial F_t\|(\mathbb{R}^n)\, dt = \|Df\|(\mathbb{R}^n) < \infty,$$

and so

$$\liminf_{t \to \infty} \|\partial F_t\|(\mathbb{R}^n) = 0. \tag{$\star$}$$

Since $\mathrm{spt}(f)$ is compact, there exists $d > 0$ such that

$$\mathcal{L}^n(\mathrm{spt}(f) \cap B(x,r)) \leq \frac{1}{8}\alpha(n)r^n \quad (x \in \mathrm{spt}(f), r \geq d). \tag{$\star\star$}$$

Fix $t > 0$. By the definitions of λ and F_t,

$$\lim_{r \to 0} \frac{\mathcal{L}^n(B(x,r) \cap F_t)}{\alpha(n)r^n} = 1 \text{ for } x \in F_t.$$

Thus for each $x \in F_t$, there exists $r > 0$ such that

$$\frac{\mathcal{L}^n(B(x,r) \cap F_t)}{\alpha(n)r^n} = \frac{1}{4}. \tag{$\star\star\star$}$$

According to $(\star\star)$, it follows that $r \leq d$.

We apply Vitali's Covering Theorem to find a countable disjoint collection $\{B(x_i, r_i)\}_{i=1}^{\infty}$ of balls satisfying $(\star\star\star)$ for $x = x_i$ and $r = r_i \leq d$, such that

$$F_t \subseteq \bigcup_{i=1}^{\infty} B(x_i, 5r_i).$$

Now $(\star\star\star)$ and the relative isoperimetric inequality imply

$$\left(\frac{\alpha(n)}{4}\right)^{\frac{n-1}{n}} \leq \frac{C\|\partial F_t\|(B(x_i, r_i))}{r_i^{n-1}};$$

that is,
$$r_i^{n-1} \le C\|\partial F_t\|(B(x_i, r_i)) \qquad (i = 1, 2, \dots).$$

Thus we may calculate

$$\mathcal{H}_{10d}^{n-1}(F_t) \le \sum_{i=1}^{\infty} \alpha(n-1)(5r_i)^{n-1}$$

$$\le C \sum_{i=1}^{\infty} \|\partial F_t\|(B(x_i, r_i))$$

$$\le C\|\partial F_t\|(\mathbb{R}^n).$$

In view of $(\star)$,
$$\mathcal{H}_{10d}^{n-1}(\{x \mid \lambda(x) = +\infty\}) = 0,$$

and so
$$\mathcal{H}^{n-1}(\{x \mid \lambda(x) = +\infty\}) = 0.$$

The proof that $\mathcal{H}^{n-1}(\{x \mid \mu(x) = -\infty\}) = 0$ is similar.

2. *Claim #2*: $\mathcal{H}^{n-1} \left(\{x \mid \mu(x) - \lambda(x) = \infty\}\right) = 0.$

Proof of claim: By Theorem 5.17, J is σ-finite with respect to $\mathcal{H}^{n-1}$ in $\mathbb{R}^n$, and thus $\{(x, t) \mid x \in J, \lambda(x) < t < \mu(x)\}$ is σ-finite with respect to $\mathcal{H}^{n-1} \times \mathcal{L}^1$ in $\mathbb{R}^{n+1}$. Consequently, Fubini's Theorem implies

$$\int_{-\infty}^{\infty} \mathcal{H}^{n-1}(\{\lambda(x) < t < \mu(x)\}) \, dt = \int_{\mathbb{R}^n} \mu(x) - \lambda(x) \, d\mathcal{H}^{n-1}.$$

But by statement $(\star)$ in the proof of Theorem 5.17 and the theory developed in Section 5.7,

$$\int_{-\infty}^{\infty} \mathcal{H}^{n-1}(\{\lambda(x) < t < \mu(x)\}) \, dt \le \int_{-\infty}^{\infty} \mathcal{H}^{n-1}(\partial_* E_t) dt$$

$$= \int_{-\infty}^{\infty} \|\partial E_t\|(\mathbb{R}^n) dt$$

$$= \|Df\|(\mathbb{R}^n) < \infty.$$

Consequently, $\mathcal{H}^{n-1}(\{x \mid \mu(x) - \lambda(x) = \infty\}) = 0.$ □

NOTATION We hereafter write

$$F(x) := \frac{\lambda(x) + \mu(x)}{2}.$$

DEFINITION 5.10. *Let ν be a unit vector in $\mathbb{R}^n$, $x \in \mathbb{R}^n$. We define the hyperplane*

$$H_\nu := \{y \in \mathbb{R}^n \mid \nu \cdot (y - x) = 0\}$$

and the half-spaces

$$H_\nu^+ := \{y \in \mathbb{R}^n \mid \nu \cdot (y - x) \geq 0\},$$
$$H_\nu^- := \{y \in \mathbb{R}^n \mid \nu \cdot (y - x) \leq 0\}.$$

THEOREM 5.19 (Fine properties of BV functions). *Assume $f \in BV(\mathbb{R}^n)$.*

(i) *Then for $\mathcal{H}^{n-1}$-a.e. $x \in \mathbb{R}^n - J$, we have*

$$\lim_{r \to 0} \fint_{B(x,r)} |f - F(x)|^{\frac{n}{n-1}} \, dy = 0.$$

(ii) *Furthermore, for $\mathcal{H}^{n-1}$-a.e. $x \in J$, there exists a unit vector $\nu = \nu(x)$ such that*

$$\lim_{r \to 0} \fint_{B(x,r) \cap H_\nu^-} |f - \mu(x)|^{\frac{n}{n-1}} \, dy = 0$$

and

$$\lim_{r \to 0} \fint_{B(x,r) \cap H_\nu^+} |f - \lambda(x)|^{\frac{n}{n-1}} \, dy = 0.$$

(iii) *In particular,*

$$\mu(x) = \operatorname*{ap\,lim}_{y \to x, y \in H_\nu^+} f(y), \quad \lambda(x) = \operatorname*{ap\,lim}_{y \to x, y \in H_\nu^-} f(y).$$

Remark. Thus we see that for $\mathcal{H}^{n-1}$-a.e. $x \in J$, f has a "measure theoretic jump" across the hyperplane $H_{\nu(x)}$.   □

Proof. We will prove only the second part of assertion (ii), as the other statements follow similarly.

1. For $\mathcal{H}^{n-1}$-a.e. point $x \in J$, there exists a unit vector ν such that ν is the measure theoretic exterior unit normal to $E_t = \{f > t\}$ at x

for $\lambda(x) < t < \mu(x)$. Thus for each $\epsilon > 0$,

$$\begin{cases} \dfrac{\mathcal{L}^n(B(x,r) \cap \{f > \lambda(x) + \epsilon\} \cap H_\nu^+)}{r^n} = 0, \\[4mm] \dfrac{\mathcal{L}^n(B(x,r) \cap \{f < \lambda(x) - \epsilon\})}{r^n} = 0. \end{cases} \quad (\star)$$

Hence if $0 < \epsilon < 1$,

$$\frac{1}{r^n} \int_{B(x,r) \cap H_\nu^+} |f - \lambda(x)|^{\frac{n}{n-1}} \, dy$$

$$\leq \frac{1}{2} \alpha(n) \epsilon^{\frac{n}{n-1}} + \frac{1}{r^n} \int_{B(x,r) \cap H_\nu^+ \cap \{f > \lambda(x) + \epsilon\}} |f - \lambda(x)|^{\frac{n}{n-1}} \, dy$$

$$+ \frac{1}{r^n} \int_{B(x,r) \cap H_\nu^+ \cap \{f < \lambda(x) - \epsilon\}} |f - \lambda(x)|^{\frac{n}{n-1}} \, dy. \quad (\star\star)$$

Now fix $M > \lambda(x) + \epsilon$. Then

$$\frac{1}{r^n} \int_{B(x,r) \cap H_\nu^+ \cap \{f > \lambda(x) + \epsilon\}} |f - \lambda(x)|^{\frac{n}{n-1}} \, dy$$

$$\leq (M - \lambda(x))^{\frac{n}{n-1}} \frac{\mathcal{L}^n(B(x,r) \cap H_\nu^+ \cap \{f > \lambda(x) + \epsilon\})}{r^n}$$

$$+ \frac{1}{r^n} \int_{B(x,r) \cap \{f > M\}} |f - \lambda(x)|^{\frac{n}{n-1}} \, dy.$$

Similarly, if $-M < \lambda(x) - \epsilon$, we have

$$\frac{1}{r^n} \int_{B(x,r) \cap \{f < \lambda(x) - \epsilon\}} |f - \lambda(x)|^{\frac{n}{n-1}} \, dy$$

$$\leq (M + \lambda(x))^{\frac{n}{n-1}} \frac{\mathcal{L}^n(B(x,r) \cap \{f < \lambda(x) - \epsilon\})}{r^n}$$

$$+ \frac{1}{r^n} \int_{B(x,r) \cap \{f < -M\}} |f - \lambda(x)|^{\frac{n}{n-1}} \, dy.$$

We employ the two previous calculations in $(\star\star)$ and then recall $(\star)$ to compute

$$\limsup_{r \to 0} \frac{1}{r^n} \int_{B(x,r) \cap H_\nu^+} |f - \lambda(x)|^{\frac{n}{n-1}} \, dy$$

$$\leq \limsup_{r \to 0} \frac{1}{r^n} \int_{B(x,r) \cap \{|f| > M\}} |f - \lambda(x)|^{\frac{n}{n-1}} \, dy \quad (\star\star\star)$$

for all sufficiently large $M > 0$.

2. Now

$$\frac{1}{r^n} \int_{B(x,r) \cap \{f > M\}} |f - \lambda(x)|^{\frac{n}{n-1}} \, dy \leq \frac{C}{r^n} \int_{B(x,r)} (f - M)^{\frac{n}{n-1}} \, dy$$
$$+ (M - \lambda(x))^{\frac{n}{n-1}} \frac{\mathcal{L}^n(B(x,r) \cap \{f < M\})}{r^n}.$$

If $M > \mu(x)$, the second term on the right-hand side of this inequality goes to zero as $r \to 0$. Furthermore, for sufficiently small $r > 0$,

$$\frac{\mathcal{L}^n(B(x,r) \cap \{f > M\})}{\mathcal{L}^n(B(x,r))} \leq \frac{1}{2};$$

and hence by Theorem 5.17, (iii) we have

$$\left(\fint_{B(x,r)} (f - M)^{\frac{n}{n-1}} \, dy \right)^{\frac{n-1}{n}} \leq \frac{C}{r^{n-1}} \|D(f - M)^+\|(B(x,r)).$$

This estimate and the analogous one over the set $\{f < -M\}$ combine with $(\star \star \star)$ to prove

$$\limsup_{r \to 0} \left(\fint_{B(x,r) \cap H_\nu^+} |f - \lambda(x)|^{\frac{n}{n-1}} \, dy \right)^{\frac{n-1}{n}}$$
$$\leq C \limsup_{r \to 0} \frac{\|D(f - M)^+\|(B(x,r))}{r^{n-1}}$$
$$+ C \limsup_{r \to 0} \frac{\|D(-M - f)^+\|(B(x,r))}{r^{n-1}} \qquad (\star \star \star \star)$$

for all sufficiently large $M > 0$.

3. Fix $\epsilon > 0$, $N > 0$, and define

$$A_\epsilon^N :=$$
$$\left\{ x \in \mathbb{R}^n \mid \limsup_{r \to 0} \frac{\|D(f - M)^+\|(B(x,r))}{r^{n-1}} > \epsilon \text{ for all } M \geq N \right\}.$$

Then

$$\epsilon \mathcal{H}^{n-1}(A_\epsilon^N) \leq C \|D(f - M)^+\|(\mathbb{R}^n) = C \int_M^\infty \|\partial E_t\|(\mathbb{R}^n) \, dt$$

for all $M \geq N$. Consequently,

$$\mathcal{H}^{n-1}(A_\epsilon^N) = 0,$$

and so

$$\lim_{M\to\infty} \limsup_{r\to 0} \frac{\|D(f-M)^+\|(B(x,r))}{r^{n-1}} = 0$$

for $\mathcal{H}^{n-1}$-a.e. $x \in J$. Similarly,

$$\lim_{M\to\infty} \limsup_{r\to 0} \frac{\|D(-M-f)^+\|(B(x,r))}{r^{n-1}} = 0.$$

These estimates and $(\star)$ prove

$$\lim_{r\to 0} \fint_{B(x,r)\cap H_\nu^+} |f - \lambda(x)|^{1^*} \, dy = 0. \qquad \square$$

THEOREM 5.20 (BV and mollifiers).

(i) *If $f \in BV(\mathbb{R}^n)$, then*

$$f^*(x) := \lim_{r\to 0} (f)_{x,r} = F(x)$$

exists for $\mathcal{H}^{n-1}$-a.e. $x \in \mathbb{R}^n$.

(ii) *Furthermore, if η_ϵ is the standard mollifier and $f^\epsilon := \eta_\epsilon * f$, then*

$$f^*(x) = \lim_{\epsilon\to 0} f^\epsilon(x)$$

for $\mathcal{H}^{n-1}$-a.e. $x \in \mathbb{R}^n$.

Proof. This is a corollary of the foregoing theorem. $\qquad \square$

5.10   Essential variation on lines

We now investigate the behavior of a BV function restricted to lines.

5.10.1 BV functions of one variable

We first study BV functions of one variable. Suppose $f : \mathbb{R} \to \mathbb{R}$ is $\mathcal{L}^1$-measurable, and $-\infty \le a < b \le \infty$.

DEFINITION 5.11. *The* **essential variation** *of f on the interval* (a, b) *is*

$$\operatorname{ess} V_a^b f := \sup \left\{ \sum_{j=1}^m |f(t_{j+1}) - f(t_j)| \right\},$$

the supremum taken over all finite partitions $\{a < t_1 < \cdots < t_{m+1} < b\}$ such that each t_i is a point of approximate continuity of f.

Remark. The **variation** of f on (a, b) is similarly defined, but without the proviso that each partition point t_j be a point of approximate continuity. Since we demand that a function remain BV even after being redefined on a set of $\mathcal{L}^1$ measure zero, we see that essential variation is the proper notion here.

In particular, if $f = g$ $\mathcal{L}^1$-a.e. on (a, b), then

$$\operatorname{ess} V_a^b f = \operatorname{ess} V_a^b g.$$

$\square$

THEOREM 5.21 (BV functions of one variable). *Suppose $f \in L^1(a, b)$. Then*

$$\|Df\|(a, b) = \operatorname{ess} V_a^b f;$$

and thus

$$f \in BV(a, b) \text{ if and only if } \operatorname{ess} V_a^b f < \infty.$$

Proof. 1. Consider first $\operatorname{ess} V_a^b f$. Fix $\epsilon > 0$ and let $f^\epsilon := \eta_\epsilon * f$ denote the usual smoothing of f. Choose any $a + \epsilon < t_1 < \cdots < t_{m+1} < b - \epsilon$. Since $\mathcal{L}^1$-a.e. point is a point of approximate continuity of f, $t_j - s$ is a point of approximate continuity of f for $\mathcal{L}^1$-a.e. s. Hence

$$\sum_{j=1}^m |f^\epsilon(t_{j+1}) - f^\epsilon(t_j)| = \sum_{j=1}^m \left| \int_{-\epsilon}^\epsilon \eta_\epsilon(s)(f(t_{j+1} - s) - f(t_j - s))\, ds \right|$$

$$\le \int_{-\epsilon}^\epsilon \eta_\epsilon(s) \sum_{j=1}^m |f(t_{j+1} - s) - f(t_j - s)|\, ds$$

$$\le \operatorname{ess} V_a^b f.$$

It follows that

$$\int_{a+\epsilon}^{b-\epsilon} |(f^\epsilon)'| \, dx = \sup \left\{ \sum_{j=1}^{m} |f^\epsilon(t_{j+1}) - f^\epsilon(t_j)| \right\} \leq \operatorname{ess} V_a^b f.$$

Thus if $\phi \in C_c^1(a, b)$ and $|\phi| \leq 1$, we have

$$\int_a^b f^\epsilon \phi' \, dx = -\int_a^b (f^\epsilon)' \phi \, dx \leq \int_{a+\epsilon}^{b-\epsilon} |(f^\epsilon)'| \, dx \leq \operatorname{ess} V_a^b f$$

for ϵ sufficiently small. Let $\epsilon \to 0$ to find

$$\|Df\|(a, b) = \sup \left\{ \int_a^b f\phi' \, dx \mid \phi \in C_c^1(a, b), |\phi| \leq 1| \right\}$$
$$\leq \operatorname{ess} V_a^b f \leq \infty.$$

In particular, if $f \notin BV(a, b)$, then $\operatorname{ess} V_a^b f = \infty$.

2. Now suppose $f \in BV(a, b)$ and choose $a < c < d < b$. Then for each $\phi \in C_c^1(c, d)$, with $|\phi| \leq 1$, and each small $\epsilon > 0$, we calculate

$$\int_c^d (f^\epsilon)' \phi \, dx = -\int_c^d f^\epsilon \, \phi' \, dx$$
$$= -\int_c^d (\eta_\epsilon * f)\phi' \, dx$$
$$= -\int_c^b f(\eta_\epsilon * \phi)' \, dx$$
$$\leq \|Df\|(a, b).$$

Thus $\int_c^d |(f^\epsilon)'| dx \leq \|Df\|(a, b)$.

3. *Claim:* $f \in L^\infty(a, b)$.

Proof of claim: Choose $\{f_j\}_{j=1}^\infty \subset BV(a, b) \cap C^\infty(a, b)$ so that

$$f_j \to f \text{ in } L^1(a, b), \quad f_j \to f \ \mathcal{L}^n\text{-a.e.}$$

and

$$\int_a^b |f_j'| \, dx \to \|Df\|(a, b).$$

For each $y, z \in (a, b)$,

$$f_j(z) = f_j(y) + \int_y^z f_j' \, dx.$$

Averaging with respect to $y \in (a, b)$, we obtain

$$|f_j(z)| \le \fint_a^b |f_j| \, dy + \int_a^b |f_j'| \, dx,$$

and so

$$\sup_j \|f_j\|_{L^\infty(a,b)} < \infty.$$

Since $f_j \to f$ $\mathcal{L}^n$-a.e., we deduce that $\|f\|_{L^\infty(a,b)} < \infty$.

4. It follows from the claim that each point of approximate continuity of f is a Lebesgue point and hence

$$f^\epsilon(t) \to f(t) \qquad\qquad (\star)$$

as $\epsilon \to 0$ for each point of approximate continuity of f. Consequently, for each partition $\{a < t_1 < \cdots < t_{m+1} < b\}$, with each t_j a point of approximate continuity of f, we have

$$\sum_{j=1}^m |f(t_{j+1}) - f(t_j)| = \lim_{\epsilon \to 0} \sum_{j=1}^m |f^\epsilon(t_{j+1}) - f^\epsilon(t_j)|$$

$$\le \limsup_{\epsilon \to 0} \int_a^b |(f^\epsilon)'| \, dx$$

$$\le \|Df\|(a, b).$$

Thus

$$\operatorname{ess} V_a^b f \le \|Df\|(a, b) < \infty. \qquad \square$$

5.10.2 Essential variation on almost all lines

We next extend our analysis to BV functions on $\mathbb{R}^n$.

NOTATION Suppose $f \colon \mathbb{R}^n \to \mathbb{R}$. Then for $k = 1, \ldots, n$, set

$$x' = (x_1, \ldots, x_{k-1}, x_{k+1}, \ldots x_n) \in \mathbb{R}^{n-1}.$$

If $t \in \mathbb{R}$, write

$$f_k(x', t) := f(\ldots, x_{k-1}, t, x_{k+1}, \ldots).$$

Thus $\operatorname{ess} V_a^b f_k$ means the essential variation of f_k as a function of $t \in (a, b)$, for each fixed x'.

LEMMA 5.7. *Assume* $f \in L^1_{\mathrm{loc}}(\mathbb{R}^n)$, $k \in \{1, \ldots, n\}$, *and* $-\infty \le a < b \le \infty$. *Then the mapping*

$$x' \mapsto \mathrm{ess}\, V_a^b f_k$$

is $\mathcal{L}^{n-1}$-*measurable.*

Proof. According to Theorem 5.21, for $\mathcal{L}^{n-1}$-a.e. $x' \in \mathbb{R}^{n-1}$

$$\mathrm{ess}\, V_a^b f_k = \|Df_k\|(a, b)$$
$$= \sup \left\{ \int_a^b f_k(x', t)\phi'(t)\, dt \mid \phi \in C_c^1(a, b), |\phi| \le 1 \right\}.$$

Take $\{\phi_j\}_{j=1}^\infty$ to be a countable, dense subset of $C_c^1(a, b) \cap \{|\phi| \le 1\}$. Then

$$x' \mapsto \int_a^b f_k(x', t)\phi_j'(t)\, dt$$

is $\mathcal{L}^{n-1}$-measurable for $j = 1, \ldots$ and so

$$x' \mapsto \sup_j \left\{ \int_a^b f_k(x', t)\phi_j'(t)\, dt \right\} = \mathrm{ess}\, V_a^b f_k$$

is $\mathcal{L}^{n-1}$-measurable. $\square$

THEOREM 5.22 (Essential variation on lines). *Assume that* $f \in L^1_{\mathrm{loc}}(\mathbb{R}^n)$. *Then* $f \in BV_{\mathrm{loc}}(\mathbb{R}^n)$ *if and only if*

$$\int_K \mathrm{ess}\, V_a^b f_k\, dx' < \infty \quad (k = 1, \ldots, n)$$

for all $-\infty < a < b < \infty$ *and all compact sets* $K \subset \mathbb{R}^{n-1}$.

Proof. 1. First suppose $f \in BV_{\mathrm{loc}}(\mathbb{R}^n)$. Choose k, a, b, K as above. Set

$$C := \{x \mid a \le x_k \le b, (x_1, \ldots, x_{k-1}, x_{k+1}, \ldots, x_n) \in K\}.$$

Let $f^\epsilon := \eta_\epsilon * f$, as before. Then

$$\lim_{\epsilon \to 0} \int_C |f^\epsilon - f|\, dx = 0, \quad \limsup_{\epsilon \to 0} \int_C |Df^\epsilon|\, dx < \infty.$$

Thus for $\mathcal{H}^{n-1}$-a.e. $x' \in K$,

$$f_k^\epsilon \to f_k \quad \text{in } L^1(a, b),$$

where

$$f_k^\epsilon(x', t) := f^\epsilon(\ldots, x_{k-1}, t, x_{k+1}, \ldots).$$

Hence

$$\mathrm{ess}\, V_a^b f_k \le \liminf_{\epsilon \to 0} \mathrm{ess}\, V_a^b f_k^\epsilon$$

for $\mathcal{H}^{n-1}$-a.e. $x' \in K$. Thus Fatou's Lemma implies

$$\int_K \mathrm{ess}\, V_a^b f_k \, dx' \le \liminf_{\epsilon \to 0} \int_K \mathrm{ess}\, V_a^b f_k^\epsilon \, dx'$$

$$= \liminf_{\epsilon \to 0} \int_C |f_{x_k}^\epsilon| \, dx$$

$$\le \limsup_{\epsilon \to 0} \int_C |Df^\epsilon| \, dx < \infty.$$

2. Now suppose $f \in L^1_{\mathrm{loc}}(\mathbb{R}^n)$ and

$$\int_K \mathrm{ess}\, V_a^b f_k \, dx' < \infty$$

for all $k = 1, \ldots, n$, $a < b$ and compact sets $K \subset \mathbb{R}^{n-1}$. Fix $\phi \in C_c^\infty(\mathbb{R}^n)$, with $|\phi| \le 1$, and choose a, b, and k such that

$$\mathrm{spt}(\phi) \subset \{x \mid a < x_k < b\}.$$

Then Theorem 5.21 implies

$$\int_{\mathbb{R}^n} f \phi_{x_k} \, dx \le \int_K \mathrm{ess} V_a^b f_k \, dx' < \infty,$$

for

$$K :=$$
$$\{x' \in \mathbb{R}^{n-1} \mid (\ldots, x_{k-1}, t, x_{k+1}, \ldots) \in \mathrm{spt}(\phi) \text{ for some } t \in \mathbb{R}\}.$$

As this estimate holds for $k = 1, \ldots, n$, we deduce $f \in BV_{\mathrm{loc}}(\mathbb{R}^n)$. $\quad\square$

5.11   A criterion for finite perimeter

We conclude this chapter by establishing a relatively simple criterion for a set E to have locally finite perimeter.

NOTATION We will write the point $x \in \mathbb{R}^n$ as $x = (x', t)$, for $x' = (x_1, \ldots, x_{n-1}) \in \mathbb{R}^{n-1}$, $t = x_n \in \mathbb{R}$.

DEFINITION 5.12.

(i) *The* **projection** $P : \mathbb{R}^n \to \mathbb{R}^{n-1}$ *is*

$$P(x) = x' \qquad (x = (x', x_n) \in \mathbb{R}^n).$$

(ii) *The* **multiplicity function** *is*

$$N(P \mid A, x') := \mathcal{H}^0(A \cap P^{-1}\{x'\})$$

for Borel sets $A \subseteq \mathbb{R}^n$ and $x' \in \mathbb{R}^{n-1}$.

LEMMA 5.8.

(i) *The mapping $x' \mapsto N(P \mid A, x')$ is $\mathcal{L}^{n-1}$-measurable.*

(ii) $\int_{\mathbb{R}^{n-1}} N(P \mid A, x') \, dx' \le \mathcal{H}^{n-1}(A).$

Proof. Assertions (i) and (ii) follow as in the proof of Lemma 3.5, Section 3.4; see also the remark in Section 3.4. □

DEFINITION 5.13. *Let $E \subseteq \mathbb{R}^n$ be $\mathcal{L}^n$-measurable. We define*

$$I := \left\{ x \in \mathbb{R}^n \mid \lim_{r \to 0} \frac{\mathcal{L}^n(B(x,r) - E)}{r^n} = 0 \right\}$$

to be the **measure theoretic interior** *of E and*

$$O := \left\{ x \in \mathbb{R}^n \mid \lim_{r \to 0} \frac{\mathcal{L}^n(B(x,r) \cap E)}{r^n} = 0 \right\}$$

to be the **measure theoretic exterior** *of E.*

Remark. Note $\partial_* E = \mathbb{R}^n - (I \cup O)$. Think of I as denoting the "inside" and O as denoting the "outside" of E. □

LEMMA 5.9.

(i) *I, O, and $\partial_* E$ are Borel measurable sets.*

(ii) *$\mathcal{L}^n((I - E) \cup (E - I)) = 0.$*

Proof. 1. There exists a Borel set $C \subseteq \mathbb{R}^n - E$ such that $\mathcal{L}^n(C \cap T) = \mathcal{L}^n(T - E)$ for all $\mathcal{L}^n$-measurable sets T. Thus

$$I = \left\{ x \mid \lim_{r \to 0} \frac{\mathcal{L}^n(B(x,r) \cap C)}{r^n} = 0 \right\},$$

and so is Borel measurable. The proof for O is similar.

2. Assertion (ii) follows from Theorem 1.35. □

THEOREM 5.23 (Criterion for finite perimeter). *Let $E \subseteq \mathbb{R}^n$ be $\mathcal{L}^n$-measurable. Then E has locally finite perimeter if and only if*

$$\mathcal{H}^{n-1}(K \cap \partial_* E) < \infty \qquad (\star)$$

for each compact set $K \subset \mathbb{R}^n$.

Proof. 1. Assume first $(\star)$ holds, fix $a > 0$, and set

$$U := (-a, a)^n \subset \mathbb{R}^n.$$

To simplify notation slightly, let us write $z = x' \in \mathbb{R}^{n-1}$, $t = x_n \in \mathbb{R}$. Note from Lemma 5.8 and hypothesis $(\star)$ that

$$\int_{\mathbb{R}^{n-1}} N(P \mid U \cap \partial_* E, z)\, dz \leq \mathcal{H}^{n-1}(U \cap \partial_* E) < \infty. \qquad (\star\star)$$

Define for each $z \in \mathbb{R}^{n-1}$

$$f^z(t) := \chi_I(z, t) \quad (t \in \mathbb{R}).$$

Select $\phi \in C_c^1(U)$, with $|\phi| \leq 1$, and then compute

$$\int_E \operatorname{div}(\phi e_n)\, dx = \int_I \operatorname{div}(\phi e_n)\, dx = \int_I \phi_{x_n}\, dx$$
$$= \int_{\mathbb{R}^{n-1}} \left[\int_{\mathbb{R}} f^z(t) \phi_{x_n}(z, t)\, dt \right] dz \qquad (\star\star\star)$$
$$\leq \int_V \operatorname{ess} V_{-a}^a f^z\, dz$$

where

$$V := (-a, a)^{n-1} \subseteq \mathbb{R}^{n-1}.$$

2. For positive integers k and m, define

$$G(k) := \left\{ x \in \mathbb{R}^n \mid \mathcal{L}^n(B(x, r) \cap O) \leq \frac{a(n-1)}{3^{n+1}} r^n \text{ for } 0 < r < \frac{3}{k} \right\},$$

$$H(k) := \left\{ x \in \mathbb{R}^n \mid \mathcal{L}^n(B(x, r) \cap I) \leq \frac{a(n-1)}{3^{n+1}} r^n \text{ for } 0 < r < \frac{3}{k} \right\},$$

and

$$G^+(k,m) := G(k) \cap \left\{ x \mid x + se_n \in O \text{ for } 0 < s < \frac{3}{m} \right\},$$

$$G^-(k,m) := G(k) \cap \left\{ x \mid x - se_n \in O \text{ for } 0 < s < \frac{3}{m} \right\},$$

$$H^+(k,m) := H(k) \cap \left\{ x \mid x + se_n \in I \text{ for } 0 < s < \frac{3}{m} \right\},$$

$$H^-(k,m) := H(k) \cap \left\{ x \mid x - se_n \in I \text{ for } 0 < s < \frac{3}{m} \right\}.$$

3. *Claim #1:* $\mathcal{L}^{n-1}(P(G^{\pm}(k,m))) = \mathcal{L}^{n-1}(P(H^{\pm}(k,m))) = 0$ for $k, m = 1, 2, \ldots$.

Proof of claim: For fixed k, m, write

$$G^+(k,m) = \bigcup_{j=-\infty}^{\infty} G_j,$$

where

$$G_j := G^+(k,m) \cap \left\{ x \mid \frac{j-1}{m} \le x_n < \frac{j}{m} \right\}.$$

Assume $z \in \mathbb{R}^{n-1}$, $0 < r < \min\{\frac{1}{k}, \frac{1}{m}\}$, and $B(z,r) \cap P(G_j) \ne \emptyset$. Then there exists a point $b \in G_j \cap P^{-1}(B(z,r)) \subseteq G(k)$ such that

$$b_n + \frac{r}{2} > \sup\{x_n \mid x \in G_j \cap P^{-1}(B(x,r))\}.$$

Thus, by the definition of $G^+(k,m)$, we have

$$\left\{ y \mid b_n + \frac{r}{2} \le y_n \le b_n + r \right\} \cap P^{-1}(P(G_j) \cap B(z,r)) \subset O \cap B(b,3r).$$

Take the $\mathcal{L}^n$ measure of each side above to calculate

$$\frac{r}{2} \mathcal{L}^{n-1}(P(G_j) \cap B(z,r)) \le \mathcal{L}^n(O \cap B(b,3r)) \le \frac{\alpha(n-1)}{3^{n+1}}(3r)^n,$$

since $b \in G(k)$. Then

$$\limsup_{r \to 0} \frac{\mathcal{L}^{n-1}(P(G_j) \cap B(z,r))}{\alpha(n-1)r^{n-1}} \le \frac{2}{3}$$

for all $z \in \mathbb{R}^{n-1}$. This implies

$$\mathcal{L}^{n-1}(P(G_j)) = 0 \quad (j = 0 \pm 1, \pm 2, \ldots);$$

and consequently
$$\mathcal{L}^{n-1}(P(G^+(k,m))) = 0.$$

Similar arguments imply
$$\mathcal{L}^{n-1}(P(G^-(k,m))) = \mathcal{L}^{n-1}(P(H^\pm(k,m))) = 0$$

for all k, m.

4. Now suppose
$$z \in V - \bigcup_{k,m=1}^{\infty} P[G^+(k,m) \cup G^-(k,m) \cup H^+(k,m) \cup H^-(k,m)]$$

$$(\star\star\star)$$

and
$$N(P \mid U \cap \partial_* E, z) < \infty.$$

Assume $-a < t_1 < \cdots < t_{m+1} < a$ are points of approximate continuity of f^z. Notice that $|f^z(t_{j+1}) - f^z(t_j)| \neq 0$ if and only if $|f^z(t_{j+1}) - f^z(t_j)| = 1$. In the latter case we may, for definiteness, suppose $(z, t_j) \in I$, but $(z, t_{j+1}) \notin I$. Since t_{j+1} is a point of approximate continuity of f^z and since $\mathbb{R}^n - (O \cup I) = \partial_* E$, it follows from the finiteness of $N(P \mid U \cap \partial_* E, z)$ that every neighborhood of t_{j+1} must contain points s such that $(z, s) \in O$ and f^z is approximately continuous at s. Consequently,

$$\operatorname{ess} V_{-a}^a f^z = \sup \left\{ \sum_{j=1}^{m} |f^z(t_{j+1}) - f^z(t_j)| \right\},$$

the supremum taken over points $-a < t_1 < \cdots < t_{m+1} < a$ such that $(z, t_j) \in (O \cup I)$ and f^z is approximately continuous at each t_j.

5. *Claim #2*: If $(z, u) \in I$ and $(z, v) \in O$, with $u < v$, then there exists $u < t < v$ such that $(z, t) \in \partial_* E$.

Proof of claim: Suppose not; then $(z, t) \in (O \cup I)$ for all $u < t < v$. We observe that
$$I \subset \bigcup_{k=1}^{\infty} G(k), \quad O \subseteq \bigcup_{k=1}^{\infty} H(k),$$

and that the sets $G(k)$, $H(k)$ are increasing and closed. Hence there exists k_0 such that $(z, u) \in G(k_0)$, $(z, v) \in H(k_0)$. Now $H(k_0) \cap G(k_0) = \emptyset$, and so
$$u_0 := \sup\{t \mid (z, t) \in G(k_0), t < v\} < v.$$

Set

$$v_0 := \inf\{t \mid (z,t) \in H(k_0), t > u_0\}.$$

Then

$$(z, u_0) \in G(k_0), \quad (z, v_0) \in H(k_0), \quad u \le u_0 < v_0 \le v,$$

and

$$\{(z,t) \mid u_0 < t < v_0\} \cap [H(k_0) \cup G(k_0)] = \emptyset.$$

Next, there exist

$$u_0 < s_1 < t_1 < v_0$$

with $(z, s_1) \in I$, and $(z, t_1) \in O$; this is a consequence of $(\star\star\star\star)$. Arguing as above, we find $k_1 > k_0$ and numbers u_1, v_1 such that

$$u_0 < u_1 < v_1 < v_0, \quad (z, u_1) \in G(k_1), \quad (z, v_1) \in H(k_1),$$

and $(z,t) \notin H(k_1) \cup G(k_1)$ if $u_1 < t < v_1$.

Continuing, we see that there exist $k_j \to \infty$ and sequences $\{u_j\}_{j=1}^\infty$, $\{v_j\}_{j=1}^\infty$ such that

$$\begin{cases} u_0 < u_1 < \ldots, \; v_0 > v_1 > v_2 \ldots, \\ u_j < v_j \text{ for all } j = 1, 2, \ldots, \\ (z, u_j) \in G(k_j), \; (z, v_j) \in H(k_j), \\ (z,t) \notin G(k_j) \cup H(k_j) \text{ if } u_j < t < v_j. \end{cases}$$

Choose

$$\lim_{j \to \infty} u_j \le t \le \lim_{j \to \infty} v_j.$$

Then

$$y := (z,t) \notin \bigcup_{j=1}^\infty [G(k_j) \cup H(k_j)];$$

hence

$$\limsup_{r \to 0} \frac{\mathcal{L}^n(B(y,r) \cap E)}{r^n} \ge \frac{\alpha(n-1)}{3^{n+1}}$$

and

$$\limsup_{r \to 0} \frac{\mathcal{L}^n(B(y,r) - E)}{r^n} \ge \frac{\alpha(n-1)}{3^{n+1}}.$$

Thus $y \in \partial_* E$.

6. Now, by Claim #2, if z satisfies $(\star\star\star\star)$, then

$$\text{ess } V_{-a}^a f^z \le \text{Card}\{t \mid -a < t < a, (z,t) \in \partial_* E\}$$
$$= N(P \mid U \cap \partial_* E, z).$$

Thus $(\star\star\star)$ implies

$$\int_V \operatorname{ess} V_{-a}^a f^z \, dz \leq \int_V N(P \mid U \cap \partial_* E, z) \, dz$$
$$\leq \mathcal{H}^{n-1}(U \cap \partial_* E) < \infty,$$

and analogous inequalities hold for the other coordinate directions. According to Theorem 5.22, E therefore has locally finite perimeter.

7. The necessity of $(\star)$ was established in Theorem 5.16. $\qquad\square$

5.12   References and notes

We principally used Giusti [G] and Federer [F, Section 4.5] for BV theory, and also Simon [S, Section 6]. The Structure Theorem is stated, for instance, in [S, Section 6.1]. The Lower Semicontinuity Theorem in Section 5.2 is [G, Section 1.9], and the Local Approximation Theorem is [G, Theorem 1.17]. (This result is due to Anzellotti and Giaquinta). The compactness assertion in Section 5.2 follows [G, Theorem 1.19]. The discussion of traces in Section 5.3 follows [G, Chapter 2]. Our treatment of extensions in Section 5.4 is an elaboration of [G, Remark 2.13].

The coarea formula for BV functions, due to Fleming and Rishel [Fl-R], is proved as in [G, Theorem 1.23]. For the isoperimetric inequalities, consult [G, Theorem 1.28 and Corollary 1.29]. The remark in Section 5.6 is related to [F, Section 4.5.9(18)]. Theorem 5.12 is due to Fleming; we followed [F-Z]. The results in Sections 5.7 and 5.8 on the reduced and measure-theoretic boundaries are from [G, Chapters 3 and 4]; these assertions were originally established by De Giorgi.

Federer [F, Section 4.5.9] presents a long list of properties of BV functions, from which we extracted the theory set forth in Section 5.9. Essential variation occurs in [F, Section 4.5.10] and the criterion for finite perimeter described in Section 5.11 is [F, Section 4.5.11].

L. Ambrosio and E. De Giorgi [A-DG] have introduced the class of "special" functions of bounded variation, denoted SBV, for which the singular part of the gradient is supported on the jump set J. See also Ambrosio [A].

Chapter 6

Differentiability, Approximation by C^1 Functions

In this final chapter we examine more carefully the differentiability properties of BV, Sobolev, and Lipschitz continuous functions. We will see that such functions are differentiable in various senses for $\mathcal{L}^n$-a.e. point in $\mathbb{R}^n$, and as a consequence are equal to C^1 functions except on small sets.

Section 6.1 investigates differentiability $\mathcal{L}^n$-a.e. in certain L^p-senses, and Section 6.2 extends these ideas to show functions in $W^{1,p}$ for $p > n$ are in fact $\mathcal{L}^n$-a.e. differentiable in the classical sense. Section 6.3 recounts the elementary properties of convex functions. In Section 6.4 we prove Aleksandrov's Theorem, asserting a convex function is twice differentiable $\mathcal{L}^n$-a.e. Whitney's Extension Theorem, ensuring the existence of C^1 extensions, is proved in Section 6.5 and is utilized in Section 6.6 to show that a BV or Sobolev function equals a C^1 function except on a small set.

6.1 L^p differentiability, approximate differentiability

6.1.1 L^{1^*} differentiability for BV

Assume $f \in BV_{\mathrm{loc}}(\mathbb{R}^n)$.

NOTATION We recall from Section 5.1 the notation

$$[Df] = [Df]_{\mathrm{ac}} + [Df]_{\mathrm{s}} = \mathcal{L}^n \llcorner Df + [Df]_{\mathrm{s}},$$

where $Df \in L^1_{\mathrm{loc}}(\mathbb{R}^n; \mathbb{R}^n)$ is the density of the absolutely continuous part $[Df]_{\mathrm{ac}}$ of $[Df]$, and $[Df]_{\mathrm{s}}$ is the singular part.

We first demonstrate that near $\mathcal{L}^n$-a.e. point x, f can be approximated in an integral norm by a linear mapping.

THEOREM 6.1 (Differentiability for BV functions). *Assume that $f \in BV_{\text{loc}}(\mathbb{R}^n)$. Then for $\mathcal{L}^n$-a.e. $x \in \mathbb{R}^n$,*

$$\left(\fint_{B(x,r)} |f(y) - f(x) - Df(x) \cdot (y-x)|^{1^*} \, dy \right)^{\frac{1}{1^*}} = o(r)$$

as $r \to 0$.

Proof. 1. $\mathcal{L}^n$-a.e. point $x \in \mathbb{R}^n$ satisfies these conditions:

(a) $\lim_{r \to 0} \fint_{B(x,r)} |f(y) - f(x)| \, dy = 0$.

(b) $\lim_{r \to 0} \fint_{B(x,r)} |Df(y) - Df(x)| \, dy = 0$.

(c) $\lim_{r \to 0} \frac{|[Df]_s|(B(x,r))}{r^n} = 0$.

2. Fix such a point x; we may as well assume $x = 0$. Choose $r > 0$ and let $f^\epsilon := \eta_\epsilon * f$. We write $B(r) = B(0,r)$ and select $y \in B(r)$. Define $g(t) := f^\epsilon(ty)$. Then

$$g(1) = g(0) + \int_0^1 g'(s) \, ds;$$

that is,

$$f^\epsilon(y) = f^\epsilon(0) + \int_0^1 Df^\epsilon(sy) \cdot y \, ds$$

$$= f^\epsilon(0) + Df(0) \cdot y + \int_0^1 [Df^\epsilon(sy) - Df(0)] \cdot y \, ds.$$

3. Choose any function $\phi \in C_c^1(B(r))$ with $|\phi| \leq 1$, multiply by ϕ, and average over $B(r)$:

$$\fint_{B(r)} \phi(y)(f^\epsilon(y) - f^\epsilon(0) - Df(0) \cdot y) \, dy$$

$$= \int_0^1 \left(\fint_{B(r)} \phi(y)[Df^\epsilon(sy) - Df(0)] \cdot y \, dy \right) ds$$

$$= \int_0^1 \frac{1}{s} \left(\fint_{B(rs)} \phi\left(\frac{z}{s}\right) [Df^\epsilon(z) - Df(0)] \cdot z \, dz \right) ds. \qquad (\star)$$

Now

$$g_\epsilon(s) := \int_{B(rs)} \phi\left(\frac{z}{s}\right) Df^\epsilon(z) \cdot z \, dz$$

$$= -\int_{B(rs)} f^\epsilon(z) \, \mathrm{div}\left(\phi\left(\frac{z}{s}\right) z\right) dz$$

$$\to -\int_{B(rs)} f(z) \, \mathrm{div}\left(\phi\left(\frac{z}{s}\right) z\right) dz \quad \text{as } \epsilon \to 0$$

$$= \int_{B(rs)} \phi\left(\frac{z}{s}\right) z \cdot d[Df]$$

$$= \int_{B(rs)} \phi\left(\frac{z}{s}\right) Df(z) \cdot z \, dz + \int_{B(rs)} \phi\left(\frac{z}{s}\right) z \cdot d[Df]_\mathrm{s}.$$

Furthermore,

$$\frac{|g_\epsilon(s)|}{s^{n+1}} \leq \frac{r}{s^n} \int_{B(rs)} |Df^\epsilon(z)| dz$$

$$= \frac{r}{s^n} \int_{B(rs)} \left| \int_{\mathbb{R}^n} D\eta_\epsilon(z-y) f(y) \, dy \right| dz$$

$$= \frac{r}{s^n} \int_{B(rs)} \left| \int_{\mathbb{R}^n} \eta_\epsilon(z-y) \, d[Df] \right| dz$$

$$\leq \frac{r}{s^n} \int_{B(rs)} \int_{\mathbb{R}^n} \eta_\epsilon(z-y) \, d\|Df\| \, dz$$

$$= \frac{r}{s^n} \int_{\mathbb{R}^n} \int_{B(rs)} \eta_\epsilon(z-y) \, dz \, d\|Df\|$$

$$\leq \frac{C}{s^n \epsilon^n} \int_{B(rs+\epsilon)} \int_{B(rs)\cap B(y,\epsilon)} dz \, d\|Df\|$$

$$\leq C \frac{\min((rs)^n, \epsilon^n)}{s^n \epsilon^n} \|Df\|(B(rs+\epsilon))$$

$$\leq C \frac{\min((rs)^n, \epsilon^n)(rs+\epsilon)^n}{s^n \epsilon^n}$$

$$\leq C \quad \text{for } 0 < \epsilon, s \leq 1.$$

4. Therefore, applying the Dominated Convergence Theorem to $(\star)$, we find

$$\fint_{B(r)} \phi(y)(f(y) - f(0) - Df(0) \cdot y) \, dy$$

$$\leq Cr \int_0^1 \fint_{B(rs)} |Df(z) - Df(0)| \, dz \, ds + Cr \int_0^1 \frac{\|[Df]\|_s|(B(rs))}{(rs)^n} \, ds$$

$$= o(r)$$

as $r \to 0$. Take the supremum over all ϕ as above to find

$$\fint_{B(r)} |f(y) - f(0) - Df(0) \cdot y| \, dy = o(r) \qquad (\star\star)$$

as $r \to 0$.

5. Finally, observe from Theorem 5.10, (ii) in Section 5.6 that

$$\left(\fint_{B(r)} |f(y) - f(0) - Df(0) \cdot y|^{\frac{n}{n-1}} \, dy \right)^{\frac{n-1}{n}}$$

$$\leq C \frac{\|D(f - f(0) - Df(0) \cdot y)\|(B(r))}{r^{n-1}}$$

$$+ C \fint_{B(r)} |f(y) - f(0) - Df(0) \cdot y| \, dy$$

$$= o(r)$$

as $r \to 0$, according to $(\star\star)$, (b), and (c). $\square$

6.1.2 L^{p*} differentiability a.e. for $W^{1,p}$

We can improve the local approximation by tangent planes if f is a Sobolev function.

THEOREM 6.2 (Differentiability for Sobolev functions). *As-sume that $f \in W^{1,p}_{\mathrm{loc}}(\mathbb{R}^n)$ for some*

$$1 \leq p < n.$$

Then for $\mathcal{L}^n$-a.e. $x \in \mathbb{R}^n$,

$$\left(\fint_{B(x,r)} |f(y) - f(x) - Df(x) \cdot (y - x)|^{p^*} \, dy \right)^{\frac{1}{p^*}} = o(r)$$

as $r \to 0$.

Proof. 1. $\mathcal{L}^n$-a.e. point $x \in \mathbb{R}^n$ satisfies

(a) $\lim_{r \to 0} \fint_{B(x,r)} |f(x) - f(y)|^p \, dy = 0,$

(b) $\lim_{r \to 0} \fint_{B(x,r)} |Df(x) - Df(y)|^p \, dy = 0.$

2. Fix such a point x; we may as well assume $x = 0$. Select $\phi \in C^1_c(B(r))$ with $\|\phi\|_{L^q(B(r))} \leq 1$, where $\frac{1}{p} + \frac{1}{q} = 1$. Then, as in the previous proof, we calculate

$$\fint_{B(r)} \phi(y)(f(y) - f(0) - Df(0) \cdot y) \, dy$$

$$= \int_0^1 \frac{1}{s} \fint_{B(rs)} \phi\left(\frac{z}{s}\right) [Df(z) - Df(0)] \cdot z \, dz \, ds$$

$$\leq r \int_0^1 \left(\fint_{B(rs)} \left| \phi\left(\frac{z}{s}\right) \right|^q dz \right)^{\frac{1}{q}} \left(\fint_{B(r)} |Df(z) - Df(0)|^p \, dz \right)^{\frac{1}{p}} ds.$$

Since

$$\fint_{B(rs)} \left| \phi\left(\frac{z}{s}\right) \right|^q dz = \fint_{B(r)} |\phi(y)|^q \, dy \leq \frac{1}{\alpha(n) r^n},$$

we obtain

$$\fint_{B(r)} \phi(y)(f(y) - f(0) - Df(0) \cdot y) \, dy = o(r^{1 - \frac{n}{q}}) \quad \text{as } r \to 0.$$

Taking the supremum over all functions ϕ as above gives

$$\frac{1}{r^n} \left(\int_{B(r)} |f(y) - f(0) - Df(0) \cdot y|^p \, dy \right)^{\frac{1}{p}} = o(r^{1 - \frac{n}{q}}).$$

Hence

$$\left(\fint_{B(r)} |f(y) - f(0) - Df(0) \cdot y|^p \, dy \right)^{\frac{1}{p}} = o(r) \quad \text{as } r \to 0. \quad (\star)$$

3. Thus Theorem 4.9,(ii) in Section 4.5 implies

$$\left(\fint_{B(r)} |f(y) - f(0) - Df(0) \cdot y|^{p^*} \, dy \right)^{\frac{1}{p^*}}$$

$$\leq Cr \left(\fint_{B(r)} |Df(y) - Df(0)|^p \, dy \right)^{\frac{1}{p}}$$

$$+ C \left(\fint_{B(r)} |f(y) - f(0) - Df(0) \cdot y|^p \, dy \right)^{\frac{1}{p}}$$

$$= o(r) \quad \text{as } r \to 0,$$

according to ($\star$) and (b). $\square$

6.1.3   Approximate differentiability

DEFINITION 6.1. *Let $f : \mathbb{R}^n \to \mathbb{R}^m$. We say f is **approximately differentiable** at $x \in \mathbb{R}^n$ if there exists a linear mapping*

$$L : \mathbb{R}^n \to \mathbb{R}^m$$

such that

$$\operatorname*{ap\,lim}_{y \to x} \frac{|f(y) - f(x) - L(y - x)|}{|y - x|} = 0.$$

(See Section 1.7 for the definition of the approximate limit.)

NOTATION As proved below, such an L, if it exists, is unique. We write

$$\operatorname{ap} Df(x)$$

for L and call $\operatorname{ap} Df(x)$ the **approximate derivative** of f at x.

THEOREM 6.3 (Approximate differentiability). *An approximate derivative is unique and, in particular,*

$$\operatorname{ap} Df = 0 \quad \mathcal{L}^n\text{-a.e. on } \{f = 0\}.$$

Proof. Suppose

$$\operatorname*{ap\,lim}_{y \to x} \frac{|f(y) - f(x) - L(y - x)|}{|y - x|} = 0$$

and

$$\operatorname*{ap\,lim}_{y \to x} \frac{|f(y) - f(x) - L'(y - x)|}{|y - x|} = 0.$$

Then for each $\epsilon > 0$,

$$\lim_{r \to 0} \frac{\mathcal{L}^n \left(B(x, r) \cap \left\{ y \mid \frac{|f(y) - f(x) - L(y - x)|}{|y - x|} > \epsilon \right\} \right)}{\mathcal{L}^n(B(x, r))} = 0 \qquad (\star)$$

and

$$\lim_{r \to 0} \frac{\mathcal{L}^n \left(B(x,r) \cap \left\{ y \mid \frac{|f(y) - f(x) - L'(y - x)|}{|y - x|} > \epsilon \right\} \right)}{\mathcal{L}^n(B(x,r))} = 0. \qquad (\star\star)$$

If $L \neq L'$, set

$$\|L - L'\| := \max_{|z|=1} |(L - L')(z)| > 0.$$

and put

$$\epsilon = \frac{1}{6} \|L - L'\|.$$

Consider then the sector

$$S := \left\{ y \mid |(L - L') \cdot (y - x)| \geq \frac{\|L - L'\| |y - x|}{2} \right\}.$$

Note

$$\frac{\mathcal{L}^n(B(x,r) \cap S)}{\mathcal{L}^n(B(x,r))} := a > 0 \qquad (\star\star\star)$$

for all $r > 0$. But if $y \in S$,

$$3\epsilon |y - x| = \frac{\|L - L'\| |y - x|}{2}$$
$$\leq |(L - L')(y - x)|$$
$$\leq |f(y) - f(x) - L(y - x)| + |f(y) - f(x) - L'(y - x)|;$$

so that

$$S \subseteq \left\{ \frac{|f(y) - f(x) - L(y - x)|}{|y - x|} > \epsilon \right\}$$
$$\cup \left\{ \frac{|f(y) - f(x) - L'(y - x)|}{|y - x|} > \epsilon \right\}.$$

Thus $(\star)$ and $(\star\star)$ imply

$$\lim_{r \to 0} \frac{\mathcal{L}^n(B(x,r) \cap S)}{\mathcal{L}^n(B(x,r))} = 0,$$

a contradiction to $(\star\star\star)$. $\qquad\square$

THEOREM 6.4 (BV and approximate differentiability). *Assume $f \in BV_{\text{loc}}(\mathbb{R}^n)$. Then f is approximately differentiable $\mathcal{L}^n$-a.e.*

Remark.

(i) We show in addition that

$$\text{ap}\, Df = Df \quad \mathcal{L}^n\text{-a.e.,}$$

the function on the right defined in Section 5.1.

(ii) Since $W^{1,p}_{\text{loc}}(\mathbb{R}^n) \subset BV_{\text{loc}}(\mathbb{R}^n)$ for ($1 \leq p \leq \infty$, we see that *each Sobolev function is approximately differentiable $\mathcal{L}^n$-a.e. and its approximate derivative equals its weak derivative $\mathcal{L}^n$-a.e.*

$\square$

Proof. Choose a point $x \in \mathbb{R}^n$ such that

$$\fint_{B(x,r)} |f(y) - f(x) - Df(x) \cdot (y - x)|\, dy = o(r) \qquad (\star)$$

as $r \to 0$; $\mathcal{L}^n$-a.e. x will do according to Theorem 6.1.
Suppose

$$\text{ap}\, \limsup_{y \to x} \frac{|f(y) - f(x) - Df(x) \cdot (y - x)|}{|y - x|} > \theta > 0.$$

Then there exist $r_j \to 0$ and $\gamma > 0$ such that

$$\mathcal{L}^n(\{y \in B(x, r_j) \mid$$
$$|f(y) - f(x) - Df(x) \cdot (y - x)| > \theta|y - x|\}) \geq \gamma\alpha(n)r_j^n > 0.$$

Hence there exists $\sigma > 0$ such that

$$\mathcal{L}^n(\{y \in B(x,r_j) - B(x, \sigma r_j) \mid$$
$$|f(y) - f(x) - Df(x) \cdot (y - x)| > \theta|y - x|\}) \geq \frac{\gamma\alpha(n)r_j^n}{2}$$

for $j = 1, 2, \ldots$. Since $|y - x| > \sigma r_j$ for $y \in B(x, r_j) - B(x, \sigma r_j)$, it follows that

$$\frac{\mathcal{L}^n(\{y \in B(x, r_j) \mid |f(y) - f(x) - Df(x) \cdot (y - x)| > \theta\sigma r_j\})}{\alpha(n)r_j^n} \geq \frac{\gamma}{2}$$

$$(\star\star)$$

for $j = 1, \ldots$ But by $(\star)$, the expression on the left-hand side of $(\star\star)$ is less than or equal to

$$\frac{o(r_j)}{\theta \sigma r_j} = o(1)$$

as $r_j \to 0$, a contradiction to $(\star\star)$.

Thus

$$\operatorname{ap\,lim\,sup}_{y \to x} \frac{|f(y) - f(x) - Df(x) \cdot (y - x)|}{|y - x|} = 0,$$

and so

$$\operatorname{ap} Df(x) = Df(x). \qquad \square$$

6.2 Differentiability a.e. for $W^{1,p}$ $(p > n)$

Recall from Section 3.1 the

DEFINITION 6.2. *A function $f : \mathbb{R}^n \to \mathbb{R}^m$ is **differentiable** at $x \in \mathbb{R}^n$ if there exists a linear mapping*

$$L : \mathbb{R}^n \to \mathbb{R}^m$$

such that
$$\lim_{y \to x} \frac{|f(y) - f(x) - L(x - y)|}{|x - y|} = 0.$$

NOTATION If such a linear mapping L exists at x, it is clearly unique, and we write

$$Df(x)$$

for L. We call $Df(x)$ the **derivative** of f at x.

THEOREM 6.5 (Almost everywhere differentiability). *Assume that $f \in W^{1,p}_{\text{loc}}(\mathbb{R}^n)$ for some*

$$n < p \leq \infty.$$

Then f is differentiable $\mathcal{L}^n$-a.e., and its derivative equals its weak derivative $\mathcal{L}^n$-a.e.

Proof. Since $W_{loc}^{1,\infty}(\mathbb{R}^n) \subset W_{loc}^{1,p}(\mathbb{R}^n)$, we may as well assume $n < p < \infty$. For $\mathcal{L}^n$-a.e. $x \in \mathbb{R}^n$, we have

$$\lim_{r \to 0} \fint_{B(x,r)} |Df(z) - Df(x)|^p \, dz = 0. \qquad (\star)$$

Choose such a point x, and write

$$g(y) := f(y) - f(x) - Df(x) \cdot (y - x) \quad (y \in B(x,r)).$$

Employing Morrey's estimate from Section 4.5, we deduce

$$|g(y) - g(x)| \le Cr \left(\fint_{B(x,r)} |Dg|^p \, dz \right)^{\frac{1}{p}}$$

for $r := |x - y|$. Since $g(x) = 0$ and $Dg = Df - Df(x)$, this reads

$$\frac{|f(y) - f(x) - Df(x) \cdot (y - x)|}{|y - x|}$$

$$\le C \left(\fint_{B(x,r)} |Df(z) - Df(x)|^p \, dz \right)^{\frac{1}{p}}$$

$$= o(1) \quad \text{as } y \to x$$

according to $(\star)$. $\qquad\qquad \square$

As an application we have a new proof of

THEOREM 6.6 (Rademacher's Theorem again). *Let $f : \mathbb{R}^n \to \mathbb{R}$ be a locally Lipschitz continuous function. Then f is differentiable $\mathcal{L}^n$-a.e.*

Proof. According to Theorem 4.5, $f \in W_{loc}^{1,\infty}(\mathbb{R}^n)$. $\qquad\qquad \square$

6.3   Convex functions

DEFINITION 6.3. *A function $f : \mathbb{R}^n \to \mathbb{R}$ is called* **convex** *if*

$$f(\lambda x + (1 - \lambda)y) \le \lambda f(x) + (1 - \lambda)f(y)$$

for all $0 \le \lambda \le 1$, $x, y \in \mathbb{R}^n$.

THEOREM 6.7 (Properties of convex functions). *Assume that* $f : \mathbb{R}^n \to \mathbb{R}$ *is convex.*

(i) *Then f is locally Lipschitz continuous on $\mathbb{R}^n$.*

(ii) *Furthermore, there exists a constant C, depending only on n, such that*

$$\sup_{B(x,\frac{r}{2})} |f| \leq C \fint_{B(x,r)} |f| \, dy$$

and

$$\operatorname*{ess\,sup}_{B(x,\frac{r}{2})} |Df| \leq \frac{C}{r} \fint_{B(x,r)} |f| \, dy$$

for each ball $B(x,r) \subset \mathbb{R}^n$.

(iii) *If, in addition, $f \in C^2(\mathbb{R}^n)$, then*

$$D^2 f \geq 0 \quad \text{on } \mathbb{R}^n;$$

that is, for each $x \in \mathbb{R}^n$, $D^2 f(x)$ is a nonnegative definite symmetric matrix.

Proof. 1. Let $Q := [-L, L]^n$ be a cube, with vertices $V = \{v_k\}_{k=1}^{2^n}$. We can write any point $x \in Q$ as a convex combination of the vertices: $x = \sum_{k=1}^{2^n} \lambda_k v_k$, where $0 \leq \lambda_k \leq 1$ and $\sum \lambda_k = 1$. Hence

$$f(x) \leq \sum_{k=1}^{2^n} \lambda_k f(v_k) \leq \max_{v_k \in V} f(v_k) < \infty,$$

and thus $M := \sup_Q f < \infty$. To derive a lower bound, again select any point $x \in Q$ and write

$$0 = \frac{1}{2} x + \frac{1}{2}(-x).$$

Then

$$f(0) \leq \frac{1}{2} f(x) + \frac{1}{2} f(-x) \leq \frac{1}{2} f(x) + \frac{1}{2} M;$$

and so

$$f(x) \geq 2f(0) - M.$$

Therefore $\inf_Q f \geq 2f(0) - M$. These estimates are valid for each cube Q as above, and hence f is locally bounded.

2. If $x, y \in B(r)$ and $x \neq y$, select $\mu > 0$ so that

$$z := x + \mu(y - x) \in \partial B(2r).$$

Then $\mu = \frac{|z-x|}{|y-x|} > 1$ and $y = \frac{1}{\mu}z + (1 - \frac{1}{\mu})x$. Hence

$$f(y) \leq \frac{1}{\mu}f(z) + (1 - \frac{1}{\mu})f(x)$$

$$= f(x) + \frac{1}{\mu}(f(z) - f(x))$$

$$\leq f(x) + C|y - x|$$

for $C := \frac{2}{r}\sup_{B(2r)}|f|$, since $|z - x| \geq r$. Interchanging x, y, we find that

$$|f(y) - f(x)| \leq C|y - x| \qquad (x, y \in B(r)).$$

This proves assertion (i).

3. Suppose next that $f \in C^2(\mathbb{R}^n)$ and is convex. Fix $x \in \mathbb{R}^n$. Then for each $y \in \mathbb{R}^n$ and $\lambda \in (0, 1)$,

$$f(x + \lambda(y - x)) \leq f(x) + \lambda(f(y) - f(x)).$$

Thus

$$\frac{f(x + \lambda(y - x)) - f(x)}{\lambda} \leq f(y) - f(x).$$

Let $\lambda \to 0$ to obtain

$$f(y) \geq f(x) + Df(x) \cdot (y - x) \qquad (\star)$$

for all $x, y \in \mathbb{R}^n$.

4. Given now $B(x, r) \subset \mathbb{R}^n$, we fix a point $z \in B(x, \frac{r}{2})$. Then $(\star)$ implies

$$f(y) \geq f(z) + Df(z) \cdot (y - z).$$

We integrate this inequality with respect to y over $B(z, \frac{r}{2})$ to find

$$f(z) \leq \fint_{B(z, \frac{r}{2})} f(y)\, dy \leq C \fint_{B(x,r)} |f|\, dy \qquad (\star\star)$$

Next choose a smooth cutoff function $\zeta \in C_c^\infty(\mathbb{R}^n)$ satisfying

$$\begin{cases} 0 \leq \zeta \leq 1, \ |D\zeta| \leq \frac{C}{r}, \\ \zeta \equiv 1 \text{ on } B(x, \frac{r}{2}), \ \zeta \equiv 0 \text{ on } \mathbb{R}^n - B(x, r). \end{cases}$$

Now $(\star)$ implies

$$f(z) \geq f(y) + Df(y) \cdot (z - y).$$

Multiply this inequality by $\zeta(y)$ and integrate with respect to y over $B(x, r)$:

$$f(z) \int_{B(x,r)} \zeta(y)\, dy \geq \int_{B(x,r)} f(y)\zeta(y)\, dy + \int_{B(x,r)} \zeta(y)Df(y) \cdot (z - y)\, dy$$

$$= \int_{B(x,r)} f(y)[\zeta(y) - \mathrm{div}(\zeta(y)\,(z - y))]\, dy$$

$$\geq -C \int_{B(x,r)} |f|\, dy.$$

This inequality implies

$$f(z) \geq -C \fint_{B(x,r)} |f|\, dy,$$

which estimate together with $(\star\star)$ proves

$$|f(z)| \leq C \fint_{B(x,r)} |f|\, dy. \qquad\qquad (\star\star\star)$$

5. For z as above, define

$$S_z := \left\{ y \mid \frac{r}{4} \leq |y - x| \leq \frac{r}{2}, Df(z) \cdot (y - z) \geq \frac{1}{2}|Df(z)||y - z| \right\},$$

and observe

$$\mathcal{L}^n(S_z) \geq Cr^n$$

where C depends only on n. Use $(\star)$ to write

$$f(y) \geq f(z) + \frac{r}{8}|Df(z)|$$

for all $y \in S_z$. Integrating over S_z gives

$$|Df(z)| \leq \frac{C}{r} \fint_{B(x,\frac{r}{2})} |f(y) - f(z)|\, dy.$$

This inequality and $(\star\star\star)$ complete the proof of assertion (i) for C^2 convex functions f.

6. If f is merely convex, define $f^\epsilon := \eta_\epsilon * f$, where $\epsilon > 0$ and η_ϵ is the standard mollifier.

Claim #2: f^ϵ is convex.

Proof of claim: Fix $x, y \in \mathbb{R}^n$, $0 \le \lambda \le 1$. Then for each $z \in \mathbb{R}^n$,

$$f(z - (\lambda x + (1 - \lambda)y)) = f(\lambda(z - x) + (1 - \lambda)(z - y))$$
$$\le \lambda f(z - x) + (1 - \lambda)f(z - y).$$

Multiply this estimate by $\eta_\epsilon(z) \ge 0$ and integrate over $\mathbb{R}^n$:

$$f^\epsilon(\lambda x + (1 - \lambda)y) = \int_{\mathbb{R}^n} f(z - (\lambda x + (1 - \lambda)y))\eta_\epsilon(z)\, dz$$
$$\le \lambda \int_{\mathbb{R}^n} f(z - x)\eta_\epsilon(z)\, dz$$
$$+ (1 - \lambda) \int_{\mathbb{R}^n} f(z - y)\eta_\epsilon(z)\, dz$$
$$= \lambda f^\epsilon(x) + (1 - \lambda)f^\epsilon(y).$$

7. According to the estimate proved above for smooth convex functions, we have

$$\sup_{B(x, \frac{r}{2})} (|f^\epsilon| + r|Df^\epsilon|) \le C \fint_{B(x,r)} |f^\epsilon|\, dy$$

for each ball $B(x, r) \subset \mathbb{R}^n$. Letting $\epsilon \to 0$, we obtain in the limit the same estimates for f. This proves assertion (i).

8. To prove assertion (ii), recall from Taylor's Theorem that

$$f(y) = f(x) + Df(x) \cdot (y - x)$$
$$+ (y - x)^T \cdot \int_0^1 (1 - s)D^2 f(x + s(y - x))\, ds \cdot (y - x).$$

This equality and $(\star)$ yield

$$(y - x)^T \cdot \int_0^1 (1 - s)D^2 f(x + s(y - x))\, ds \cdot (y - x) \ge 0$$

for all $x, y \in \mathbb{R}^n$. Thus, given any vector ξ, we can set $y = x + t\xi$ above for $t > 0$, to compute:

$$\xi^T \cdot \int_0^1 (1 - s)D^2 f(x + st\xi)\, ds \cdot \xi \ge 0.$$

Send $t \to 0$:

$$\xi^T \cdot D^2 f(x) \cdot \xi \ge 0. \qquad \square$$

THEOREM 6.8 (Second derivatives as measures). *Let $f : \mathbb{R}^n \to \mathbb{R}$ be convex.*

(i) *There exist signed Radon measures $\mu^{ij} = \mu^{ji}$ such that*

$$\int_{\mathbb{R}^n} f \phi_{x_i x_j} \, dx = \int_{\mathbb{R}^n} \phi \, d\mu^{ij} \quad (i, j = 1, \dots, n)$$

for all $\phi \in C_c^2(\mathbb{R}^n)$. Furthermore, the measures μ^{ii} are nonnegative $(i = 1, \dots, n)$.

(ii) *Furthermore,*

$$f_{x_1}, \dots, f_{x_n} \in BV_{\text{loc}}(\mathbb{R}^n).$$

Proof. 1. Fix any vector $\xi \in \mathbb{R}^n$, $\xi = (\xi_1, \dots, \xi_n)$, with $|\xi| = 1$. Let η_ϵ be the standard mollifier. Write $f^\epsilon := \eta_\epsilon * f$. Then f^ϵ is smooth and convex, whence

$$D^2 f^\epsilon \geq 0.$$

Thus for all $\phi \in C_c^2(\mathbb{R}^n)$ with $\phi \geq 0$,

$$\sum_{i,j=1}^{n} \int_{\mathbb{R}^n} f^\epsilon \phi_{x_i x_j} \xi_i \xi_j \, dx = \int_{\mathbb{R}^n} \phi \sum_{i,j=1}^{n} f^\epsilon_{x_i x_j} \xi_i \xi_j \, dx \geq 0.$$

Let $\epsilon \to 0$ to conclude

$$L(\phi) := \sum_{i,j=1}^{n} \int_{\mathbb{R}^n} f \phi_{x_i x_j} \xi_i \xi_j \, dx \geq 0.$$

Then Theorem 1.39 implies the existence of a Radon measure μ^ξ such that

$$L(\phi) = \int_{\mathbb{R}^n} \phi \, d\mu^\xi$$

for all $\phi \in C_c^2(\mathbb{R}^n)$.

2. Let $\mu^{ii} := \mu^{e_i}$ for $i = 1, \dots, n$. If $i \neq j$, set $\xi := \frac{e_i + e_j}{\sqrt{2}}$. Note that then

$$\sum_{k,l=1}^{n} \phi_{x_k x_l} \xi_k \xi_l = \frac{1}{2} (\phi_{x_i x_i} + 2\phi_{x_i x_j} + \phi_{x_j x_j}).$$

Thus

$$\int_{\mathbb{R}^n} f \phi_{x_i x_j} \, dx = \int_{\mathbb{R}^n} f \sum_{k,l=1}^{n} \phi_{x_k x_l} \xi_k \xi_l \, dx$$

$$-\frac{1}{2}\left[\int_{\mathbb{R}^n} f\phi_{x_i x_i}\,dx + \int_{\mathbb{R}^n} f\phi_{x_j x_j}\,dx\right]$$

$$=\int_{\mathbb{R}^n}\phi\,d\mu^\xi - \frac{1}{2}\int_{\mathbb{R}^n}\phi\,d\mu^{ii} - \frac{1}{2}\int_{\mathbb{R}^n}\phi\,d\mu^{jj}$$

$$=\int_{\mathbb{R}^n}\phi\,d\mu^{ij},$$

where

$$\mu^{ij} := \mu^\xi - \frac{1}{2}\mu^{ii} - \frac{1}{2}\mu^{jj}. \qquad\qquad \square$$

3. Let $V \subset\subset \mathbb{R}^n, \phi \in C_c^2(V,\mathbb{R}^n), |\phi| \le 1$. Then for $k = 1,\dots,n$,

$$\int_{\mathbb{R}^n} f_{x_k}\operatorname{div}\phi\,dx = -\int_{\mathbb{R}^n} f\sum_{i=1}^n \phi_{x_i x_k}^i\,dx$$

$$=\sum_{i=1}^n\int_{\mathbb{R}^n}\phi^i\,d\mu^{ik} \le \sum_{i=1}^n \mu^{ik}(V) < \infty.$$

NOTATION By analogy with the notation introduced in Section 5.1, let us write for a convex function f:

$$[D^2 f] := \begin{pmatrix} \mu^{11} & \cdots & \mu^{1n} \\ \vdots & \ddots & \vdots \\ \mu^{n1} & \cdots & \mu^{nn} \end{pmatrix} = \|D^2 f\|\,\llcorner\Sigma,$$

where $\Sigma : \mathbb{R}^n \to \mathbb{M}^{n\times n}$ is $\|D^2 f\|$-measurable, with $|\Sigma| = 1$ $\|D^2 f\|$-a.e. (Recall that $\mathbb{M}^{n\times n}$ denotes the space of real $n \times n$ matrices.) We also write

$$[f_{x_i x_j}] = \mu^{ij}\quad (i,j = 1,\dots,n).$$

By Lebesgue's Decomposition Theorem, we may further set

$$\mu^{ij} = \mu_{\text{ac}}^{ij} + \mu_{\text{s}}^{ij},$$

where

$$\mu_{\text{ac}}^{ij} << \mathcal{L}^n, \quad \mu_{\text{s}}^{ij} \perp \mathcal{L}^n.$$

But then

$$\mu_{\text{ac}}^{ij} = \mathcal{L}^n\,\llcorner f_{ij}$$

for some $f_{ij} \in L_{\text{loc}}^1(\mathbb{R}^n)$. Set

$$f_{x_i x_j} := f_{ij}\quad (i,j = 1,\dots,n),$$

$$D^2 f := \begin{pmatrix} f_{x_1 x_1} & \cdots & f_{x_1 x_n} \\ \vdots & \ddots & \vdots \\ f_{x_n x_1} & \cdots & f_{x_n x_n} \end{pmatrix},$$

$$[D^2 f]_{\mathrm{ac}} := \begin{pmatrix} \mu_{\mathrm{ac}}^{11} & \cdots & \mu_{\mathrm{ac}}^{1n} \\ \vdots & \ddots & \vdots \\ \mu_{\mathrm{ac}}^{n1} & \cdots & \mu_{\mathrm{ac}}^{nn} \end{pmatrix} = \mathcal{L}^n \, \llcorner \, D^2 f,$$

$$[D^2 f]_{\mathrm{s}} := \begin{pmatrix} \mu_{\mathrm{s}}^{11} & \cdots & \mu_{\mathrm{s}}^{1n} \\ \vdots & \ddots & \vdots \\ \mu_{\mathrm{s}}^{n1} & \cdots & \mu_{\mathrm{s}}^{nn} \end{pmatrix}.$$

Thus $[D^2 f] = [D^2 f]_{\mathrm{ac}} + [D^2 f]_{\mathrm{s}} = \mathcal{L}^n \, \llcorner \, D^2 f + [D^2 f]_{\mathrm{s}}$. Therefore $D^2 f \in L_{\mathrm{loc}}^1(\mathbb{R}^n; \mathbb{M}^{n \times n})$ is the density of the absolutely continuous part $[D^2 f]_{\mathrm{ac}}$ of $[D^2 f]$.

6.4 Second derivatives a.e. for convex functions

Next we show that a convex function is twice differentiable-a.e. This assertion is in the same spirit as Rademacher's Theorem, but is perhaps even more remarkable in that we have only "one-sided control" on the second derivatives.

THEOREM 6.9 (Aleksandrov's Theorem). *Let* $f : \mathbb{R}^n \to \mathbb{R}$ *be convex. Then* f *has second derivatives* $\mathcal{L}^n$*-a.e.*

More precisely, for $\mathcal{L}^n$*-a.e.* x,

$$\left| f(y) - f(x) - Df(x) \cdot (y - x) - \frac{1}{2}(y - x)^T \cdot D^2 f(x) \cdot (y - x) \right|$$
$$= o(|y - x|^2) \quad \textit{as } y \to x. \tag{$\star$}$$

Proof. 1. $\mathcal{L}^n$-a.e. point x satisfies these conditions:

(a) $Df(x)$ exists and $\lim_{r \to 0} \fint_{B(x,r)} |Df(y) - Df(x)| \, dy = 0$.

(b) $\lim_{r \to 0} \fint_{B(x,r)} |D^2 f(y) - D^2 f(x)| \, dy = 0$. $\tag{$\star\star$}$

(c) $\lim_{r \to 0} \frac{|[D^2 f]_{\mathrm{s}}|(B(x,r))}{r^n} = 0$.

2. Fix such a point x; we may as well assume $x = 0$. Choose $r > 0$ and let $f^\epsilon := \eta_\epsilon * f$. Fix $y \in B(r)$. By Taylor's Theorem,

$$f^\epsilon(y) = f^\epsilon(0) + Df^\epsilon(0) \cdot y + \int_0^1 (1-s)y^T \cdot D^2 f^\epsilon(sy) \cdot y \, ds.$$

Therefore

$$f^\epsilon(y) = f^\epsilon(0) + Df^\epsilon(0) \cdot y + \frac{1}{2}y^T \cdot D^2 f(0) \cdot y$$
$$+ \int_0^1 (1-s)y^T \cdot \left[D^2 f^\epsilon(sy) - D^2 f(0)\right] \cdot y \, ds.$$

3. Fix any function $\phi \in C_c^2(B(r))$ with $|\phi| \leq 1$, multiply the equation above by ϕ, and average over $B(r)$:

$$\fint_{B(r)} \phi(y)(f^\epsilon(y) - f^\epsilon(0) - Df^\epsilon(0) \cdot y - \frac{1}{2}y^T \cdot D^2 f(0) \cdot y) \, dy$$
$$= \int_0^1 (1-s) \left(\fint_{B(r)} \phi(y)y^T \cdot [D^2 f^\epsilon(sy) - D^2 f(0)] \cdot y \, dy \right) ds \quad (\star\star\star)$$
$$= \int_0^1 \frac{(1-s)}{s^2} \left(\fint_{B(rs)} \phi\left(\frac{z}{s}\right) z^T \cdot [D^2 f^\epsilon(z) - D^2 f(0)] \cdot z \, dz \right) ds.$$

Now

$$g_\epsilon(s) := \int_{B(rs)} \phi\left(\frac{z}{s}\right) z^T \cdot D^2 f^\epsilon(z) \cdot z \, dz$$
$$= \int_{B(rs)} f^\epsilon(z) \sum_{i,j=1}^n \left(\phi\left(\frac{z}{s}\right) z_i z_j\right)_{z_i z_j} dz$$
$$\to \int_{B(rs)} f(z) \sum_{i,j=1}^n \left(\phi\left(\frac{z}{s}\right) z_i z_j\right)_{z_i z_j} dz \quad \text{as } \epsilon \to 0$$
$$= \sum_{i,j=1}^n \int_{B(rs)} \phi\left(\frac{z}{s}\right) z_i z_j \, d\mu^{ij}$$
$$= \int_{B(rs)} \phi\left(\frac{z}{s}\right) z^T \cdot D^2 f(z) \cdot z \, dz + \sum_{i,j=1}^n \int_{B(rs)} \phi\left(\frac{z}{s}\right) z_i z_j \, d\mu_s^{ij}.$$

Furthermore, we can calculate

$$\frac{|g_\epsilon(s)|}{s^{n+2}} \leq \frac{r^2}{s^n} \int_{B(rs)} |D^2 f^\epsilon(z)| dz$$

$$= \frac{r^2}{s^n} \int_{B(rs)} \left| \int_{\mathbb{R}^n} D^2 \eta_\epsilon (z-y) f(y) \, dy \right| dz$$

$$\leq \frac{r^2}{s^n} \int_{B(rs)} \left| \int_{\mathbb{R}^n} \eta_\epsilon (z-y) \, d[D^2 f] \right| dz$$

$$\leq \frac{C}{s^n \epsilon^n} \int_{B(rs+\epsilon)} \left(\int_{B(rs) \cap B(y,\epsilon)} dz \right) d\|D^2 f\|$$

$$\leq C \frac{\min((rs)^n, \epsilon^n)}{s^n \epsilon^n} \|D^2 f\|(B(rs+\epsilon))$$

$$\leq C \frac{\min((rs)^n, \epsilon^n)(rs+\epsilon)^n}{s^n \epsilon^n}$$

$$\leq C$$

for $0 < \epsilon, s \leq 1$ by $(\star\star)$.

4. Hence we may apply the Dominated Convergence Theorem to let $\epsilon \to 0$ in $(\star\star\star)$:

$$\int_{B(r)} \phi(y) \left[f(y) - f(0) - Df(0) \cdot y - \frac{1}{2} y^T \cdot D^2 f(0) \cdot y \right] dy$$

$$\leq Cr^2 \int_0^1 \int_{B(rs)} |D^2 f(z) - D^2 f(0)| \, dz \, ds$$

$$+ Cr^2 \int_0^1 \frac{|[D^2 f]_s|(B(rs))}{(sr)^n} \, ds$$

$$= o(r^2) \quad \text{as } r \to 0,$$

according to $(\star\star)$ with $x = 0$. Take the supremum over all ϕ as above to obtain

$$\int_{B(r)} |h(y)| \, dy = o(r^2) \quad \text{as } r \to 0 \qquad (\star\star\star\star)$$

for

$$h(y) := f(y) - f(0) - Df(0) \cdot y - \frac{1}{2} y^T \cdot D^2 f(0) \cdot y.$$

5. *Claim #1*: There exists a constant C such that

$$\sup_{B(\frac{r}{2})} |Dh| \leq \frac{C}{r} \int_{B(r)} |h| \, dy + Cr \quad (r > 0).$$

Proof of claim: Let $\Lambda := |D^2 f(0)|$. Then $g := h + \frac{\Lambda}{2} |y|^2$ is convex. Apply Theorem 6.7.

6. *Claim #2*: $\sup_{B(\frac{r}{2})} |h| = o(r^2)$ as $r \to 0$.

Proof of claim: Fix $0 < \epsilon, \eta < 1, \eta^{\frac{1}{n}} \le \frac{1}{2}$. Then $(\star\star\star\star)$ implies

$$\mathcal{L}^n\{z \in B(r) \mid |h(z)| \ge \epsilon r^2\} \le \frac{1}{\epsilon r^2} \int_{B(r)} |h| \, dz$$
$$= o(r^n)$$
$$< \eta \mathcal{L}^n(B(r))$$

for $0 < r < r_0 := r_0(\epsilon, \eta)$. Thus for each $y \in B(\frac{r}{2})$ there exists $z \in B(r)$ such that

$$|h(z)| \le \epsilon r^2$$

and

$$|y - z| \le \sigma := \eta^{\frac{1}{n}} r.$$

To see this, observe that if not, then

$$\mathcal{L}^n\{z \in B(r) \mid |h(z)| \ge \epsilon r^2\}$$
$$\ge \mathcal{L}^n(B(y, \sigma)) = \alpha(n)\eta r^n = \eta \mathcal{L}^n(B(r)).$$

Consequently,

$$|h(y)| \le |h(z)| + |h(y) - h(z)| \le \epsilon r^2 + \sigma \sup_{B(r)} |Dh| \le \epsilon r^2 + C\eta^{\frac{1}{n}} r^2$$

by Claim # 1 and $(\star\star\star\star)$, provided we fix η such that $C\eta^{\frac{1}{n}} = \epsilon$ and then choose $0 < r < r_0$.

7. According to Claim #2,

$$\sup_{B(\frac{r}{2})} \left| f(y) - f(0) - Df(0) \cdot y - \frac{1}{2} y^{\mathrm{T}} \cdot D^2 f(0) \cdot y \right| = o(r^2)$$

as $r \to 0$. This proves $(\star)$ for $x = 0$. $\qquad\qquad\square$

6.5   Whitney's Extension Theorem

We next identify conditions ensuring the existence of a C^1 extension $\bar{f}$ of a given function f defined on a closed subset C of $\mathbb{R}^n$.

Let $C \subset \mathbb{R}^n$ be a closed set and assume $f : C \to \mathbb{R}, d : C \to \mathbb{R}^n$ are given functions.

NOTATION

(i)

$$R(y, x) := \frac{f(y) - f(x) - d(x) \cdot (y - x)}{|x - y|} \quad (x, y \in C, x \neq y).$$

(ii) Let $K \subseteq C$ be compact, and for $\delta > 0$ set

$$\rho_K(\delta) := \sup\{|R(y, x)| \mid 0 < |x - y| \leq \delta, \, x, y \in K\}.$$

THEOREM 6.10 (Whitney's Extension Theorem). *Assume that f, d are continuous, and for each compact set $K \subseteq C$,*

$$\rho_K(\delta) \to 0 \quad as \, \delta \to 0. \tag{$\star$}$$

Then there exists a function $\bar{f} : \mathbb{R}^n \to \mathbb{R}$ such that

(i) *$\bar{f}$ is C^1.*

(ii) *$\bar{f} = f, \, D\bar{f} = d$ on C.*

The proof is a sort of "C^1-version" of the proof of the extension Theorem 1.13 in Section 1.2.

Proof. 1. Let $U := \mathbb{R}^n - C$; U is open. Define

$$r(x) := \frac{1}{20} \min\{1, \text{dist}(x, C)\}.$$

By Vitali's Covering Theorem, there exists a countable set $\{x_j\}_{j=1}^\infty \subset U$ such that

$$U = \bigcup_{j=1}^\infty B(x_j, 5r(x_j))$$

and the balls $\{B(x_j, r(x_j))\}_{j=1}^\infty$ are disjoint. For each $x \in U$, define

$$S_x := \{x_j \mid B(x, 10r(x)) \cap B(x_j, 10r(x_j)) \neq \emptyset\}.$$

2. *Claim #1:* $\text{Card}(S_x) \leq (129)^n$ and

$$\frac{1}{3} \leq \frac{r(x)}{r(x_j)} \leq 3$$

if $x_j \in S_x$.

Proof of claim: If $x_j \in S_x$, then

$$|r(x) - r(x_j)| \leq \frac{1}{20}|x - x_j|$$
$$\leq \frac{1}{20}(10(r(x) + r(x_j))) = \frac{1}{2}(r(x) + r(x_j)).$$

Hence

$$r(x) \leq 3r(x_j), \ r(x_j) \leq 3r(x).$$

In addition, we have

$$|x - x_j| + r(x_j) \leq 10(r(x) + r(x_j)) + r(x_j)$$
$$= 10r(x) + 11r(x_j) \leq 43r(x);$$

consequently,

$$B(x_j, r(x_j)) \subset B(x, 43r(x)).$$

Since the balls $\{B(x_j, r(x_j))\}_{j=1}^{\infty}$ are disjoint, we have $r(x_j) \geq \frac{r(x)}{3}$,

$$\mathrm{Card}(S_x)\alpha(n)\left(\frac{r(x)}{3}\right)^n \leq \alpha(n)(43r(x))^n.$$

Therefore

$$\mathrm{Card}(S_x) \leq (129)^n.$$

3. Now choose $\mu : \mathbb{R} \to \mathbb{R}$ such that

$$\mu \in C^{\infty}, \ 0 \leq \mu \leq 1, \ \mu(t) \equiv 1 \text{ for } t \leq 1, \ \mu(t) \equiv 0 \text{ for } t \geq 2.$$

For each $j = 1, \ldots,$ define

$$u_j(x) := \mu\left(\frac{|x - x_j|}{5r(x_j)}\right) \quad (x \in \mathbb{R}^n).$$

Then

$$\begin{cases} u_j \in C^{\infty}, \ 0 \leq u_j \leq 1, \\ u_j \equiv 1 \text{ on } B(x_j, 5r(x_j)), \\ u_j \equiv 0 \text{ on } \mathbb{R}^n - B(x_j, 10r(x_j)). \end{cases}$$

Also

$$|Du_j(x)| \leq \frac{C}{r(x_j)} \leq \frac{C_1}{r(x)} \quad \text{if } x_j \in S_x \qquad (\star\star)$$

and
$$u_j = 0 \quad \text{on } B(x, 10r(x)), \text{ if } x_j \notin S_x.$$

Define
$$\sigma(x) := \sum_{j=1}^{\infty} u_j(x) \quad (x \in \mathbb{R}^n).$$

Since $u_j = 0$ on $B(x, 10r(x))$ if $x_j \notin S_x$, we see that
$$\sigma(y) = \sum_{x_j \in S_x} u_j(y) \quad \text{if } y \in B(x, 10r(x)).$$

By Claim #1, Card $(S_x) \leq (129)^n$; this and $(\star\star)$ imply
$$\sigma \in C^\infty(U), \ \sigma \geq 1 \text{ on } U, \ |D\sigma(x)| \leq \frac{C_2}{r(x)} \quad (x \in U).$$

Now for each $j = 1, \ldots$, define
$$v_j(x) := \frac{u_j(x)}{\sigma(x)} \quad (x \in U).$$

Notice
$$Dv_j = \frac{Du_j}{\sigma} - \frac{u_j D\sigma}{\sigma^2}.$$

Thus
$$\begin{cases} \sum_{j=1}^{\infty} v_j(x) = 1 \\ \sum_{j=1}^{\infty} Dv_j(x) = 0 \quad (x \in U) \\ |Dv_j(x)| \leq \frac{C_3}{r(x)}. \end{cases}$$

The functions $\{v_j\}_{j=1}^{\infty}$ are thus a smooth partition of unity in U.

4. Now for each $j = 1, \ldots$, choose any point $s_j \in C$ such that
$$|x_j - s_j| = \text{dist}(x_j, C).$$

Finally, define $\bar{f} : \mathbb{R}^n \to \mathbb{R}$ this way:
$$\bar{f}(x) := \begin{cases} f(x) & \text{if } x \in C \\ \sum_{j=1}^{\infty} v_j(x)[f(s_j) + d(s_j) \cdot (x - s_j)] & \text{if } x \in U. \end{cases}$$

Observe that $\bar{f} \in C^\infty(U)$ and
$$D\bar{f}(x) = \sum_{x_j \in S_x} \{[f(s_j) + d(s_j) \cdot (x - s_j)]Dv_j(x) + v_j(x)d(s_j)\}$$

for $x \in U$.

5. *Claim #2:* $D\bar{f}(a) = d(a)$ for all $a \in C$.

Proof of claim: Fix $a \in C$ and let $K := C \cap B(a, 1); K$ is compact. Define

$$\phi(\delta) := \sup \{|R(x, y)| \mid x, y \in K, 0 < |x - y| \le \delta\}$$
$$+ \sup \{|d(x) - d(y)| \mid x, y \in K, |x - y| \le \delta\}.$$

Since $d : C \to \mathbb{R}^n$ is continuous and $(\star)$ holds,

$$\phi(\delta) \to 0 \quad \text{as } \delta \to 0. \qquad (\star\star\star)$$

If $x \in C$ and $|x - a| \le 1$, then

$$\begin{aligned}
|\bar{f}(x) - \bar{f}(a) - d(a) \cdot (x - a)| &= |f(x) - f(a) - d(a) \cdot (x - a)| \\
&= |R(x, a)||x - a| \\
&\le \phi(|x - a|)|x - a|
\end{aligned}$$

and

$$|d(x) - d(a)| \le \phi(|x - a|).$$

Now suppose $x \in U$, $|x - a| \le \frac{1}{6}$. We calculate

$$\begin{aligned}
&|f(x) - f(a) - d(a) \cdot (x - a)| \\
&= |\bar{f}(x) - f(a) - d(a) \cdot (x - a)| \\
&\le \sum_{x_j \in S_x} |v_j(x)[f(s_j) - f(a) + d(s_j) \cdot (x - s_j) - d(a) \cdot (x - a)]| \\
&\le \sum_{x_j \in S_x} v_j(x)|f(s_j) - f(a) + d(s_j) \cdot (a - s_j)| \\
&\quad + \sum_{x_j \in S_x} v_j(x)|(d(s_j) - d(a)) \cdot (x - a)|.
\end{aligned}$$

Now $|x - a| \le \frac{1}{6}$ implies $r(x) \le \frac{1}{20}|x - a|$. Thus for $x_j \in S_x$,

$$\begin{aligned}
|a - s_j| &\le |a - x_j| + |x_j - s_j| \\
&\le 2|a - x_j| \\
&\le 2(|x - a| + |x - x_j|) \\
&\le 2(|x - a| + 10(r(x) + r(x_j))) \\
&\le 2(|x - a| + 40r(x)) \\
&\le 6|x - a|.
\end{aligned}$$

Hence the calculation above and Claim #1 show

$$|\bar{f}(x) - \bar{f}(a) - d(a) \cdot (x - a)| < C\phi(6|x - a|)|x - a|.$$

In view of $(\star\star\star)$, the calculations above imply that for each $a \in C$,

$$|\bar{f}(x) - \bar{f}(a) - d(a) \cdot (x - a)| = o(|x - a|) \quad \text{as } x \to a.$$

Thus $D\bar{f}(a)$ exists and equals $d(a)$.

6. *Claim #3*: $\bar{f} \in C^1(\mathbb{R}^n)$.

Proof of claim: Fix $a \in C, x \in \mathbb{R}^n, |x - a| \le \frac{1}{6}$. If $x \in C$, then

$$|D\bar{f}(x) - D\bar{f}(a)| = |d(x) - d(a)| \le \phi(|x - a|).$$

If $x \in U$, choose $b \in C$ such that

$$|x - b| = \text{dist}(x, C).$$

Then

$$|D\bar{f}(x) - D\bar{f}(a)| = |D\bar{f}(x) - d(a)| \le |D\bar{f}(x) - d(b)| + |d(b) - d(a)|.$$

Since

$$|b - a| \le |b - x| + |x - a| \le 2|x - a|,$$

we have

$$|d(b) - d(a)| \le \phi(2|x - a|).$$

We thus must estimate:

$$|D\bar{f}(x) - d(b)|$$

$$= \left| \sum_{x_j \in S_x} [f(s_j) + d(s_j) \cdot (x - s_j)]Dv_j(x) + v_j(x)[d(s_j) - d(b)] \right|$$

$$\le \left| \sum_{x_j \in S_x} [-f(b) + f(s_j) + d(s_j) \cdot (b - s_j)]Dv_j(x) \right|$$

$$+ \left| \sum_{x_j \in S_x} [(d(s_j) - d(b)) \cdot (x - b)]Dv_j(x) \right|$$

$$+ \left| \sum_{x_j \in S_x} v_j(x)[d(s_j) - d(b)] \right| \qquad (\star\star\star\star)$$

$$\leq \frac{C}{r(x)} \sum_{x_j \in S_x} \phi(|b - s_j|)|b - s_j| + \frac{C}{r(x)} \sum_{x_j \in S_x} \phi(|b - s_j|)|x - b|$$
$$+ \sum_{x_j \in S_x} \phi(|b - s_j|).$$

Now

$$|x - b| \leq |x - a| \leq \frac{1}{6},$$

and therefore

$$r(x) = \frac{1}{20}|x - b| \leq \frac{1}{120}.$$

If $x_j \in S_x$,

$$r(x_j) \leq 3r(x) \leq \frac{1}{40} < \frac{1}{20}.$$

Hence

$$r(x_j) = \frac{1}{20}|x_j - s_j| \quad (x_j \in S_x).$$

Accordingly, if $x_j \in S_x$,

$$|b - s_j| \leq |b - x| + |x - x_j| + |x_j - s_j|$$
$$\leq 20r(x) + 10(r(x) + r(x_j)) + 20r(x_j)$$
$$\leq 120r(x) = 6|x - b| \leq 6|x - a|.$$

Consequently $(\star\star\star\star)$ implies

$$|D\bar{f}(x) - d(b)| \leq C\phi(6|x - a|).$$

This estimate and the calculations before show

$$|D\bar{f}(x) - D\bar{f}(a)| \leq C\phi(6|x - a|). \qquad \square$$

6.6   Approximation by C^1 functions

We now make use of Whitney's Extension Theorem to show that if f is a Lipschitz continuous, BV or Sobolev function, then f actually equals a C^1 function $\bar{f}$, except on a small set. In addition, $Df = D\bar{f}$, except on a small set.

6.6.1 Approximation of Lipschitz continuous functions

THEOREM 6.11 (Approximating Lipschitz functions). *Suppose $f : \mathbb{R}^n \to \mathbb{R}$ is Lipschitz continuous. Then for each $\epsilon > 0$, there exists a C^1 function $\bar{f} : \mathbb{R}^n \to \mathbb{R}$ such that*

$$\mathcal{L}^n(\{x \mid \bar{f}(x) \neq f(x) \text{ or } D\bar{f}(x) \neq Df(x)\}) \leq \epsilon.$$

In addition,

$$\sup_{\mathbb{R}^n} |D\bar{f}| \leq C \operatorname{Lip}(f)$$

for some constant C depending only on n.

Proof. By Rademacher's Theorem, f is differentiable on a set $A \subseteq \mathbb{R}^n$, with $\mathcal{L}^n(\mathbb{R}^n - A) = 0$. Using Lusin's Theorem, we see that there exists a closed set $B \subseteq A$ such that $Df|_B$ is continuous and $\mathcal{L}^n(\mathbb{R}^n - B) < \frac{\epsilon}{2}$. Set

$$d(x) := Df(x)$$

and

$$R(y, x) := \frac{f(y) - f(x) - d(x) \cdot (y - x)}{|x - y|} \quad (x \neq y).$$

Define also

$$\eta_k(x) := \sup\left\{ |R(y, x)| \;\middle|\; y \in B, 0 < |x - y| \leq \frac{1}{k} \right\}.$$

Then $\eta_k(x) \to 0$ as $k \to \infty$, for all $x \in B$. By Egoroff's Theorem, there exists a closed set $C \subseteq B$ such that $\eta_k \to 0$ uniformly on compact subsets of C, and

$$\mathcal{L}^n(B - C) \leq \frac{\epsilon}{2}.$$

This implies hypothesis $(\star)$ of Whitney's Extension Theorem.

The stated estimate on $\sup_{\mathbb{R}^n} |D\bar{f}|$ follows from the construction of $\bar{f}$ in the proof in Section 6.5, since $\sup_C |d| \leq \operatorname{Lip}(f)$ and thus

$$|R| \leq C \operatorname{Lip}(f). \qquad \square$$

6.6.2 Approximation of BV functions

THEOREM 6.12 (Approximating BV functions). *Let $f \in BV(\mathbb{R}^n)$. Then for each $\epsilon > 0$, there exists a Lipschitz continuous function $\bar{f} : \mathbb{R}^n \to \mathbb{R}$ such that*

$$\mathcal{L}^n(\{x \mid \bar{f}(x) \neq f(x)\}) \leq \epsilon.$$

Proof. 1. Define for $\lambda > 0$

$$R^\lambda := \left\{ x \in \mathbb{R}^n \mid \frac{\|Df\|(B(x,r))}{r^n} \leq \lambda \text{ for all } r > 0 \right\}.$$

2. *Claim #1:*

$$\mathcal{L}^n(\mathbb{R}^n - R^\lambda) \leq \frac{\alpha(n)5^n}{\lambda} \|Df\|(\mathbb{R}^n).$$

Proof of claim: According to Vitali's Covering Theorem, there exist disjoint balls $\{B(x_i, r_i)\}_{i=1}^\infty$ such that

$$\mathbb{R}^n - R^\lambda \subset \bigcup_{i=1}^\infty B(x_i, 5r_i)$$

and

$$\frac{\|Df\|(B(x_i, r_i))}{r_i^n} > \lambda.$$

Thus

$$\mathcal{L}^n(\mathbb{R}^n - R^\lambda) \leq 5^n \alpha(n) \sum_{i=1}^\infty r_i^n \leq \frac{5^n \alpha(n)}{\lambda} \|Df\|(\mathbb{R}^n).$$

3. *Claim #2:* There exists a constant C, depending only on n, such that

$$|f(x) - f(y)| \leq C\lambda|x - y|$$

for $\mathcal{L}^n$-a.e. $x, y \in R^\lambda$.

Proof of claim: Let $x \in R^\lambda$, $r > 0$. By Poincaré's inequality, Theorem 5.10,(ii) in Section 5.6,

$$\fint_{B(x,r)} |f - (f)_{x,r}| \, dy \leq \frac{C\|Df\|(B(x,r))}{r^{n-1}} \leq C\lambda r.$$

Thus, in particular,

$$|(f)_{x,\frac{r}{2^{k+1}}} - (f)_{x,\frac{r}{2^k}}| \leq \fint_{B(x,\frac{r}{2^{k+1}})} |f - (f)_{x,\frac{r}{2^k}}| \, dy$$

$$\leq 2^n \fint_{B(x,\frac{r}{2^k})} |f - (f)_{x,\frac{r}{2^k}}| \, dy$$

$$\leq \frac{C\lambda r}{2^k}.$$

Since

$$f(x) = \lim_{r \to 0} (f)_{x,r}$$

for $\mathcal{L}^n$-a.e. $x \in R^\lambda$, we have

$$|f(x) - (f)_{x,r}| \le \sum_{k=1}^{\infty} |(f)_{x, \frac{r}{2^{k+1}}} - (f)_{x, \frac{r}{2^k}}| \le C\lambda r.$$

Now for $x, y \in \mathbb{R}^\lambda$, $x \ne y$, set $r = |x - y|$. Then

$$
\begin{aligned}
|(f)_{x,r} &- (f)_{y,r}| \\
&\le \fint_{B(x,r) \cap B(y,r)} |(f)_{x,r} - f(z)| + |f(z) - (f)_{y,r}| \, dz \\
&\le C \left(\fint_{B(x,r)} |f(z) - (f)_{x,r}| \, dz + \fint_{B(y,r)} |f(z) - (f)_{y,r}| \, dz \right) \\
&\le C\lambda r.
\end{aligned}
$$

We combine the inequalities above, to estimate

$$|f(x) - f(y)| \le C\lambda r = C\lambda |x - y|$$

for $\mathcal{L}^n$-a.e. $x, y \in \mathbb{R}^\lambda$.

4. In view of Claim #2, there exists a Lipschitz continuous mapping $\bar{f} : R^\lambda \to \mathbb{R}$ such that $\bar{f} = f$ $\mathcal{L}^n$-a.e. on R^λ. Now recall Theorem 3.1 and extend $\bar{f}$ to a Lipschitz continuous mapping $\bar{f} : \mathbb{R}^n \to \mathbb{R}$. □

THEOREM 6.13 (Pointwise approximations for BV functions). *Let $f \in BV(\mathbb{R}^n)$. Then for each $\epsilon > 0$ there exists a C^1-function $\bar{f} : \mathbb{R}^n \to \mathbb{R}$ such that*

$$\mathcal{L}^n(\{x \mid f(x) \ne \bar{f}(x) \text{ or } Df(x) \ne D\bar{f}(x)\}) \le \epsilon.$$

Proof. According to Theorems 6.11 and 6.12, there exists $\bar{f} \in C^1(\mathbb{R}^n)$ such that

$$\mathcal{L}^n(\{\bar{f} \ne f\}) < \epsilon.$$

Furthermore,

$$D\bar{f}(x) = Df(x)$$

$\mathcal{L}^n$-a.e. on $\{f = \bar{f}\}$, according to Theorem 6.3. □

6.6.3   Approximation of Sobolev functions

THEOREM 6.14 (Pointwise approximations for Sobolev functions I). *Let $f \in W^{1,p}(\mathbb{R}^n)$ for some $1 \leq p < \infty$. Then for each $\epsilon > 0$ there exists a Lipschitz continuous function $\bar{f} : \mathbb{R}^n \to \mathbb{R}$ such that*

$$\mathcal{L}^n(\{x \mid f(x) \neq \bar{f}(x)\}) \leq \epsilon$$

and

$$\|f - \bar{f}\|_{W^{1,p}(\mathbb{R}^n)} \leq \epsilon.$$

Proof. 1. Write $g := |f| + |Df|$, and define for $\lambda > 0$

$$R^\lambda := \left\{ x \in \mathbb{R}^n \mid \fint_{B(x,r)} g \, dy \leq \lambda \text{ for all } r > 0 \right\}.$$

2. *Claim #1*: $\mathcal{L}^n(\mathbb{R}^n - R^\lambda) = o(\frac{1}{\lambda^p})$ as $\lambda \to \infty$.

Proof of claim: By Vitali's Covering Theorem, there exist disjoint balls $\{B(x_i, r_i)\}_{i=1}^\infty$ such that

$$\mathbb{R}^n - R^\lambda \subseteq \bigcup_{i=1}^\infty B(x_i, 5r_i) \tag{$\star$}$$

and

$$\fint_{B(x_i, r_i)} g \, dy > \lambda \quad (i = 1, \dots).$$

Hence

$$\lambda \leq \frac{1}{\mathcal{L}^n(B(x_i, r_i))} \int_{B(x_i, r_i) \cap \{g > \frac{\lambda}{2}\}} g \, dy$$

$$+ \frac{1}{\mathcal{L}^n(B(x_i, r_i))} \int_{B(x_i, r_i) \cap \{g \leq \frac{\lambda}{2}\}} g \, dy$$

$$\leq \frac{1}{\mathcal{L}^n(B(x_i, r_i))} \int_{B(x_i, r_i) \cap \{g > \frac{\lambda}{2}\}} g \, dy + \frac{\lambda}{2}$$

and so

$$\alpha(n) r_i^n \leq \frac{2}{\lambda} \int_{B(x_i, r_i) \cap \{g > \frac{\lambda}{2}\}} g \, dy \quad (i = 1, \dots).$$

Using $(\star)$ therefore, we see

$$\mathcal{L}^n(\mathbb{R}^n - R^\lambda) \leq 5^n \alpha(n) \sum_{i=1}^\infty r_i^n$$

$$\leq \frac{2 \cdot 5^n}{\lambda} \int_{\{g>\frac{\lambda}{2}\}} g \, dy$$

$$\leq \frac{2 \cdot 5^n}{\lambda} \left(\int_{\{g>\frac{\lambda}{2}\}} g^p \, dy \right)^{\frac{1}{p}} \left(\mathcal{L}^n \{g > \frac{\lambda}{2}\} \right)^{1-\frac{1}{p}}$$

$$\leq \frac{C}{\lambda^p} \int_{\{|f|+|Df|>\frac{\lambda}{2}\}} |Df|^p + |f|^p \, dy$$

$$= o(\lambda^{-p})$$

as $\lambda \to \infty$, since $\mathcal{L}^n \{g > \frac{\lambda}{2}\} \leq \frac{2^p}{\lambda^p} \int_{\{g>\frac{\lambda}{2}\}} g^p \, dy$.

3. *Claim #2*: There exists a constant C, depending only on n, such that

$$|f(x)| \leq \lambda, \quad |f(x) - f(y)| \leq C\lambda|x - y|$$

for $\mathcal{L}^n$-a.e. $x, y \in R^\lambda$.

Proof of claim: This is almost exactly like the proof of Claim #2 in the proof of Theorem 6.12.

4. In view of Claim #2 we may extend f using Theorem 3.1 to a Lipschitz continuous mapping $\bar{f} : \mathbb{R}^n \to \mathbb{R}$, with

$$|\bar{f}| \leq \lambda, \ \text{Lip}(\bar{f}) \leq C\lambda, \ \bar{f} = f \ \mathcal{L}^n\text{-a.e. on } R^\lambda.$$

5. *Claim #3*: $\|f - \bar{f}\|_{W^{1,p}(\mathbb{R}^n)} = o(1)$ as $\lambda \to \infty$.

Proof of claim: Since $f = \bar{f}$ on R^λ, we have

$$\int_{\mathbb{R}^n} |f - \bar{f}|^p \, dx = \int_{\mathbb{R}^n - R^\lambda} |f - \bar{f}|^p \, dx$$

$$\leq C \int_{\mathbb{R}^n - R^\lambda} |f|^p \, dx + C\lambda^p \mathcal{L}^n(\mathbb{R}^n - R^\lambda)$$

$$= o(1) \quad \text{as } \lambda \to \infty,$$

according to Claim #1.

Similarly, $Df = D\bar{f}$ $\mathcal{L}^n$-a.e. on R^λ, and so

$$\int_{\mathbb{R}^n} |Df - D\bar{f}|^p \, dx \leq C \int_{\mathbb{R}^n - R^\lambda} |Df|^p \, dx + C\lambda^p \mathcal{L}^n(\mathbb{R}^n - R^\lambda)$$

$$= o(1)$$

as $\lambda \to \infty$. $\qquad \square$

THEOREM 6.15 (Pointwise approximations for Sobolev functions II). *Let $f \in W^{1,p}(\mathbb{R}^n)$ for some $1 \leq p < \infty$. Then for each $\epsilon > 0$, there exists a C^1-function $\bar{f} : \mathbb{R}^n \to \mathbb{R}$ such that*

$$\mathcal{L}^n(\{x \mid f(x) \neq \bar{f}(x) \ or \ Df(x) \neq D\bar{f}(x)\}) \leq \epsilon$$

and

$$\|f - \bar{f}\|_{W^{1,p}(\mathbb{R}^n)} \leq \epsilon.$$

Proof. This follows from Theorems 6.12 and 6.14.   □

6.7   References and notes

The principal sources for this chapter are Federer [F], Liu [L], Reshetnjak [R], and Stein [St]. Our treatment of L^p-differentiability utilizes ideas from [St, Section 8.1]. Approximate differentiability is discussed in [F, Sections 3.1.2–3.1.5]. D. Adams showed us the proof of Theorem 6.5 in Section 6.2.

We followed [R] for the proof of Aleksandrov's Theorem, and we took Whitney's Extension Theorem from [F, Sections 3.1.13–3.1.14]. The approximation of Lipschitz continuous function by C^1 functions is from Simon [S, Section 5.3]. See also [F, Section 3.1.15]. We relied upon Liu [L] for the approximation of Sobolev functions. Fefferman [Ff] has established a refined version of Whitney's extension theorem.

Bibliography

[A]   L. Ambrosio, "A new proof of the SBV compactness theorem",
        Calc. Var. PDE 3 (1995), 127–137.

[A-DG] L. Ambrosio and E. De Giorgi, "New functionals in the calcu-
        lus of variations" (Italian), Atti Accad. Naz. Lincei Rend. Cl.
        Sci. Fis. Mat. Natur. 82 (1988), 199–210.

[B-M]   J. Ball and F. Murat, "Remarks on Chacon's Biting Lemma",
        Proc. Amer. Math. Soc. 107 (1989), 655–663.

[B-C]   J. K. Brooks and R. Chacon, "Continuity and compactness of
        measures", Adv. in Math. 37 (1980), 16–26.

[C]   F. Clarke, *Optimization and Nonsmooth Analysis*, Wiley–
        Interscience, 1983.

[DB]   E. DiBenedetto, *Real Analysis*, Birkhäuser, 2002.

[D]   R. Durrett, *Probability: Theory and Examples* (3rd edition),
        Thomson, Brooks/Cole, 2005.

[E1]   L. C. Evans, *Weak Convergence Methods for Nonlinear Partial
        Differential Equations*, CBMS Lectures # 74, Amer. Math.
        Soc., 1990.

[E2]   L. C. Evans, *Partial Differential Equations* (2nd edition),
        Amer. Math. Soc., 2010.

[Fa1]   K. Falconer, *The Geometry of Fractal Sets*, Cambridge U.
        Press, 1985.

[Fa2]   K. Falconer, *Fractal Geometry*, Wiley, New York, 1990.

[F]   H. Federer, *Geometric Measure Theory*, Springer, 1969.

[F-Z]   H. Federer and W. Ziemer, "The Lebesgue set of a function whose distribution derivatives are p-th power summable", Indiana U. Math J. 22 (1972), 139–158.

[Ff]   C. Fefferman, "A sharp form of Whitney's extension theorem", Ann. Math. 161 (2005), 509–577.

[Fg]   A. Figalli, "A simple proof of the Morse-Sard theorem in Sobolev spaces", Proc. Amer. Math. Soc. 136 (2008) 3675–3681.

[F-R]   P. M. Fitzpatrick and H.L. Royden, *Real Analysis* (4th edition), Prentice Hall, 2010.

[Fl-R] W. Fleming and R. Rishel, "An integral formula for the total gradient variation", Arch. Math. 11, 218–222 (1960).

[Fo]   G. Folland, *Real Analysis,* Wiley-Interscience, 1984.

[F-L]   Z. Furedi and P. A. Loeb, "On the best constant for the Besicovitch covering theorem", Proc. Amer. Math. Soc. 121 (1994) 1063–1073.

[Ga]   F. R. Gantmacher, *The Theory of Matrices,* Vol. I. Chelsea Publishing Co., 1977.

[G-T]   D. Gilbarg and N. Trudinger, *Elliptic Partial Differential Equations of Second Order* (2nd edition), Springer, 1983.

[G]   E. Giusti, *Minimal Surfaces and Functions of Bounded Variation,* Birkhäuser, 1984.

[H]   R. Hardt, *An Introduction to Geometric Measure Theory,* Lecture notes, Melbourne University, 1979.

[K-P]   S. Krantz and H. Parks, *The Geometry of Domains in Space,* Birkhäuser, 1999.

[L-Y]   F. Lin and X. Yang, *Geometric Theory: An Introduction,* International Press, 2002.

[L]   F.-C. Liu, "A Lusin property of Sobolev functions", Indiana U. Math. J. 26, 645–651 (1977).

[M-Z] J. Maly and W. Ziemer, *Fine Regularity of Solutions of Elliptic Partial Differential Equations*, American Math Society, 1997.

[M-S-Z] J. Maly, D. Swanson and W. Ziemer, "The co-area formula for Sobolev mappings", Trans. Amer. Math. Soc. 355 (2003), 477–492.

[Ma] P. Mattila, *Geometry of Sets and Measures in Euclidean Spaces*, Cambridge U Press, 1995.

[M] V. Maz'ja, *Sobolev Spaces*, Springer, 1985.

[Mo] F. Morgan, *Geometric Measure Theory: A Beginner's Guide* (4th edition), Academic Press, 2009.

[My] C. B. Morrey, Jr., *Multiple Integrals in the Calculus of Variations*, Springer, 1966.

[O] J. C. Oxtoby, *Measure and Category*, Springer, 1980.

[R] J. G. Reshetnjak, "Generalized derivatives and differentiability almost everywhere", Math USSR-Sbornik 4, 293–302 (1968).

[S] L. Simon, *Lectures on Geometric Measure Theory*, Centre for Mathematical Analysis, National University, 1984.

[St] E. Stein, *Singular Integrals and Differentiability Properties of Functions*, Princeton University Press, 1970.

[Sk] D. Stroock, *A Concise Introduction to the Theory of Integration* (3rd edition), Birkhäuser, 1999.

[T] L. Tartar, "Compensated compactness and applications to partial differential equations", 136–212 in *Nonlinear Analysis and Mechanics: Heriot-Watt Symposium*, Vol. IV, Res. Notes in Math 39, Pitman, 1979.

[W-Z] R. L. Wheeden and A. Zygmund, *Measure and Integral, An Introduction to Real Analysis*, Dekker, 1977.

[Z] W. Ziemer, *Weakly Differentiable Functions*, Springer, 1989.

Notation

A. Set and geometric notation

$\mathbb{R}^n$	n-dimensional real Euclidean space
$\mathbb{Z}$	set of integers
$\mathbb{Z}^+$	set of nonnegative integers
$\mathbb{M}^{m \times n}$	space of real $m \times n$ matrices
e_i	$(0, \ldots, 1, \ldots, 0)$, with 1 in the ith slot
$x = (x_1, x_2, \ldots, x_n)$	typical point in $\mathbb{R}^n$
$|x|$	$(x_1^2 + x_2^2 + \cdots + x_n^2)^{\frac{1}{2}}$
$x \cdot y$	$x_1 y_1 + x_2 y_2 + \cdots + x_n y_n$
$x^T \cdot Ay$	bilinear form $\sum_{i,j=1}^n a_{ij} x_i y_j$, where $x, y \in \mathbb{R}^n$ and $A = ((a_{ij}))$ is an $n \times n$ matrix
$B(x, r)$	$\{y \in \mathbb{R}^n \mid |x - y| \leq r\}$ = closed ball with center x, radius r
$B(r)$	$B(0, r)$ = closed ball with center 0, radius r
$B^0(x, r)$	$\{y \in \mathbb{R}^n \mid |x - y| < r\}$ = open ball with center x, radius r
$C(x, r, h)$	$\{y \in \mathbb{R}^n \mid |y' - x'| < r, |y_n - x_n| < h\}$ = open cylinder with center x, radius r, height $2h$
$Q(x, r)$	$\{y \in \mathbb{R}^n \mid |x_i - y_i| < r, i = 1, \ldots, n\}$ = open cube with center x, side length $2r$
$\alpha(s)$	$\dfrac{\pi^{\frac{s}{2}}}{\Gamma\left(\frac{s}{2} + 1\right)} \quad (0 \leq s < \infty)$
$\alpha(n)$	volume of the unit ball in $\mathbb{R}^n$
$\mathrm{dist}(A, B)$	distance between the sets $A, B \subset \mathbb{R}^n$
U, V, W	open sets, usually in $\mathbb{R}^n$
$V \subset\subset U$	V is compactly contained in U; that is, $\bar{V}$ is compact and $\bar{V} \subset U$
K	compact set, usually in $\mathbb{R}^n$
χ_E	indicator function of the set E
$\bar{E}$	closure of E
E^0	interior of E
$S_a(E)$	Steiner symmetrization of a set E; Section 2.3
∂E	topological boundary of E

$\partial^* E$	reduced boundary of E; Section 5.7
$\partial_* E$	measure theoretic boundary of E; Section 5.8
$\|\partial E\|$	perimeter measure of E; Section 5.1

B. Functional notation

$\fint_E f \, d\mu$ or $(f)_E$	$\frac{1}{\mu(E)} \int_E f \, d\mu$ = average of f over E with respect to the measure μ
$(f)_{x,r}$	$\fint_{B(x,r)} f \, dx$ = average of f over $B(x,r)$ with respect to Lebesgue measure
spt(f)	support of f
f^+, f^-	$\max(f,0), \max(-f,0)$
f^*	precise representative of f; Section 1.7
$f\|_E$	f restricted to the set E
$\bar{f}$ or Ef	an extension of f; cf. Sections 1.2, 3.1, 4.4, 5.4, 6.5
Tf	trace of f; Sections 4.3, 5.3
Df	derivative of f
$[Df]$	(vector-valued) measure for gradient of $f \in BV$; Section 5.1
$[Df]_{\mathrm{ac}}, [Df]_{\mathrm{s}}$	absolutely continuous, singular parts of $[Df]$; Section 5.1
ap Df	approximate derivative of f; Section 6.1
$Jf = [\![Df]\!]$	Jacobian of f; Section 3.2
Lip(f)	Lipschitz constant of f; Sections 2.4, 3.1
$D^2 f$	Hessian matrix of f
$[D^2 f]$	(matrix-valued) measure for Hessian of convex f; Section 6.3
$[D^2 f]_{\mathrm{ac}}, [D^2 f]_{\mathrm{s}}$	absolutely continuous, singular parts of $[D^2 f]$; Section 6.3
$G(f, A)$	graph of f over the set A; Section 2.4

C. Function spaces

Let $U \subseteq \mathbb{R}^n$ be an open set.

$C(U)$	$\{f : U \to \mathbb{R} \mid f \text{ continuous}\}$
$C(\bar{U})$	$\{f \in C(U) \mid f \text{ locally uniformly continuous}\}$
$C^k(U)$	$\{f : U \to \mathbb{R} \mid f \text{ is } k\text{-times continuously differentiable }\}$
$C^k(\bar{U})$	$\{f \in C^k(U) \mid D^\alpha f \text{ locally uniformly continuous on } U \text{ for } |\alpha| \leq k\}$

$C_c(U), C_c(\bar{U})$, etc.	functions in $C(U), C(\bar{U})$, etc. with compact support
$C(U; \mathbb{R}^m)$	functions $f : U \to \mathbb{R}^m, f = \{f^1, f^2, \ldots, f^m\}$, with $f^i \in C(U)$ for $i = 1, \ldots, m$
$C(\bar{U}; \mathbb{R}^m)$	functions $f : U \to \mathbb{R}^m, f = \{f^1, f^2, \ldots, f^m\}$, with $f^i \in C(\bar{U})$, for $i = 1, \ldots, m$
$L^p(U)$	$\{f : U \to \mathbb{R} \mid (\int_U |f|^p \, dx)^{\frac{1}{p}} < \infty, f$ Lebesgue measurable$\}$ $(1 \le p < \infty)$
$L^\infty(U)$	$\{f : U \to \mathbb{R} \mid \operatorname{ess\,sup}_U |f| < \infty, f$ Lebesgue measurable $\}$
$L^p_{\text{loc}}(U)$	$\{f : U \to \mathbb{R} \mid f \in L^p(V)$ for each open set $V \subset\subset U \}$
$L^p(U; \mu)$	$\{f : U \to \mathbb{R} \mid (\int_U |f|^p \, d\mu)^{\frac{1}{p}} < \infty, f$ μ-measurable $\}$ $(1 \le p < \infty)$
$L^\infty(U; \mu)$	$\{f : U \to \mathbb{R} \mid f$ is μ-measurable, $\mu - \operatorname{ess\,sup}_U |f| < \infty\}$
$W^{1,p}(U)$	Sobolev space; Section 4.1
K^p	$\{f : \mathbb{R}^n \to \mathbb{R} \mid f \ge 0, f \in L^{p^*}, Df \in L^p\}$; Section 4.7
$BV(U)$	space of functions of bounded variation; Section 5.1

D. Measures and capacity

$\mathcal{L}^n$	n-dimensional Lebesgue measure
$\mathcal{H}^s_\delta$	approximate s-dimensional Hausdorff measure; Section 2.1
$\mathcal{H}^s$	s-dimensional Hausdorff measures; Section 2.1
H_{dim}	Hausdorff dimension; Section 2.1
Cap_p	p-capacity; Section 4.7

E. Other notation

$\mu \llcorner A$	μ restricted to the set A; Section 1.1
$\mu \llcorner f$	(signed) measure with density f with respect to μ; Section 1.3
$D_\mu \nu$	derivative of ν with respect to μ; Section 1.6
$\nu << \mu$	ν is absolutely continuous with respect to μ; Section 1.6

$\nu \perp \mu$	ν and μ are mutually singular; Section 1.6
$\operatorname{ap\,lim}_{y \to x} f$	approximate limit; Section 1.7
$\operatorname{ap\,lim\,sup}_{y \to x} f$	approximate lim sup; Section 1.7
$\operatorname{ap\,lim\,inf}_{y \to x} f$	approximate lim inf; Section 1.7
$\rightharpoonup$	weak convergence; Section 1.9
S	symmetric linear mapping; Section 3.2
O	orthogonal linear mapping; Section 3.2
L^*	adjoint of L; Section 3.2
$[\![L]\!]$	Jacobian of linear mapping L; Section 3.2
$\Lambda(m,n)$	$\{\lambda : \{l,\dots,n\} \to \{1,\dots,m\} \mid \lambda$ increasing $\}$; Section 3.2
P_λ	projection associated with $\lambda \in \Lambda(m,n)$; Section 3.2
η, η_ϵ	mollifiers; Section 4.2
p^*	$\frac{np}{n-p}$ = Sobolev conjugate of p; Section 4.5
H, H^+, H^-	hyperplane, half spaces; Section 5.7
μ, λ	approximate lim sup, lim inf for BV function; Section 5.9
J	set of "measure theoretic jumps" for BV function; Section 5.9
$\operatorname{ess} V_a^b f$	essential variation; Section 5.10

Index

λ-system, 7, 9
π-λ Theorem, 7, 8, 79
π-system, 7–9
σ-algebra, 5, 6, 8, 9
σ-finite, 30, 31

absolutely continuous, 50
    part, 52, 257
approximate
    continuity, 58, 59, 245, 247,
        253
    derivative, 262
    differentiability, 262, 263
    lim inf, 57
    lim sup, 57
    limit, 56, 57
approximation
    by C^1 functions, 257–288
    by compact sets, 9–14
    by continuous functions, 22
    by open sets, 9–14
    by smooth functions,
        145–152, 199–203, 244
    in K^p, 171
area formula, 98, 101, 114, 119,
    206
average, 53

Binet–Cauchy formula, 112, 121
Biting Lemma, 72
Borel
    σ-algebra, 6

measure, 7, 8, 10, 11, 14, 21,
    22, 45, 50, 60, 82, 174
    regular measure, 9

capacity, 56, 170–183, 185
Caratheodory's criterion, 14, 60,
    83
chain rule, 153
changing variables, 122–123
coarea formula, 31, 101, 134, 210
    for BV functions, 141, 212,
        238, 239
convex functions, 266–276
cover, 36
    fine, 36
Covering Theorem
    Besicovitch, 39–46
    Vitali, 35–39, 45, 93, 99, 209,
        220, 231, 239, 277, 284,
        286

density, 47, 55, 56, 92, 93, 197,
    257, 273
derivative, 47, 50–53, 103, 202,
    265
    weak, 143
differentiable, 47, 103, 265, 266
distance function, 141
Dominated Convergence
    Theorem, 28, 33, 105,
        147, 259, 275
Dunford–Pettis Theorem, 70, 72

essential variation, 245, 247–249

extension, 20
  C^1 functions, 234, 276–282
  BV functions, 210–211
  by reflection, 160
  continuous functions, 19
  Lipschitz functions, 102
  Sobolev functions, 158–162

Fatou's Lemma, 26–28, 49, 128,
      215, 216, 249
Fubini's Theorem, 29–34, 89, 101,
      104, 105, 126, 127, 134,
      189, 240
fundamental theorem of calculus,
      50

Gagliardo–Nirenberg–Sobolev
      inequality, 162–164, 172,
      181, 219
Gauss–Green Theorem, 157, 158,
      235–236
gradient matrix, 114
graph, 87, 97, 98, 124

Hausdorff
  dimension, 86, 97, 179–180
  measure, 56, 59, 81–100, 179,
      181
Hessian, 294
hypersurface, 124, 232, 238

Implicit Function Theorem, 234
integral
  lower, 25
  upper, 24
inverse images, 16
isodiametric inequality, 89, 91,
      128
isoperimetric inequality, 218–220,
      224

Jacobian, 114

Lebesgue
  –Besicovitch Differentiation
      Theorem, 53, 56
  Decomposition Theorem, 52,
      272
  measure, 34, 87, 91
  point, 54, 56, 59, 146
level sets, 140, 141
Lipschitz
  boundary, 150
  constant, 96
  continuous function, 96–98,
      101–108
Lusin's Theorem, 21, 22, 58, 232,
      283

measurable
  function, 16–19
  set, 2, 3, 6, 9, 10, 13, 16
measure, 1–5
  signed, 25
measure theoretic
  boundary, 235
  exterior, 56, 250
  interior, 56, 250
  jump, 238, 241
mollifier, 145, 146, 244
Monotone Convergence Theorem,
      27, 29, 34, 116, 120, 123,
      139
Morrey's inequality, 167, 266
Morse–Sard Theorem, 134, 191
multiplicity function, 115, 250
mutually singular, 50

orthogonal
  linear mapping, 108–111, 127,
      136

perimeter, 197, 212
  finite, 194, 211, 212, 217

locally finite, 194, 198
polar coordinates, 140
precise representative, 56, 185, 187, 188
product measure, 30
product rule, 153
projection, 75, 112, 250

quasicontinuity, 183–187

Rademacher's Theorem, 103, 108, 114, 120, 151, 204, 266, 273, 283
Radon measure, 9–11, 13, 25, 39, 47–54, 59, 64–66, 75, 82, 194, 195, 202, 271
Radon–Nikodym Theorem, 50
reduced boundary, 221–230, 232
relative isoperimetric inequality, 218, 220, 225, 235, 239
restriction of measure, 10
Riesz Representation Theorem, 59–64, 67, 76, 174, 195, 197

simple function, 24
singular part, 52, 257
slicing measures, 75
Steiner symmetrization, 87, 88
subadditivity, 2–4, 14
summable, 25
symmetric
    linear mapping, 109–111, 117, 127, 132, 133, 135, 136
    matrix, 110, 267

trace
    BV function, 204–210
    Sobolev function, 156–158

uniform integrability, 70, 72

variation, 245
variation measure, 197

weak convergence
    in L^1, 70–75
    in L^p, 68–70
    of measures, 65–67
Whitney's Extension Theorem, 234, 277, 282, 283

Young measures, 77

Printed in the United States
by Baker & Taylor Publisher Services